The Latin American Studies Book Series

The Latin American Studies Book Series promotes quality scientific research focusing on Latin American countries. The series accepts disciplinary and interdisciplinary titles related to geographical, environmental, cultural, economic, political and urban research dedicated to Latin America. The series publishes comprehensive monographs, edited volumes and textbooks refereed by a region or country expert specialized in Latin American studies.

The series aims to raise the profile of Latin American studies, showcasing important works developed focusing on the region. It is aimed at researchers, students, and everyone interested in Latin American topics.

Submit a proposal: Proposals for the series will be considered by the Series Advisory Board. A book proposal form can be obtained from the Publisher, Juliana Pitanguy (juliana.pitanguy@springer.com).

More information about this series at http://www.springer.com/series/15104

John E. Staller
Editor

Andean Foodways

Pre-Columbian, Colonial, and Contemporary Food and Culture

Springer

Editor
John E. Staller
Botanical Research Institute of Texas
Fort Worth, TX, USA

ISSN 2366-3421 ISSN 2366-343X (electronic)
The Latin American Studies Book Series
ISBN 978-3-030-51628-4 ISBN 978-3-030-51629-1 (eBook)
https://doi.org/10.1007/978-3-030-51629-1

This Springer imprint is published by the registered company Springer Nature Switzerland AG
The registered company address is: Gewerbestrasse 11, 6330 Cham, Switzerland

Contents

Chapter 1
Andean Foodways: Interdisciplinary Approaches to Pre-Columbian, Colonial, and Contemporary Food and Culture

John E. Staller

Abstract Pre-Columbian Andean cultures have generally been characterized as having strong cultural and religious ties to their surrounding landscape and the natural world. Plants and animals associated with this sacred landscape have had, and in some societies continue to have, a particular cultural and religious meaning. Food crops and cultigens that sustained life and their associated preparation were often seen as a sacred act, with strong cultural associations to ethnic identity. Other plants have had direct associations to ritual and religious practices and were seen as sacred, reaffirming the diversity and complexity of Andean foodways and local cuisines. Anthropologists and archaeologists have documented the symbolic complexity of the natural world and the social importance of feasting, rituals and rites in contemporary and historical societies. Cultural perceptions and beliefs regarding the natural world and their associated plants and cuisines were subsequently modified to varying degrees by the Spanish conquest and introduction of foreign plants and animals. Contributions in this volume explore the art history and history to examine the roles of food through particular interdisciplinary lenses. Research from diverse regions of the cordillera further emphasize the diversity of Andean cultures. Contributions explore and analyze these topics in the context of historic and contemporary Andean culture, with examples of how domesticates, cuisines, their preparations, and basic ingredients continue to influence present day foodways and regional tastes.

Introduction

Contributions to this volume provide evidence of the diversity and variability of Andean food and culture and how traditional foodways were transformed by the Spanish conquest and introduction of native food crops and domesticated biota. The volume is primarily organized chronologically. Part I *Pre-Columbian Food and Cultures: Andean Culinary and Ritual Practices* consists of archaeological research

J. E. Staller (✉)
Botanical Research Institute of Texas, Fort Worth, USA
e-mail: jstaller@earthlink.net

J. E. Staller (ed.), *Andean Foodways*, The Latin American Studies Book Series,
https://doi.org/10.1007/978-3-030-51629-1_1

and studies of changing patterns of consumption during pre-Columbian times in different regions of the coast and highlands. Several chapters also incorporate interdisciplinary evidence in order to support their interpretations. In Part II, *Andean Foodways, Indigenous Customs, and Transformations among Colonial and Contemporary Andean Cultures* contributors explore the changes and transformations which occurred after the Spanish conquest with the introduction of Eurasian domesticated food crops and animals. Several contributions in this part of the volume use ethnographic evidence from contemporary indigenous Andean cultures to document their assertions with regard to preparations of drink and traditional cuisines in the highlands.

The Andes are geographically and culturally diverse with many environmentally challenging habitats and represent the second highest and the longest mountain range in the world. The Andes extend from Colombia and Venezuela in the north to Argentina and Tierra del Fuego in the south. Consequently, ecologies are extremely complex and environmentally distinct and diverse due in large part to the fact that a major part of the cordillera is near or on the equator. The multiple habitats and ecological settings, their associated culinary, ritual, and religious traditions are highly varied through time and space among Andean cultures (Murra 1972; Ugent and Ochoa 2006; Dillehay and Kaulicke 2011). In the central highlands, food and cuisines are not only phenomena within culture, but they are also referents through which many behavioral and performative aspects of culture may be better understood. Cultural, religious, and culinary traditions in combination with striking geographic diversity and a complex colonial history make the Central Andes a unique, important, and interesting region of the world. This volume on Andean food and culture presents evidence that analyzes and documents the various ways that both indigenous and Spanish colonial cultures past and present incorporated plants, animals, and natural resources into their lives. Contributors examine topics related to food from pre-Colombian archaeology and colonial accounts, as well as ethnographic evidence, and contemporary international economies. These contributions provide important evidence on how cultural and economic associations were linked to culinary traditions, cultural and ethnic identities, as well as political authority during pre-Columbian times, as well as after the Spanish conquest in the context of the Colonial, and contemporary economies. Many of the modern Andean culinary practices have their origins in the pre-Columbian past, and some of the food practices are syncretic in that they reflect a combination of ancient Andean food habits with practices and behaviors which emerged after the Spanish conquest and introduction of exotic flora and fauna. These contributions document how food practices are directly tied to Andean social organization, spirituality, and cultural perceptions regarding what is generally referred to as a sacred landscape.

Indigenous Andes cultures consume and prepare distinct foods, beverages, and cuisines to a large degree to create their own ethnic identities and foster social alliances. This has been documented archaeologically, mentions in numerous colonial accounts, and continues among contemporary Andean cultures. Contributions to this volume on foodways present research which indicate pre-Columbian, Colonial, and contemporary culinary traditions are the result from the union of a cornucopia of native foodstuffs which were subsequently transformed with introduced non-native plants, animals, and associated foodways (Lau, Toohey, Covey, Martinez, McDonell, Sayre and Rosenfeld, this volume). Anthropological and historical research on how the conquest of the New World changed culinary traditions and methods of preparation among contemporary cultures is also presented that are linked to long-term themes in anthropological query (e.g., Coe 1994; Schiebinger 2004; Schiebinger and Swan 2005; Villavicencio 2007; Staller 2010a, b; Earle and Costin 1989; Dietler and Hayden 2001; Jennings and Bowser 2009). In many cases, the preparation and ingredients of different cuisines were preserved archaeologically or combined in interesting and innovative ways with introduced species among contemporary indigenous cultures (Staller, Covey, Martinez Borrero, Bonzani, Dozier and Jennings, Sayre and Rosenfeld, this volume).

Andean Culture and the Natural World

In the Andes, cultural behaviors and associations that accompany the cultivation, production, and consumption of food are interrelated to religious rituals and practices commonly related to calculating the annual solar and lunar cycles ((Veintimilla and Garcia Caputi, Scher, Jackson, this volume). Pre-Columbian Andean religious ideologies were inherently telluric, that is, naturalistic and spatial, essentially representing a veneration of the natural world and celestial cycles (Sullivan 1984, 1988; Sharon 2001; Staller 2006, 2008a, b). Western epistemological distinctions between the natural world and culture, of humanity as distinct and separate from nature, or acting in a certain way upon the social and natural environment contrasts with traditional indigenous Andean concepts, where this dichotomy is essentially reciprocal rather than oppositional (Sullivan 1988; Sharon 2001; Staller 2008a, b). Anthropologists, archaeologists, and ethnobotanists have in recent decades applied their knowledge of Andean culture and ritual practices to investigate the significance of food and cuisines to culture, ethnic identity, political economies, and ancient religious ideologies (Prieto, Tantaleán and Rodriquez, Bonzani, Staller, Toohey, this volume). The natural landscape, its cycles and rhythms, continues to be perceived as interrelated and dynamic expressions of mythology, history, and economics and intrinsically related to their ethnic identities (Sullivan 1987, 1988; Staller 2008b).

In the Andes, certain plants and food fulfill spiritual needs and have sacred meanings that go beyond purely economic subsistence or requirements for sustenance (Scher, Jackson, Staller, Lambaré et al., this volume). Cultural, religious, and culinary traditions among Andean cultures reflect the integration of native biota and introduced maritime resources, domesticated plants, animals, and their associated cuisines. Foods and beverages among ancient and contemporary Andean cultures have historically played major roles in defining cultural and ethnic identities to their surrounding landscapes (Tantaleán and Rodríguez, Toohey, Sayre and Rosenfeld, this volume). Andean foods and cuisines frequently are referents through which people conduct many behavioral and performative aspects of their rites and rituals, as well as articulate individual identities and maintain social obligations. These performances of ritual practice and rites associated with food and cuisine take place along the coast as well as on mountainous landscapes punctuated by powerful natural landmarks: snow-capped mountain peaks, active volcanoes, hot water springs, and rushing waterfalls. Landscapes are viewed as animate, life-giving forces among indigenous and many mestizo cultures. Beliefs in the dynamic power of the natural environment and respect to these natural forces are enduring features of the Andean worldview. A common theme in several of the contributions is how people integrate their food practices with beliefs regarding veneration of the surrounding landscape (Tantaleán and Rodríguez, Veintimilla and Garcia Caputi, Scher, Bonzani, Staller, Sayre and Rosenfeld, this volume). In addition to linking food practices to surrounding landscapes, Andean cultures use specific plants, animals, and cuisines to convey complex symbolic and social meanings. Foods and culinary practices provide insights into regional interactions, social complexity, and cultural change among different peoples and between regions (e.g., Custer 2003; Dietler and Hayden 2001; Earle and Costin 1989; Gálvez and Runcio 2009; Jackson 2008; Wust and Cabieses 2004; Staller 2008b, 2010a, b, Staller and Carrasco 2010). Such connections and interactions are addressed by several contributions (e.g., Prieto, Lau, Toohey, McDonnell, this volume). Given the ecological and environmental diversity of the Andean region, the foodways and cultural interactions with the natural world which evolved there were and continue to be highly complex and have considerable variability through different regions along the coast and the cordillera.

Andean Political Economies, Resources, and Adaptation

Many Andean civilizations sustained their political authority through the maintenance of large irrigation networks all along the eastern and western cordillera along with systems of production that guaranteed a surplus of foodstuffs (Morris 1979; Murra 1972; Staller 2005; Jennings and Bowser 2009). These civilizations filled local markets and storehouses with economic staples, textiles, and an incredibly diverse set of foodstuffs that sustained populations throughout the annual cycle (Goody 1982; Montanari 1994; Dietler and Hayen 2001). Such commodities were dispersed in the context of ritual feasting and rites as forms of tribute. They were particularly critical to sustaining societies before the coming rainy seasons during the austral summer and fall (December to April). Various lines of evidence in this volume focus upon the role of foodways in ancient highland agricultural economies of scale (Bonzani, Sayre and Rosenfeld, this volume). Such commodities were also critical to coastal societies and formed the basis of long-distance interaction with societies in the cordillera (Murra 1978, 1980; Morris 1979; Staller 2010a).

The eastern Pacific coast is one of the most ecologically complex regions of the world with regard to maritime resources (Prieto, Tantaleán and Rodríguez, Veintimilla and Garcia Caputi, this volume). Along the Pacific Ocean, the complex coastal ecology is generated in part by the Humboldt Current that channels its cold current from Antarctica north along the eastern Pacific Ocean to the equator where it streams west to the Galapagos Islands. This cold-water current makes the maritime ecology the richest and most diverse in the western hemisphere (Moseley 1975; Moseley and Feldman 1988; Staller 2010a; Prieto 2020). Numerous fish species and mollusks are adapted to its cold waters. Consequently, such resources have sustained Pre-Columbian as well as contemporary cultures throughout Chile, Peru, and southern coastal Ecuador. However, the Humboldt Current is periodically disrupted by cycles of the El Niño Southern Oscillation (ENSO) (Keefer et al. 2003). Particularly, intense mega El Niño events have created extreme disruptions in the maritime ecology, along with widespread flooding from torrential rains as well as widespread destruction of the natural and human landscapes by inundation and mudslides (Glynn 1988; Keefer et al. 2003; Staller 2013, 2015; Prieto 2020). Such meteorological events result in the widespread death of marine life upon which most coastal Andean communities are dependent for their sustenance and survival (Moseley 1975; Prieto 2020). During ENSO events, highland habitats often experience drought resulting in detrimental consequences for agro-pastoral communities (Staller 2013, 2015).

As documented in these chapters, the Andean region provided the essential ingredients for distinct Andean communities and cultures to create foods and drinks that fulfilled the social needs of both individuals and communities. Some chapters highlight how certain cuisines and beverages conveyed important symbolic information regarding class, ethnicity, and social rank as well as access to the spiritual world (Staller, Covey, Martinez, Scher, this volume).

Contributions in Part II document that after the Spanish conquest some Andean communities developed local food habits that conformed to Western cognitive perceptions of proper and moral behavior (Lambaré et al., Covey, Martinez, Sayre and Rosenfeld, this volume). Colonial and contemporary changes in Andean food and cuisine highlight the persistence of cultural and religious tensions that have existed among traditional Andean communities with regard to the conquest and the modern world. The research presented here complements the growing body of literature related to food across the globe (e.g., Hastorf 2016; Hayden 2014, Peres and Deter-Wolf 2018; Twiss 2007; Van Derwarker and Wilson 2016; Wilkins and Nadeau 2015).

Prior to the Spanish conquest, most of the Andean cordillera was politically and linguistically united under control of the Inca Empire (Murra 1973, 1976, 1981; Morris 1979; Staller 2006, 2008b, 2010a, b). When the Inca Empire dominated and ruled much of the Andes, there existed a host of local ethnic variations in cuisines and beverages and preparation practices, which existed alongside traditions and behaviors shared by Andean peoples across diverse geographic spaces (Staller, Covey, Martinez, this volume). Diachronic changes in foodways, culinary traditions, and modern consumption practices reflect the repercussions of a complex indigenous, colonial, and modern history. After the changes in food habits that accompanied Spanish colonization, influxes of immigrants from many regions of the globe contributed to the emergence of contemporary food practices (Lambaré et al., Dozier and Jennings, this volume). Among contemporary cultures variation in foodways are generally a result of or reflect of deep historical roots related to traditional forms of social organization, gender divisions, communal labor, and reciprocity and not solely colonial and contemporary practices. Additionally, modern global economic forces and the international marketability of some Andean food crops have had a profound impact on local culinary customs and patterns of consumption (Martinez, McDonnell, this volume).

In addition to linking food practices to landscapes, people use specific plants, animals, and cuisines to convey complex symbolic and social meanings (Lau, Tantaleán and Rodriquez, Toohey, this volume). Foods and culinary practices provide insights into regional interactions, social complexity, and cultural change among different peoples and between regions (e.g., Custer 2003; Dietler and Hayden 2001; Earle and Costin 1989; Gálvez and Runcio 2009; Jackson 2008; Wust and Cabieses 2004; Staller 2008b; Staller and Carrasco 2010). Given the ecological and environmental diversity of the Andean region, the foodways and human interactions with the natural world that developed there were, and continue to be, complex and diverse.

Anthropological Approaches to Andean Food and Culture

The research presented in this volume builds on previous anthropological studies of Andean food; however, to varying degrees these contributions go beyond earlier literature to include more holistic analyses of the cultural meanings associated with the production and consumption of foods and beverages in different Andean settings and time periods (Prieto, Lau, Bonzani, Jackson, Sayre and Rosenfeld, this volume). Previously, scholars have addressed the cultivation and preparation of Andean food with accounts focused on particular cultigens, regions, beverages, or culinary traditions during specific time periods (e.g., Jennings and Bowser 2009; Olivas Weston 2001; Rodríguez 2007; Staller et al. 2006; Ugent and Ochoa 2006; Wust 2006). In the recent decades, the published literature surrounding Andean food and cuisines and their relation to cultural and ethnic identity have grown (e.g., Alcalde 2005; Bray 2003; Cutright 2009; Earle 2012; Hastorf 2003; Kennedy et al. 2019; Klarich 2010; Jamieson and Sayre 2010). Archaeologists traditionally have contributed long-term studies focusing on early plant and animal domestication as well as the spread of agriculture and its roles to the rise of complexity along the coast and in the highlands (e.g., Almekinders and de Boef 2000; Cowan et al. 1992; Dillehay and Kaulicke 2011; Harper 2003; Hastorf 2016; Staller et al. 2006). Some studies of contemporary food culture also consider the role of long-term historical processes in modern habits (e.g., Krögel 2011; León 2017; Weismantel 1988).

Other publications related to food focus specifically on Andean political economies and the ways in which people cultivated food commodities as tribute, generated stored foodstuffs, and ultimately redistributed food as forms of reciprocity for corvée labor (Murra 1980). A popular theme is how various pre-Columbian Andean states sponsored feasts to create status differentiation in conjunction with religious and political power (Bray 2003; Cuéllar 2013; Dillehay and Kaulicke 2011; Isbell and Groleau 2010; Nash and deFrance 2019; Rosenfeld 2012; Staller 2008a, b,

2010a, b). Previous literature has also revealed how iconographic and symbolic representations had reference to ancient ideologies, religions, and ritual practice (e.g., Bray 2003; Jackson 2008; Ramirez 2005), and several contributions in this volume expand on such cultural beliefs and traditions (Jackson, Veintimilla and Garcia Caputi, this volume).

As shown in these studies, the Andean region provided the essential ingredients for distinct Andean communities and cultures to create foods and drinks that fulfilled both individual and community social needs. Some chapters highlight how ingesting foods and beverages conveyed important symbolic information regarding class, ethnicity, and social rank (Lau, Tantaleán and Rodriquez, Covey, this volume). Analyses also show that following the Spanish conquest people developed local food habits that conformed to cognitive perceptions of proper and moral behavior. Contemporary changes in food and cuisine highlight the continuing and persistent cultural tensions that exist as traditional Andean practices and beliefs intersect with the modern world. The research presented here complements the growing body of literature related to the archaeology of food across the globe (e.g., Hastorf 2016; Hayden 2014, Peres and Deter-Wolf 2018; Twiss 2007; Van Derwarker and Wilson 2016; Wilkins and Nadeau 2015).

The extreme complexity and distinct environmental diversity of the Andean cordillera are due in part to the fact that a major part of the Andes is near or on the equator (Staller 2010a). When considered in light of the ecological dynamics of the eastern Pacific Ocean the numerous fish species and mollusks are adapted to the cold waters of the Humboldt Current, it is obvious why such resources maritime as well as highland resources sustained pre-Columbian as well as contemporary cultures particularly throughout Chile, Peru, and southern coastal Ecuador. In some cases, people created cultural adaptations to mitigate ecological and environmental instability, especially along the coastal deserts and sierra habitats of Peru and Chile through long-distance interaction with the highlands. Trade between the coast and highlands was particularly common among coastal cultures involved in large-scale cultivation along the various highland-fed rivers and streams (Murra 1973, 1976, 1980; Morris 1993; Staller 2005). The maritime resources exploited by cultures along the Pacific coast generated long-distance interaction extending back to the earliest evidence of agriculture in this region (e.g., Gose 1994; Olivas Weston 2001; Dillehay and Kaulicke 2011; Hastorf 2016; Prieto 2020; Shady and Leyva 2003).

Economic specialization along the coast and western valleys created significant ethnic variability over the millennia. The western cordillera and desert coasts were more linguistically diverse than the highlands (Heggerty and Beresford-Jones 2010). In the Andean Cordillera, the Aymara and Quechua language families spread in association with Wari and later Inca State expansion from northern Chile to northernmost highland Ecuador (Murra 1976, 1980; Staller and Stross 2013). Consequently, certain cuisines and associated terminologies for food, plants, and forms of preparation are much more distinct along the coast and western cordillera than in other regions of the Andes. Moreover, maritime resources played much larger roles in the subsistence economies and cuisines in these regions of the Andes (see Sandweiss and Prieto 2020).

Andean Foods

In conjunction with the diverse coastal resources, Andean economies relied on local food crops, potatoes, manioc, and tubers of various kinds (see Bonzani, McDonell, this volume). Colonial accounts provide significant details regarding primary staples at the time of Spanish colonization. Chroniclers state that Inca storehouses were distributed throughout the empire and for the most part held maize, however, those in the higher *puna* and *Altiplano* environmental zones also stored freeze-dried ducks and tubers such as potatoes or oca (*Oxalis tuberosa* Molina) (Murra 1980; Morris 1993). Although oca is one of the important staple crops of the Andes, the primary food staple throughout the Andes was and continues to be the potato (*Solanum tuberosum* L.) (Coe 1994; Hastorf 2003; Staller 2010a). Potatoes and quinoa were staples in the highlands and along both sides of the cordillera (Cowan et al. 1992; Langlie et al. 2011; Staller 2010a, 2016). Other native food plants in the *Altiplano* and higher parts of the Andes include numerous varieties of quinoa (*kañiwa* or *juyra*), oca (*uka* or *ulluku*), and native lupin (*tarwi*) (McDonell, Laffey and Llanque Chana, this volume). Andean food staples, particularly those in the higher altitudes, require adaptation to the extreme diurnal variations in temperatures, particularly the cold nights and generally poor soils in order to survive in such extreme environments.

Subsistence agriculture is critical to adaptation in the Andes, particularly in the higher parts of the cordillera. In such altitudes, most indigenous communities are primarily engaged in camelid herding and the cultivation of food crops (Carpio and Yupa Vereau 2014). Some food crops, particularly potatoes and quinoa, have through human and natural selection resulted in numerous varieties whose biogeography reflects this extreme biodiversity (Murra 1973, 1982; Morris 1993; Langlie et al. 2011). For example, the domestication of quinoa c. 3000 B.C. in the regions around the Lake Titicaca Basin resulted in distinct genetic varieties around the basin (Bonzani, McDonell, this volume). Quinoa and potatoes are two food crops that spread throughout the Andes as staples and generally have strong cultural associations with ethnic identity.

Other important staples that spread throughout the highlands and lowlands are the peanut and chili peppers. The inhabitants of the Andes may have domesticated peanuts in Peru as early as 7600 years ago (Dillehay et al. 2007:1890). Their cultivation spread in pre-Hispanic times to Mesoamerica where accounts indicate that people sold them in the Aztec markets in Tenochtitlan (Cieza de León 1998 [1553]).] People also domesticated chili peppers in the Central Andes and distributed them widely as an important component of both coastal and highland diets (Chiou and Hastorf 2014).

In addition to native crops of potatoes, quinoa, coca, and other agricultural products, maize (*Zea mays* L.), an exotic plant introduced from Mesoamerica, is one of the most important and widely researched cultigens in the Central Andes (Staller, Covey, Martinez, this volume). In contrast to Mesoamerican, Andean cultures primarily consumed maize as beer or *chicha* (Staller, Covey, Martinez, Veintimilla and Garcia Caputi, this volume). The Inca and other political entities redistributed *chicha* as a form of reciprocity in exchange for *mit'a* or corvée labor (i.e., obligatory, forced labor) by subject cultures (Murra 1972; Morris 1979; Staller 2006, 2008a, b, 2010a, b). Various Andean cultures linked ritualized *chicha* consumption to maize cultivation, but also to widespread cultural practices related to reciprocity, gendered behavior, and establishing political authority (Staller 2006, 2010a, b). Such cultural practices have ancient origins in this region.

In combination with the diverse plant resources cultivated in the agricultural economy, domesticated camelids (llamas and alpacas), and a pastoral economy are associated with cultures in the Andean cordillera (Lau, this volume). People used both llamas and alpacas for meat, hair, and dung as fertilizer, while llamas primarily served as pack animals in the mountainous terrain. Previously published research has emphasized their importance in long-distance interaction between highland cultures and those along the lowlands and coast (Browman 1987; Chepstow-Lusty et al. 2007; Murra 1980; Morris 1993). These camelids along with the domesticated guinea pig were both important in Andean animal sacrifices to *huacas* (sacred places) (Lau, this volume). Additionally, the Andean diet consisted of the capture and consumption of an array of wild terrestrial mammals (e.g., deer, vizcachas, guanacos, and vicuñas) and an abundance of marine finfish and shellfish. The domesticated Muscovy duck and a variety of wild birds also provided meat. For indigenous peoples and Spaniards alike, Spanish colonization resulted in economic and religious changes that resulted in the variable retention of some Andean food animals and the incorporation of non-native, Eurasian animals into Andean foodways (Staller, Covey, Martinez, Lambaré et al., this volume). Colonization also created very strong cognitive perceptions regarding either the acceptability or unsuitability of animal foods (Earle 2012). The rejection of some native Andean animals for food operated in combination with the imposition of Christian doctrine opposing animal sacrifices.

Colonial Change and Continuity in Traditional Andean Food Practices

Maize (*Zea mays* L.) is one of the most important and widely researched cultigens in the associated literature. The interdisciplinary evidence has documented that maize was an exotic plant introduced to the Andes from Mesoamerica (Staller 2006, 2010a; Staller et al. 2006; Staller and Carrasco 2010; Staller and Stross 2013). Significantly, the preparation and consumption of maize in the Andes are quite distinct from its region of origin (Staller, this volume). Its preparation and consumption as a fermented intoxicant, maize beer or *chicha,* and its most common association in traditional cuisines as a side dish have broader implications for its dietary importance among Andean cultures (Wacher 2003; Staller 2018; Tykot and Staller 2003; Katz et al. 1976; Tykot et al. 2006).

The maize was central to the rise of Andean civilization related to its consumption as maize beer or *chicha* and particularly its importance to religious rituals, rites, and as offerings to sacred places in the surrounding landscape (Morris 1979; Staller 2006, 2008b, 2010a, b, 2018), and its cultivation resulted in large-scale modifications of the landscape in the form of artificial terraces, and the spread of various pre-Columbian civilizations (Bonzani, Dozier and Jennings, Laffey et al., Staller, this volume). Contributors examine the importance of maize and maize beer or *chicha* to the rise of civilization and in political economies. The evidence also indicates that many associated rites and rituals involving *chicha* were critical to the development and spread of cultural complexity as well as the spread of various cultures throughout the highlands (Staller, Dozier and Jennings, Laffey et al., this volume). Inca territorial expansion was based in part upon large-scale maize production associated with widespread terrace construction. Such terrace constructions and maize cultivation largely involved *chicha* production. When consumed in rituals or in association with annual rites of passage, it reaffirmed status and rank in Andean cultures (Staller, Bonzani, this volume). The Inca Empire had many large-scale storehouses and administrative centers which have provided evidence of *chicha* production. Such administrative centers stored maize and various other economic staples which were redistributed to subject populations and cultures during periods of environmental disruption or instability when crop failures occurred (Staller 2006).

Research presented by contributors indicates that *chicha* was commonly prepared by females in the highlands (Bonzani, Martinez, Laffey et al., this volume). Among the Inca elite, *chicha* preparation was solely carried out by the chosen women or *mamaconas* that the wives of Inca emperors. These women were from high-ranking lineages of different ethnic groups during the Inca expansion (Coe 1994; Staller 2006, 2008b, 2010a). This is in contrast to various coastal cultures and lowland regions where it was commonly prepared by males (*chicheros*) (Morris 1979; Jennings and Bowser 2009; Staller 2010a). The anthropological and historical researches regarding how the Spanish conquest of the New World transformed culinary traditions and methods of preparation among Andean cultures are topics of interest to archaeologists and ethnohistorians (e.g., Coe 1994; Earle and Costin 1989; Ruhl 1990, 1997; Schiebinger 2004; Schiebinger and Swan 2005; Staller 2008b, 2010a; Villavicencio 2007). Changes to Andean foodways and food practices resulted from the introduction of Eurasian plants, animals, foods, and beverages, along with the establishment of control by Spanish colonial governments and the Catholic Church (Staller 2008b, 2010a).

Contributors provide methodological and analytical examples of the transformations brought on during the Colonial Period, where historical written records are available for some parts of the region (Covey, Martinez, this volume). Others research and analyze indigenous descendants of ancient cultures that, in some cases, continue to practice rituals and prepare cuisines with native plants and animals or make ritual offerings to the surrounding landscape as in ancient times (Laffey et al., Sayre and Rosenfeld, this volume). In some cases, these culinary and cultural practices are distinct from earlier times, but in every case, they speak to an intrinsic link between such cultures and their natural surroundings. It is in part these cultural, religious, and culinary traditions that make the Andes such a unique, important, and interesting region of the world.

Spaniards planted Iberian crops, particularly at lower elevations where Eurasian crops were more successful. Spaniards also encouraged native populations under their political control, to grow Iberian food crops to provide tribute to colonial overlords (Ramírez 1996). In the arable western river valleys along the Peruvian and Chile coast, colonists altered the landscape by planting grape stock in large vineyards for the production of wine and brandy (see Cushner 1980; Rice 2011, 2013). Spaniards also planted wheat throughout much of the Andes in an effort to produce familiar breadstuffs for domestic consumption as well as wheat bread for Eucharistic rituals associated with the Catholic Church (Convey, Martinez, this volume). Traditional Andean cultures generally continued to consume native food crops in their cuisines particularly in the context of feasting. For example, contributors provide evidence on culinary traditions such as *pachamanca* and how class and social differences in the preparation and consumption of traditional indigenous foods are linked to ethnic identities (Tantaleán and Rodriquez, Toohey, Sayre and Rosenfeld, this volume).

Archaeobotanical evidence indicates that the early colonial plant introductions were very successful. Wheat, which thrived in coastal Peru, highland Bolivia, and Argentina, transformed cuisine in those regions of the Andes (Capparelli et al. 2005; Kennedy et al. 2019; Munoz 2019; Petrucci et al. 2018). Colonists also introduced barley early on, and it also thrived in highland settings providing essential crops colonial cuisines (Capparelli, et al. 2005; Jameison and Sayre 2010). Sugarcane and olives were crops that necessitated large swaths of arable and low elevation land. Sugarcane came to play a major role in the consumption of fermented intoxicants in colonial and contemporary Andean culture cornucopia (Martinez, this volume). Many other Eurasian crops of more restricted economic value were cultivated, particularly citrus fruits and Eurasian vegetables. Various cultigens were grown on a much smaller scale than crops of greater commercial value (McDonell, this volume).

In addition to food crops, Spaniards also introduced a range of domestic animals for consumption and for labor and transport. Meat animals included Eurasian sheep, goats, pigs, cattle, and fowl. Horses, burros, and donkeys were central for colonial transport and trade, often replacing the native camelids. As with newly established Eurasian crops, many introduced animals (e.g., chickens, pigs, and caprids) initially thrived at lower elevations where high altitude stress was not as great (Kennedy and VanValkenburgh 2015); however, most non-native species eventually adapted to high elevation settings (Staller 2008b). In many instances, indigenous and mestizo populations adopted the use of non-native animals for food (and work) selectively incorporating the meat from various animals into recipes alongside native food animals in the process of creating criollo cuisine.

The introduction of food crops and domesticated plants and animals from the Old World dramatically changed New World economies and cuisines, and the Andes was no exception in this regard. Contributions in this volume present evidence of introduced plants, in this case a fruit from the lowlands, which becomes an important food source in the province of Jujuy in northwestern Argentina. Lambaré et al. use archaeological research and ethnohistoric accounts to document a relatively early introduction of peaches (*Prunus persica*) into the highland valley of Humahuaca by European conquistadores. As the ethnographic evidence indicates, peaches were

Isbell WH, Groleau A (2010) The Wari Brewer woman: feasting, offerings, and memory. In: Klarich E (ed) Inside ancient kitchens: new directions in the study of daily meals and feasting events. University of Colorado Press, Boulder, pp 191–220
Jackson MA (2008) Moche art and visual culture in ancient Peru. University of New Mexico Press, Albuquerque
Jamieson RW, Sayre MB (2010) Barley and identity in the Spanish colonial audiencia of quito: archaeobotany of the 18th century San Blas neighborhood in Riobamba. J Anthropol Archaeol 29:208–218
Jennings J, Bowser BJ (eds) (2009) Drink. Power and society in the Andes, University of Florida Press, Gainesville
Johannessen S, Hastorf CA (eds) (1994) Corn and culture in the prehistoric new world. Westview Press, Boulder Co
Keefer DK, Moseley ME, deFrance SD (2003) A 38,000-year record of flood and debris flows in the ilo region of southern Peru and its relation to El Niño events and great earthquakes. Palaeogeogr Palaeoclimatol Palaeoecol 194:41–77
Kennedy SA, VanValkenburgh P (2015) Zooarchaeology and changing food practices at carrizales, Peru. Int J Hist Archaeol 20(1):73–104
Kennedy SA, Chiou K, Parker VV (2019) Inside the reducción: crafting colonial foodways at carrizales and mocupe Viejo, Zaña Valley, Peru (1570–1700). Int J Histor Archaeol, published online February 6, 2019
Klarich EA (ed) (2010) Inside ancient kitchens: new directions in the study of daily meals and feasts. University of Colorado Press, Boulder
Krögel A (2011) Food, power, and resistance in the andes: exploring quechua verbal and visual narratives. Lexington Books, Lanham, MD
Langlie BS, Hastorf CA, Bruno MC, Bermann M, Bonzani RM, Condarco WC (2011) Diversity in andean chenopodium domestication: describing a new morphological type from La Barca, Bolivia 1300–1250 B.C. J Ethnobiol 31:72–88
Lavallée D, Julien M (eds) (2012) Prehistoria de la costa extremo-sur del Perú: los Pescadores arcaicos de la Quebrada de los Burros (1000–7000 a. P.). Instituto Francés de Estudios Andinos, Lima
León E (2017) 14,000 años de alimentación en el Perú. Fondo Editorial Universidad de San Martín de Porres, Lima
Marcus J, Sommer JD, Glew CP (1999) Fish and mammals in the economy of an ancient peruvian kingdom. Proceedings of the National Academy of Sciences USA 96:6564–6570
Morris C (1979) Maize beer in the economics, politics, and religion of the Inca state. In: Gastineau C, Darby W, Turner T (eds.), The role of fermented beverages in nutrition, pp 21–34. Academic Press, New York
Morris C (1993) The wealth of a native American State. value, investment, and mobilization in Inka economy. In: Henderson J, Netherly PJ (eds.), Configurations of power, pp 36–50. Cornell University Press, Ithaca
Moseley M (1975) The maritime foundations of andean civilization. Cummings Pub, Co, Menlo Park, California
Mumford JR (2012) Vertical empire: the general resettlement of Indians in the colonial andes. Duke University Press, Durham, NC
Murra JV (1972) El control vertical de un maximo de pisos ecológicos en la economia de las sociedades Andinas. In: Murra JV (ed), Visita de la Provincia de León de Huanuco en 1562, Tomo II, pp 429–476. Universidad Nacional Hermilio Valdizan, Huanuco
Murra JV (1973) Rite and crop in the Inca state. In: Gross DR (ed.), Peoples and cultures of South America, pp. 377–389. Natural History Press, Doubleday, Garden City, New Jersey
Murra JV (1976) Los limites y las limitaciones del "archipielago vertical", en los Andes. Anales de la Universidad del Norte 10:141–146
Murra JV (1980) Economic organization of the Inca state. JAI Press, Greenwich Ct

Murra JV (1982) The Mit'a obligation of ethnic groups to the Inca state. In: Collier G, Rosaldo R, Wirth J (eds.), The Inca and Aztec states, 1400–1800, pp 237–264. Academic Press, New York

Nash DJ, deFrance SD (2019) Plotting abandonment: excavating a ritual deposit at the Wari Site of Cerro Baúl. J Anthropol Archaeol 53:112–132

Olivas Weston R (2001) La Cocina de los Incas: Costumbres Gastronómicas y Técnicas Culinarias. Universidad San Martín de Porres, Escuela Profesional de Turismo y Hotelería, Lima, Perú

Orlove BS (1993) Putting race in its place: order in colonial and postcolonial Peruvian geography. Soc Res 60(2):301–336

Ortiz AI (2012) Los Maíces en la Seguridad Alimentaria de Bolivia. La Paz, Bolivia, Centro de Investigación y Promoción del Campesinado (CIPCA)

Peres TM, Deter-Wolf A (eds) (2018) Baking, bourbon, and black drink: foodways archaeology in the American Southeast. University of Alabama Press, Birmingham

Petrucci N, Lema VS, Pochettino ML, Palamarczuk V, Spano R, Tarragó M (2018) From weeds to wheat: a diachronic approach to ancient biocultural diversity in the Santa María valley, northwest Argentina. Veget Hist Archaeob 27:229–239

Prieto G (2020) The fisherman's garden. horticultural practices in a second millennium community of the north coast of Peru. In: Prieto G, Sandweiss DH (eds.), Maritime communities of the ancient Andes, pp 218–246. University Press of Florida, Gainesville, FL

Quilter J, Ojeda B, Pearsall DM, Sandweiss DH, Jones JG, Wing ES (1991) Subsistence economy of El Paraíso, an early Peruvian Site. Science 251:277–283

Ramírez SE (2005) To feed and be fed: the cosmological bases of authority and identity in the Andes. Stanford University Press, Stanford

Ramírez SE (2007) Archaeological evidence of Zea mays L. (POACEAE) in the southern argentinean Puna (Antofagasto de la Sierra, Catamarca). J Ethnobiol 27(2):256–271

Rice PM (2011) Vintage moquegua: history, wine, and archaeology on a colonial peruvian periphery. University of Texas Press, Austin

Rice PM (2013) Time-space perspectives on early colonial Moquegua. University Press of Colorado, Boulder

Rodríguez Pastor H (2007) La vida en el entorno del tamal peruano. Universidad de San Martín de Porres, Escuela Profesional de Turismo y Hotelería, Lima

Rosenfeld S (2012) Animal wealth and local power in the Huari empire. Ñawpa Pacha: J Andean Archaeol 32(1):131–164

Ruhl DL (1990) Spanish mission paleoethnobotany: an overview and some speculations for the 16th and 17th century La Florida. In: Thomas DH (ed.), Columbian consequences, archaeological and historical perspectives on the Spanish Borderlands East, vol 2, pp 560–580. Smithsonian Institution Press, Washington, DC

Ruhl DL (1997) Oranges and wheat: Spanish attempts at agriculture in La Florida. Histor Archaeol 31(1):36–45

Ryser GL (2008) Moche bean warriors and the paleobotanic record: why privilege beans? In: Castillo LJ (ed.), Arqueología Mochica: Nuevos Enfoques, Fondo Editorial. Pontificia Universidad Católica del Perú, Lima, Perú

Schiebinger L (2004) Plants and empire: colonial bioprospecting in the Atlantic world. Harvard University Press, Cambridge

Schiebinger L, Claudia S (eds.) (2005) Introduction. In: Colonial Botany: Science, Commerce, and Politics in the Early Modern World, pp 1–16. University of Pennsylvania Press, Philadelphia

Shady RS, Leyva C (eds) (2003) La ciudad sagrada del Caral-Supe: los orígenes de la civilización andina y la formación del estado prístino en el antiguo Perú. Instituto Nacional de Cultura, Lima, Peru

Staller JE (2005) La agricultura de la irrigación en el Período Formativo del Desierto de Atacama. In: Rivera MA (ed.), Arqueología del Desierto de Atacama, Chile: La Etapa Formativa en el Area de Ramaditas/Guatecondo, pp 89–102. Ediciones Universidad Bolivarana, Santiago

Staller JE (2006) The social, symbolic, and economic significance of *Zea mays* L. in the late horizon period. In: Staller JE, Tykot RH, Benz BF (eds.), Histories of Maize: multidisciplinary approaches

to the prehistory, linguistics, biogeography, domestication, and evolution of maize, pp 449–467. Routledge, Taylor & Francis, New York
Staller JE (2008a) An introduction to pre-columbian landscapes of creation and origin. In: Staller JE (ed.), Pre-Columbian landscapes of creation and origin, pp 1–9. Springer, New York
Staller JE (2008b) Dimensions of place: the significance of centers to the development of andean civilization: an examination of the *Ushnu* concept. In: Staller JE (ed.), Pre-columbian landscapes of creation and origin, pp 269–313. Springer, New York
Staller JE (2010) Maize cobs and cultures: history of Zea mays L. SpringerVerlag, Heidelberg, Berlin, p 2
Staller JE (2010b) Ethnohistoric sources on foodways, feasts and festivals in mesoamerica. In: Staller JE, Carrasco M (eds.), Pre-Columbian foodways: interdisciplinary approaches to food, culture and markets in ancient Mesoamerica, pp 23–69. Springer, New York
Staller JE (2013) Ancient El Niño Events, human adaptation, and ecological transformations: early formative period (2400–1450 B.C.) occupations in southern coastal Ecuador. Dialogo Andino 41:101–132
Staller JE (2015) Geological and archaeological evidence of El Niño Events along the Coast of El Oro province Ecuador: excavations at La Emerenciana, a Late Valdivia (ca. 2200 1450 B.C.) ceremonial center. Maskana 6(2):155–186
Staller JE (2018) Origin, evolution, biogeography, and cultivation of maize: the role of nixtamal to nutrition and economic importance in ancient and contemporary food and culture. In: Sanz N (ed.), The Origin and Evolution of Food Production and its Impact on Consumption Patterns, pp 179–195. Centre for Research and Advanced Studies (Cinvestav), México DF
Staller JE, Stross B (2013) Lightning in the Andes and Mesoamerica: Pre-Columbian, colonial, and contemporary perspectives. Oxford University Press, New York and Oxford
Staller JE, Tykot RH, Benz BF (2006) Histories of maize: multidisciplinary approaches to the prehistory, linguistics, biogeography, domestication, and evolution of Maize. Routledge, New York
Sullivan L (1987) Nature, the worship of nature. In: Eliade M (ed) Encyclopedia of religion. Macmillian Publishing Co, New York, pp 324–328
Sullivan L (1988) Icanchu's Drum. an orientation to meaning of south American religions. Macmillian Publishing Co, New York
Twiss KC (ed.) (2007) The archaeology of food and identity. Occasional Paper No. 34. Center for Archaeological Investigations, Southern Illinois University, Carbondale
Ugent D, Ochoa CM (2006) La Ethnobotánica del Perú. Universidad Nacional Mayor de San Marcus, Lima
Derwarker V, Amber M, Wilson GD (eds) (2016) The archaeology of food and warfare. Springer Press, New York
Wacher C (2003) Nixtamalization, a Mesoamerican technology to process maize at a small-scale with great potential for improving the nutritional quality of maize based foods. 2nd International Workshop: Food Based Approaches for a Healthy Nutrition 23–28:735–744
Weismantel MJ (1988) Food, gender, and poverty in the ecuadorian Andes. University of Pennsylvania Press, Philadelphia
Wilkins J, Nadeau R (eds) (2015) A Companion to food in the ancient world. Wiley Blackwell, Oxford
Wust WH (2006) La Pesca en el Antiguo Perú. Wust ediciones, Lima
Wust WH, Cabieses F (2004) Agricultura en el Perú: Cultivando la Tierra de los Incas. Neptunia, Lima, Peru

Part I
Pre-Columbian Foods and Cultures: Andean Culinary and Ritual Practices

Chapter 2
Grilling Clams and Roasting Tubers: Andean Maritime Foodways in the Second Millennium B.C.

Gabriel Prieto

Abstract This research explores the significance of marine and valley crops processed and consumed at Gramalote, a fishing community in the Moche Valley, North Coast of Peru occupied 300 years and dated to 1500–1200 cal. BC. It was an organized settlement with a permanent occupation. Its inhabitants were also part-time artisans producing a number of local products such as basketry, matting, carved bones and shells, and processing red pigment using a nearby iron oxide mine. Marine resources have received serious attention in the literature but little efforts have been made to understand their density and diversity in early Andean prehistory. This research contextualizes the importance of marine resources in a fishing community and the way people obtained, processed, and prepared these products. This research also uses accounts of ethnographic studies of contemporary maritime communities to determine how the inhabitants prepared and consumed marine foods. Finally, the presence of abundant macrobotanical remains at Gramalote suggests that the inhabitants had access to plant resources that were cultivated in the vicinity of the site and in the nearby valleys. In a broad perspective, the data presented enhance our understanding of the social dynamics and economic interactions of the early civilizations of the Central Andean region.

Introduction

Previous studies have documented a diversity of cultigens and animals consumed as food over long periods of cultural development (see Hastorf 2008; Lavallée and Julien 2012; Marcus 1987); however, a few studies have addressed how these food products were prepared and consumed (see Gumerman 2010; Goldstein and Shimada 2010; Isbell and Groleau 2010; Nash 2010). Additionally, although there has been extensive research on how marine and terrestrial foodstuffs contributed to the diet during the early stages of Andean prehistory (Moseley 1975; Osborn 1977; Pozorski

G. Prieto (✉)
University of Florida, Gainesville, FL, US
e-mail: gabrielprietobur@gmail.com; ogabriel.prietob@ufl.edu

J. E. Staller (ed.), *Andean Foodways*, The Latin American Studies Book Series,
https://doi.org/10.1007/978-3-030-51629-1_2

and Pozorski 1979; Quilter and Stocker 1983; Raymond 1981; Wilson 1981), modern approaches to paleonutrition through bioarchaeology have emphasized analysis of small group and individual diet as opposed to a collective dietary analysis of an entire region (Sutton et al. 2010: 3). Here, in an effort to document how both cultural and biological preferences affected food habits, this research details the food choices of a second millennium B.C. maritime community on Peru's North Coast in an attempt to understand how a small group of people solved the problems of food production, processing, and consumption not only for survival but as a reflection of cultural selection.

In recent years the study of food in archaeology has centered on the issue of understanding resource accessibility and status (e.g., Curet and Pestle 2010; Joyce and Henderson 2007), ethnicity and cultural preferences (e.g., La Violette 2008; Smith 2006), gender roles (e.g., Hastorf 2008; Isbell and Groleau 2010; Joyce 2010), and feasting (e.g., Norman 2010; Ben-Shlomo et al. 2009). Advances in the analysis of archaeological materials, such as the study of residues on lithic tools or ceramic fragments, the presence of phytoliths on human teeth, and the analysis of other residues using chromatography have advanced our understanding of ancient culinary behavior in various geographic settings (e.g., Seeman et al. 2008; Piperno and Dillehay 2008; Henderson et al. 2007; Olsson and Isaksson 2008; Lu et al. 2005). Indeed, archaeologists recently identified an organic mass found in a Chinese cemetery as the earliest known cheese by using mass spectrometry-based quantitative proteomic analyses (Yang et al. 2014). One conclusion from this fascinating study is that the technological improvement of making cheese provided economic benefits to the pastoral households of the Early Bronze Age eastern Eurasia population. This example shows the great potential that data and evidence on diet provides the interpretation of the past, and the necessity of starting long-term programs of specialized analyses on food remains recovered in excavations of pre-Columbian sites in the Central Andes.

This study explores the preparation and consumption of marine and inland valley products that were part of the daily diet at Gramalote, a second millennium B.C. maritime community on the Peruvian North Coast. The goal of this research is to elucidate food choices during the emergence and consolidation of early complex societies on the Central Andean coast using a bottom-up perspective (see also Lau [Chapter 2], Tantaleán, and Rodríguez [Chapter 3], Toohey [Chapter 4] for other perspectives on the roles of food and domesticates in the creation of identity, status, and rank).

This research suggests that food remains found at Gramalote are not only the result of subsistence practices but are also evidence of cultural choices deeply embedded in the ideological perspective of early fishermen (Prieto 2015). The ancient observations of natural coastal phenomena such as tides, ocean currents, winds, moon cycle, etc. along with the presence/absence of certain marine species during the annual cycle were crucial in providing the maximum harvest of particular marine resources. Since paleonutrition can benefit from ethnoarchaeological cases, which help to fill gaps archaeological materials could not provide (Sutton et al. 2010: 2), an important component of this discussion is the incorporation of ethnographic data on subsistence practices and beliefs from various modern maritime communities of the Peruvian North Coast, particularly from Huanchaco, a modern fishing town near the archaeological site of Gramalote. Ethnographic data, when understood with appropriate limitations, could help to build local models that empower archaeological data sets. In the North Coast of Peru, few local ethnographic studies have been discussed in the archaeology of coastal communities, although there is a great potential for future analysis (Prieto 2018a: 533). Ethnographic evidence presented in this research is focused upon integrating the evidence with what is recorded in the archaeological data, with the understanding that traditional-present-day cultural patterns of food preparation and consumption were to varying degrees modified by the Spanish conquest and the introduction of non-native plants and animals.

The Site of Gramalote

Gramalote covers a total of 3.5 hectares of a residential settlement on the northern end of the Moche valley, North Coast of Peru. It is located on top of a bluff overlooking the beach, about 250 m from the shoreline. It is surrounded by marshlands, fog vegetation areas, rocky, and sandy beaches (Prieto 2014, 2015). This setting epitomizes a community centered on a marine-oriented economy. Previous research at Gramalote demonstrated the site potential for revealing material goods pertaining to household studies, and the recovery of faunal and archaeobotanical remains amenable to detailed analyses (Pozorski 1976; Pozorski and Pozorski 1979; Velasquez 1987; Briceño and Billman 2008; McTavish 2013).

Since 2010, I have been excavating Gramalote for understanding its social dynamics and economic interactions. This research involved large-scale excavation units to expose domestic areas where the distribution and presence/absence of different types of artifacts and food remains could be evaluated as a window into the social dynamic of this maritime community (Prieto 2015). The excavations between 2010 and 2014 determined that the site was occupied in three occupational phases spanning 300 years, between 1500 and 1200 B.C. (see Table 2.1). Phase 2 represents the primary occupation, averaging about 300 or less individuals in two sectors: a domestic occupation located on the southwest sector and a ceremonial area on the northeast end of the settlement (Fig. 2.1).

The domestic sectors are organized around a central plaza which faces the ocean on the edge of a bluff. The central plaza is surrounded by a number of narrow corridors separating the domestic units from the open space or plaza. This large three hundred square meter plaza served as a processing area for a number of marine products. It also had a ritual function related to the use of red pigment and the carving of seabird bones (Prieto 2015; Prieto et al. 2016). Indeed, the central plaza of the domestic sector is associated with family-level ceremonies related to the preparation and consumption of marine food described below (Fig. 2.2a).

The ceremonial sector of Gramalote is defined by a 1000 square meter architectural compound that includes a large cooking facility with more than eight hearths and stone grills. Overall, the density of food remains in this sector was larger than

Table 2.1 Absolute dates and the three proposed occupational phases at Gramalote

Sample number	Material	Years (B.P)	1 Sigma calibration (ShCal 04)	2 Sigma calibration (ShCal 04)	Phases	Proposed dates
BETA-321936	*Tillandsia* sp.	3030 ± 30	Cal BC 1264–1129 (68.2%)	Cal BC 1315–1055 (94.2%) Cal BC 1369–1358 (1.2%)	Phase 3	B.C. 1300–1200
BETA-321937	*Tillandsia* sp	3140 ± 30	Cal BC 1362–1314 (34.7%) Cal BC 1411–1367 (33.5%)	Cal BC 1431–1264 (95%) Cal BC 1300–1250 (4.1%)		
BETA-321939	*Tillandsia* sp	3070 ± 30	Cal BC 1317–1212 (56.3%) Cal BC 1372–1344 (11.9%)	Cal BC 1390–1153 (91.5%) Cal BC 1146–1129 (3.9%)		
BETA-321938	*Tillandsia* sp.	3180 ± 30	Cal BC 1441–1378 (58.5%) Cal BC 1337–1321 (9.7%)	Cal BC 1464–1306 (91.3%) Cal BC 1494–1473 (4.1%)	Phase 2	B.C. 1400–1300
BETA-321940	*Tillandsia* sp.	3110 ± 30	Cal BC 1389–1292 (63.9%) Cal BC 1279–1271 (4.3%)	Cal BC 1416–1251 (90.1%) Cal BC 1243–1213 (5.3%)		
BETA-321941	*Tillandsia* sp.	3040 ± 30	Cal BC 1272–1189 (44.9%) Cal BC 1293–1278 (6%)	Cal BC 1321–1112 (88.1%) Cal BC 1377–1337 (5.3%)		
BETA-321942	*Tillandsia* sp	3140 ± 30	Cal BC 1362–1314 (34.7%) Cal BC 1411–1367 (33.5%)	Cal BC 1431–1264 (95.4%)		

(continued)

Table 2.1 (continued)

Sample number	Material	Years (B.P)	1 Sigma calibration (ShCal 04)	2 Sigma calibration (ShCal 04)	Phases	Proposed dates
BETA-321943	*Tillandsia* sp	3200 ± 30	Cal BC 1456–1390 (63.3%) Cal BC 1490–1481 (4.9%)	Cal BC 1500–1370 (87.7%) Cal BC 1346–1316 (7.7%)	Phase 1	B.C. 1500–1400
BETA-321945	*Tillandsia* sp	3130 ± 30	Cal BC 1404–1311 (68.2%)	Cal BC 1429–1260 (95.4%)		
BETA-321946	*Tillandsia* sp.	3170 ± 30	Cal BC 1433–1375 (54%) Cal BC 1339–1320 (14.2%)	Cal BC 1457–1294 (93.9%) Cal BC 1491–1479 (1.5%)		

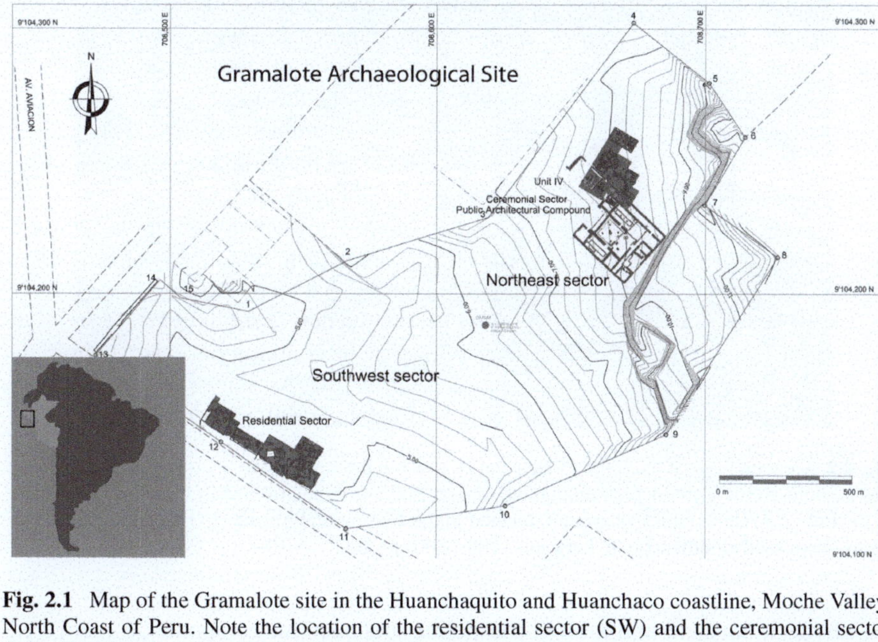

Fig. 2.1 Map of the Gramalote site in the Huanchaquito and Huanchaco coastline, Moche Valley, North Coast of Peru. Note the location of the residential sector (SW) and the ceremonial sector (NE). The areas shown in this map are the total extend of the excavations carried out between 2010 and 2014 field seasons. It should be understood that each sector must have been more extensive than what is shown on this map

Fig. 2.2 General view of the Public Plaza and House 1 in the Residential Sector of Gramalote (**2a**) and a South-North view of the rectangular sunken plaza with central hearth found in the middle of the Public Architectural Compound, located in the Ceremonial Sector of Gramalote (**2b**). Both places were used simultaneously between 1400-1300 cal. B.C

the domestic area. In and around the public architectural compound, I recovered thousands of samples of food remains, including products absent from the domestic sector, suggesting evidence of ritual feasting (Prieto 2018b) (Fig. 2.2b).

Food Consumed at Gramalote

Faunal and botanical remains are not necessarily the result of food consumption (Sobolik 1994); therefore, at times it is difficult to ascertain which remains are the result of anthropogenic activities related to food processing and consumption from remains generated by non-food activities. Asserting that the menu eaten by the ancient settlers was simply based on faunal or botanical remains is more problematic, but the variety of these products recovered allow us to infer possible dishes, as well as the variety of resources used for food consumption. Five main categories of food products are clearly identified: fish, marine mammals, seabirds, shellfish, and food plants. The most common sources of meat protein were sharks, possibly the blue shark (*Prionace glauca*) or the copper shark (*Carcharhinus brachyurus*), robalo (*Robaloscion wieneri*), corvina drum (*Cilus gilberti*), sea lion (*Otaria* sp.), pearly top shell (*Tegula atra*), mussel (*Choromytilus chorus*), dye shell (*Stramonita haemastoma*), and the purple crab (*Platyxanthus orbignyi*).

Fish

Marine resources constituted the primary consumables at Gramalote. More than twenty-one thousand fish remains comprised on thirty-two species were recovered. Of these, sharks, rays, drums, and croakers are the most common (Table 2.2). However, the most important marine resources at Gramalote consisted of possibly two species of sharks of the Carcharhinidae family (Prieto 2014, 2015). Blue sharks and copper sharks occur from pelagic to coastal waters, and they are adapted to cool temperatures. During the reproductive season in early summer, these sharks arrive usually to mate along the coastal waters. Therefore, they become vulnerable to be captured by experienced fishermen. This species is an important resource because it can yield more than fifteen to twenty kilograms of edible meat for each specimen captured.

Several other marine fishes were common. Small sharks (*Mustelus* sp.) and rays (*Myliobatis peruvianus*) were heavily exploited for consumption. Large drums such as *corvina* (*Cilus gilberti*) and *robalo* (*Robaloscion wieneri*) were also captured in large quantity (Table 2.2). All these species are adapted to coastal environments (within three miles from the shoreline) but because of their size, reed boats were needed to fish them. Temporal variation in fish consumption appears to be consistent with various occupations of the site, and one could argue that it was fairly consistent from the foundation to the abandonment of this settlement.

It is important, however, to remember that the sharks and many of the *Sciaenid* species, speckled smooth-hound (*Mustelus* sp.) and the Peruvian ray (*Myliobatis peruvianus*) are seasonal resources (according to local fishermen in Huanchaco) when limited fishing technology is available. There is a seasonal pattern that could have affected the distribution and presence or absence of certain fish species. It

Table 2.2 Total fish remains by NISP and MNI

Fish species recovered at Gramalote: family, scientific and common names in English and Spanish					
Family	Scientific name	Common name in English	Common name in Spanish	NISP	NMI
Carcharhinidae	*Prionace glauca*	Blue shark	Tiburon Azul/Tintorera / Luna	16,214	389
Carcharhinidae	*Carcharhinus brachyurus*	Copper shark	Cazon	38	3
Triakidae	*Mustelus* sp	Speckled Smooth-Hound	Tollo	648	106
Lamnidae	*Isurus oxyrinchus*	Shortfin mako	Tiburon Bonito/T. Diamante	105	31
Hexanchidae	*Notorhynchys cepedianus*	Sevengill shark	Tiburon de siete agallas/T. Gata	1	1
Myliobatidae	*Myliobatis peruvianus*	Peruvian ray	Raya	1901	155
Sciaenidae	*Robaloscion wieneri*	Stark Drum	Robalo	323	154
Sciaenidae	*Cilus gilberti*	Corvina	Corvina	392	114
Sciaenidae	*Paralonchurus peruanus*	Peruvian Banded Croaker	Coco/Suco	418	114
Sciaenidae	*Sciaena calaensis*	Callao Drum	Lorna Grande	121	49
Sciaenidae	*Cheilotrema fasciatum*	Arnillo Drum	Gallinaza	112	42
Sciaenidae	*Sciaena deliciosa*	Lorna Fish	Lorna	107	42
Sciaenidae	*Cynoscion analis*	Peruvian weakfish	Cachema	27	14
Sciaenidae	*Menticirrhus ophicephalus*	Snakehead kingcroaker	Misho/Mis-Mis	1	1
Ophidiidae	*Genypterus maculatus*	Cusk eel	Congrio	86	38
Cheilodactylidae	*Cheilodactylus variegatus*	Peruvian morwong	Pintadilla	25	15
Scombridae	*Sarda chiliensis chiliensis*	Eastern pacific bonito	Bonito	19	15
Engraulidae	*Engraulis ringens*	Anchovy	Anchoveta	5212	246
Engraulidae	*Anchoa nasus*	Longnose anchovy	Samasa	14	10
Cupleidae	*Sardinops sagax sagax*	Sardine	Sardina	231	40

(continued)

Table 2.2 (continued)

Fish species recovered at Gramalote: family, scientific and common names in English and Spanish					
Family	Scientific name	Common name in English	Common name in Spanish	NISP	NMI
Cupleidae	*Ethmidium maculatum*	Pacific menhaden	Machete	19	10
Atherinidae	*Odontesthes regia*	Silverfish	Pejerrey	81	26
Ariidae	*Galeichthys peruvianus*	Catfish	Bagre	19	17
Mugilidae	*Mugil cephalus*	Mullet	Lisa	2	1
Serranidae	Paralabrax humeralis	Peruvian rock seabass	Cabrilla	3	3
Serranidae	Hemilutjanus macrophthalmos	Grape-eye seabass	Ojo de Uva	1	1
Labrisomidae	Labrisomus phillippii	Chalapo clinid	Tramboyo/Cabezon	5	4
Haemulidae	Anisotremus scapularis	Peruvian grunt	Chita	7	4
Carangidae	Trachurus murphyi	Inca scad	Jurel	5	1
Centrolophidae	Seriolella violacea	Palm ruff	Cojinoba	1	1
Xiphioidei	Xiphias gladius	Sword fish	Pez Espada	1	1
Paralichthydae	Paralichthys adspersus	Flounder fish	Lenguado	20	3
N/I	N/I	N/I	N/I	18	12
Total				26,177	1663

Names taken from Chirichigno and Cornejo 2001. The second common name is usually what these species are called in Huanchaco

appears that Gramalote fishermen, like the present-day traditional fishing communities on the North Coast, were subject to a season of "*pesca grande*" (big game fishing) during the austral summer (November–April), and a season of "*pesca chica*" (small game fishing) during May through October. Species such as the sharks, the speckled smooth-hound, the Peruvian ray (*Myliobatis peruvianus*), anchovies (*Engraulis ringens*), as well as robalo (*Robaloscion wieneri*) and corvina drum (*Cilus gilberti*) were available during the summer season (Diaz-Alvitez 1999). On the other hand, during the winter season, when most of the previously listed fish decreased in availability, Gramalote fishermen turned their attention to species such as the small- to medium-sized drums: Peruvian banded croaker (*Paralonchurus peruanus*) or the lorna (*Sciaena deliciosa*).

Previously, scholars proposed that anchovies (*Engraulis ringens*) were crucial to the early diet of the coastal pre-Hispanic societies (Moseley 1975; Moseley and Feldman 1988; Sandweiss 2009). Certainly anchovies were important in other periods and sites of the Peruvian coast, but the five thousand plus anchovy remains found at Gramalote are literally a miniscule contribution of meat biomass when compared with the meat contribution of sharks and drums, particularly considering that almost all the deposits were screened using fine mesh size: 1/4 (6.35 mm); 1/8 (3.2 mm) and 1/16 (1.7 mm) inches (Prieto 2015: 574). And in many cases, anchovies were found whole, suggesting they were not consumed. It is possible remains of anchovies may be preserved in Late Preceramic human coprolites and well-preserved human intestines (Weir and Bonavia 1985). However, the quantity of anchovy remains at Gramalote, compared with the abundance of larger fish such as sharks and drums, suggests that, if consumed, anchovies were considered a minor complement to the diet. Today, several women from Huanchaco stated that anchovies are extremely oily and strong-tasting and usually consumed as a condiment (once ground to powder) or boiled with other fish as a flavoring. It is intriguing why so many complete anchovies appeared in the archaeological deposits at Gramalote. Interestingly a significant quantity of anchovy remains was identified as part of ritual caches in domestic contexts, suggesting their presence may be related to religious and ritual practices (Prieto 2015: 632).

Marine Mammals

Evidence of sea lions (*Otaria* sp.), dolphins (*Delphinus* spp.), bottle-nose dolphins (*Tursiups truncatus*), and porpoises (*Phocoena spinipinnis*) were identified at Gramalote (Table 2.3). After sharks and large drums, sea lions (*Otaria* sp.) had a major role in the diet. Each adult sea lion could weight over two hundred kilos, so there could be astonishing quantities of meat contributed by this species. The abundance of sea lion remains suggests they were hunted throughout the year (but most likely during the summer) and probably it was a staple food in Gramalote.

Shellfish

Only fifteen of the sixty identified species have economic or dietary importance. However, only eight of the fifteen shellfish species were widely consumed in the local diet (Table 2.4). Gathering shellfish was a systematic activity and it was focused on a narrow range of species relevant for local consumption and the economic activities at Gramalote.

The most abundant species in all the occupational phases is the pearly top shell (*Tegula atra*). Almost all the pearly top shell specimens are complete, suggesting that they were boiled and then consumed using a pointed tool of some sort to extract the meat. The turban mollusk (*Prisogaster niger*) is the second most popular gastropod at the site. Today, local fisherman use turban mollusks for food or as bait. For many

Table 2.3 Marine mammals organized by family. Total numbers by units in NISP and MNI values

		Unit I		Unit II		Unit IV		Total	
Family	Possible species	NISP	MNI	NISP	MNI	NISP	MNI	NISP	MNI
Otariidae	Sea lion; Fur seal	284	70	788	118	345	108	1417	296
Delphinidae	Dolphins	73	14	15	13	24	18	112	45
Phocoenidae	Porpoises	2	1	7	5	9	2	18	8
Physeteridae	Whales	103	8	2	2	48	8	153	18
Camelidae	Llamas, Alpacas, Vicuñas, Guanacos	1	1	0	0	4	3	5	4
Canidae	Dogs	4	1	1	1	0	0	5	2
Rodentia (order)	Mices, rats	2	1	88	11	1	1	91	13
							Total	1801	386

Table 2.4 Shellfish remains recovered at Gramalote by units and phases

	Unit I			Unit II			Unit IV			Total			Total
Bivalves	Phase 1	Phase 2	Phase 3	Phase 1	Phase 2	Phase 3	Phase 1	Phase 2	Phase 3	Phase 1	Phase 2	Phase 3	
Semele sp.	251	1874	298	110	3442	2352	37	2118	1705	398	7434	4355	12,187
Protothaca thaca	139	1041	400	1284	3147	1891	54	2048	1824	1477	6236	4115	11,828
Eurhomalea rufa	28	112	33	183	339	180	4	282	180	215	733	393	1341
Tellina sp.	15	84	24	93	192	144	7	84	59	115	360	227	702
Choromytilus chorus	33	2732	640	352	3708	1501	35	3128	2933	420	9568	5074	15,062
Donax obesulus	27	603	121	180	1110	1940	43	1861	1817	250	3574	3878	7702
Spisula adamasi	2	13		12	28	28	1	156	28	15	197	56	268
Tagelus dombeii	2	10		14	3		1	3	1	17	16	1	34
Mulinia coloradensis				16	4	2		1	2	16	5	4	25
Mulinia sp.		5	6	14	5	1		2	1	14	12	8	34
Semimytilus algosus	3	27	21	48	131	27		377	123	51	535	171	757
Perumitylus purpuratus		4		1		4		1	5	1	5	9	15
Mesodesma donacium				7	3	1		1		7	4	1	12
Argopecten purpuratus					5	5		3	3	0	8	8	16
Argopecten circularis						12				0	0	12	12
Trachycardium muricatum/mocerum				1				1		1	1	0	2
Pholas chiloensis				9	46	12		15	7	9	61	19	89

(continued)

Table 2.4 (continued)

	Unit I			Unit II			Unit IV			Total			Total
Bivalves	Phase 1	Phase 2	Phase 3	Phase 1	Phase 2	Phase 3	Phase 1	Phase 2	Phase 3	Phase 1	Phase 2	Phase 3	
Litophaga peruviana					1	1		73	4	0	74	5	79
Gastropods													
Tegula atra	772	3294	168	7445	12,048	9113	216	11,484	8292	8433	26,826	17,573	52,832
Tegula picta	2	7		41	114	115	8	217	198	51	338	313	702
Tegula tridentata	12			30	119	45		41	3	42	160	48	250
Tegula luctuosa		34				2				0	34	2	36
Tegula euryomphalu						2				0	0	2	2
Prisogaster niger	268	1038	50	2285	6751	6435	132	9666	10,741	2685	17,455	17,226	37,366
Stramonita haemastoma	285	1273	110	1061	3347	4005	155	9303	8205	1501	13,923	12,320	27,744
Stramonita chocolata	42	321	30	288	1143	1410	83	4244	4463	413	5708	5903	12,024
Mitra orientalis	12	47	4	71	158	247	20	481	379	103	686	630	1419
Polinices uber	5	37	7	288	283	249	19	480	610	312	800	866	1978
Xantochorus buxea	55	262	15	291	719	1403	76	1362	1925	422	2343	3343	6108
Xantochorus cassidiforme	25			95	305	228		214	107	120	519	335	974
Xantochorus sp.				9						9	0	0	9
Nassarius dentifer	25	104	17	308	563	810	42	596	716	375	1263	1543	3181
Nassarius gayi		13								0	13	0	13
Fissurella maxima	5	40	5	40	93	58	2	81	64	47	214	127	388

(continued)

Table 2.4 (continued)

	Unit I			Unit II			Unit IV			Total			Total
Bivalves	Phase 1	Phase 2	Phase 3	Phase 1	Phase 2	Phase 3	Phase 1	Phase 2	Phase 3	Phase 1	Phase 2	Phase 3	
Fissurella latimarginata	14	47	1	50	155	220		123	114	64	325	335	724
Fissurella peruviana	7	46	4	152	213	144	7	112	125	166	371	273	810
Fissurella limbata		7	2	4	6	19		10	7	4	23	28	55
Fissurella pulchra		18	1		5			1		0	24	1	25
Fissurella crassa		3	1	2	8	10	2	12	8	4	23	19	46
Fissurella cumingi							4	1		4	1	0	5
Crepipatella dilatata	16	66	3	121	200	180		193	157	137	459	340	936
Ceratostoma		1								0	1	0	1
Concholepas concholepas	1	7			14	16	2	23	18	3	44	34	81
Crassilabrum crassilabrum	2		1	4	7	18		12	10	6	19	29	54
Sinum cymba		4		10	11	10	1	39	36	11	54	46	111
Oliva peruviana		6			8	30		4	14	0	18	44	62
Oliva incroasata		5								0	5	0	5
Oliva columellaris						3				0	0	3	3
Prunum curtum				1					1	1	0	1	2
Familia Muricidae					1					0	1	0	1
Cancellaria sp.					4	1			2	0	4	3	7

(continued)

Table 2.4 (continued)

	Unit I			Unit II			Unit IV			Total			Total
Bivalves	Phase 1	Phase 2	Phase 3	Phase 1	Phase 2	Phase 3	Phase 1	Phase 2	Phase 3	Phase 1	Phase 2	Phase 3	
Crepidula onyx						9				0	0	9	9
Scutalus sp.		3		6	10	3	1	34	87	7	47	90	144
Familia Ascidiaceae		5	60	6	39	179		25		6	69	239	314
Crustaceans													
Platyxanthus orbignyi	110	1362	13	774	3054	4611	116	1641	2187	1000	6057	6811	13,868
Cancer sestosus		89		67	140	181	2	59	146	69	288	327	684
Emerita analoga									2	0	0	2	2
Balanaidae	1	117		2	157	428		584	274	3	858	702	1563
Chiton sp.				3	40	15		36	5	3	76	20	99
Echinoidea		53		5	42	1	6	76		11	171	1	183
Total	2159	14,814	2035	15,783	41,921	38,301	1076	51,308	47,588	19,018	108,043	87,924	214,985
Total by unit	19,008			96,005			99,972						

years it has been assumed that the presence of turban mollusks in archaeological deposits was a result of accidental gathering, but more than thirty thousand of such mollusk species in archaeological deposits could not be a result of "accidental gathering." Therefore, as suggested by evidence from locals at Huanchaco, this mollusk species was probably consumed as a condiment. Evidence of boiling suggests turban mollusks were cooked with other shellfish to flavor the dish but was not the main ingredient. This method of preparation of turban mollusks may explain why almost all of the shells of this species still have their operculum in place, suggesting the meat was never removed from the shell.

Species of bivalves are also common in the middens. The most popular bivalve was the large mussels (*Choromytilus chorus*). This species yields a great volume of meat and is no longer endemic to the Huanchaco coastline. However, it prevails along the south coast of Peru. The archaeological evidence suggests that this large mussel may have thrived around Gramalote or to the south where the sea conditions are more suitable. Razor clam species were well represented in the collections: the white clam and a smaller yellow clam (*Semele* spp.; *Protothaca thaca*).

Crab species deserve special mention. The most popular species recovered in the excavations was the purple crab (*Platyxanthus orbignyi*). The pincers are particularly diagnostic of the species more so than carapace fragments. The purple crab is abundant today along the Huanchaco coast and captured using traps or by hand during low tides. During the winter season harvesting purple crab is a major economic activity in Huanchaco, and one single reed boat can yield one hundred kilos of purple crabs per journey.

Seabirds

A large variety of seabirds also contributed to the diet. There were twenty-one species present in the middens (Table 2.5). Among these species, the guanay cormorant (*Phalacrocorax bougainvillii*) was by far the species most largely consumed at Gramalote, but Peruvian boobies (*Sula variegata*), Peruvian pelicans (*Pelecanus thagus*), sea gulls (*Larus* sp.) and the Humboldt penguin (*Spheniscus humboldtii*) were also consistently eaten. However, seabirds were not as common as shark, drums, sea lion, and shellfish.

Considering the quantity of seabird bones and their distribution by sector and occupational phase, it is clear that seabirds were a resource that was consumed only occasionally. Although there were a number of different taxa, seabirds were a very modest dietary source compared with the amount of fish or marine mammal meat.

Table 2.5 Seabird species recovered at Gramalote in the residential sector and in the ceremonial facility (Public Architectural Compound)

Family	Genera	Species	Common name	Phase I		Phase II		Phase III		Total	
				NISP	MNI	NISP	MNI	NISP	MNI	NISP	MNI
Phalacrocoracidae	*Phalacrocorax*	*Phalacrocorax bougainvilli*	Guanay Cormorant	459	79	1077	237	560	139	2090	455
Phalacrocoracidae	*Phalacrocorax*	*Phalacrocorax olivaseus*	Cushuri Cormorant	8	3	18	16	15	11	41	30
Phalacrocoracidae	*Phalacrocorax*	*Phalacrocorax gaimardi*	Chuita Cormorant	5	3	25	13	7	6	37	22
Phalacrocoracidae	*Phalacrocorax*			343	62	427	176	395	134	1165	372
Sulidae	*Sula*	*Sula variegata*	Peruvian Booby	131	32	213	91	116	69	460	192
Sulidae	*Sula*	*Sula nebouxii*	Blue Footed Booby	1	1	3	3			4	4
Sulidae	*Sula*			91	23	122	60	103	50	316	133
Pelecanidae	*Pelecanus*	*Pelecanus thagus*	Peruvian Pelican	160	38	319	25	88	25	149	56
Pelecanidae	*Pelecanus*	*Pelecanus occidentalis*	Brown Pelican			1	1	1	1	2	2
Laridae	*Larus*	*Larus pipixcan*	Franklin's Gull	6	6	55	25	88	25	149	56
Laridae	*Larus*	*Larus dominicanus*	Dominican Gull	1	1	3	2	1	1	5	4
Laridae	*Larus*	*Larus atricilla*	Laughting Gull			1	1	7	5	8	6
Laridae	*Larus*			75	18	135	42	56	28	266	88

(continued)

Table 2.5 (continued)

Family	Genera	Species	Common name	Phase I		Phase II		Phase III		Total	
				NISP	MNI	NISP	MNI	NISP	MNI	NISP	MNI
Procelaridae	*Puffinus*	*Puffinus grisseus*	*Sooty Shearwater*	12	7	50	34	31	17	93	58
Procelaridae	*Puffinus*			3	1	11	6	4	2	18	9
Spheniscidae	*Spheniscus*	*Spheniscus humboldti*	Humboldt Penguin	74	17	110	74	81	40	265	131
Procelaridae	*Pelecanoides*	*Pelecanoides garnotii*	Peruvian Diving-Petrel	39	11	49	34	53	33	141	78
Sternidae	*Thalasseus*	*Thalasseus maximus*	Royal Tern	2	1			2	2	4	3
Sternidae	*n/d*	*n/d*	Gaviotin					1	1	1	1
Sternidae	*Larosterna*	*Larosterna inca*	Inca Tern	2	1	1	1	7	6	10	8
Vulturidae	*Vultur*	*Vultur gryphus*	Peruvian Condor			3	1			3	1
Scolopacidae			Playero	15	5	14	10	8	6	37	21
Diomedeidae			Albatross	2	2	1	1	1	1	4	4
Accipitridae			Hawk			1	2			1	2
Falconidae			Eagle			1	1			1	1
Sternidae				5	2					5	2
Procelaridae				3	1	6	4	2	2	11	7
N/I				2	2					2	2

Plants

The abundance of macrobotanical remains is remarkable. Thirty-two plant species were uncovered, but only sixteen were identified as food crops (Table 2.6). Similarly, microbotanical remains (pollen, phytoliths, and starch grains) yielded food plant species not represent in the macrobotanical samples. Most food plant resources were exploited from the river valley and in the surrounding marshlands. Evidence from the marshlands suggests ancient occupants were involved in some sort of gardening activities, cultivating squash and beans (Prieto 2020). Economic food plants from the river valley are principally fruits, beans, tubers, and maize. The peanut (*Arachis hypogaea* L.) a cultivated legume, is well represented, as are sweet potatoes (*Ipomoea batatas* L.), and manioc (*Manihot esculenta*). Manioc appears to represent an important crop in the Gramalote diet. It was present throughout the samples and in all occupational levels (Prieto 2020).

A few plants that are traditionally condiments in modern Andean cuisine are present such as chili peppers or "*ají*" (*Capsicum* spp.), "*achiote*" (*Bixa orellana* L.), and "*tomatillo*" (*Solanum lycopersicon* L.). Macrobotanical remains from maize are scarce, with only seven samples recovered consisting mainly of inflorescences and a few maize cobs, but no kernels. This is in stark contrast with the microbotanical data of pollen and starch remains from soil samples and fragments of ceramic vessels in which maize is well represented. The quantity of maize starch grains in vessels fragment suggests that although it was present in the site, maize was possibly consumed in low quantities, or primarily as *chicha* or maize beer, since maize starch is mostly found in ceramic bottles and a few ceramic pots.

Finally, it appears that an algae species was important to the Gramalote diet as it was for other contemporaneous settlements during the early Initial Period and Early Horizon throughout the Andean region (Burger 1985: 276). Seaweed locally known as *mococho* (*Gracilaria chamissoi*) was intensively eaten. However, its consumption does not produce botanical residue that could be tracked in an archaeological context. The excellent preservation of organic remains at Gramalote, enabled the recovery of a significant amount of dried seaweed, which is common during the winter season along the coasts of Huanchaco. Once harvested from the rocky bottoms by shell-gatherers, it is usually eaten fresh or dried and later on consumed in stews with tubers. Similar uses of seaweed may have taken place in the ancient past.

In the following sections, I will detail the similarities and differences on food preparation and consumption at the Gramalote residential units and in the ceremonial facility of the site.

Table 2.6 Plants species at Gramalote by habitat

Family	Scientific name	Common name	NISP	Habitat	Condition	Uses
Solanaceae	*Capsicum* sp.	ají	4	Valley	Planted	Food
Convolvulaceae	*Ipomoea batatas*	camote	1	Valley	Planted	Food
Fabaceae	*Pochyrhizus ahipa*	ajipa	1	Valley	Planted	Food
Sapotaceae	*Pouteria lucuma*	lúcuma	368	Valley	mostly wild/but also planted	Food
Lauraceae	*Persea americana*	avocado	306	Valley	mostly wild/but also planted	Food
Fabaceae	*Arachis hypogaea*	maní	263	Valley	Planted	Food
Malpighiaceae	*Bunchosia armeniaca*	cansaboca	242	Valley	mostly wild/but also planted	Food
Fabaceae	*Inga feuillei*	pacae	229	Valley	mostly wild/but also planted	Food
Myrtaceae	*Psidium guajava*	guayaba	3	Valley	mostly wild/but also planted	Food
Bixaceae	*Bixa orellana*	achiote	2	Valley	planted	food/industrial
Fabaceae	*Acacia macracantha*	huarango	1	Valley	mostly wild/but also planted	industrial
Cucurbitaceae	*Cucurbita* sp.	zapallo	25	Marsh	Planted	Food
Solanaceae	*Lycopersicon* sp.	tomatillo	2	Marsh	wild	Food
Cucurbitaceae	*Lagenaria siceraria*	mate	2354	Marsh	Planted	industrial
Cyperaceae	*Scirpus californicus*	junco	101	Marsh	wild/planted	industrial
Cyperaceae	*Schoenoplectus* sp.	totora	71	Marsh	planted/wild	industrial
Typhaceae	*Typha* sp.	enea	42	Marsh	wild/planted	industrial
		grama	6	marsh	wild	industrial
Fabaceae	*Phaseolus vulgaris*	frijo	37	marsh/valley	Planted	Food
Cucurbitaceae	*Cucurbita moschata*	loche	12	marsh/valley	Planted	Food
Fabaceae	*Phaseolus lunatus*	pallar	10	marsh/valley	Planted	Food

(continued)

Table 2.6 (continued)

Family	Scientific name	Common name	NISP	Habitat	Condition	Uses
Poaceae	*Zea mays*	maíz	7	marsh/valley	Planted	Food
Malvaceae	*Gossypium barbadense*	algodón	303	marsh/valley	wild/planted	industrial
Poaceae	*Gynerium sagitatum*	caña brava	345	river banks	wild	industrial
Poaceae	*Phragmites australis*	carrizo	59	river banks	wild	industrial
Equisetaceae	*Equisetum* sp.	cola de caballo	3	river banks	wild	medicinal
Cactaceae	n/i	cactus n/i	5	lomas (fog vegetation)/desert	wild	medicinal?
Cactaceae	*Espostoa melanostele*	viejito	1	lomas (fog vegetation)/desert	wild	medicinal?
Cactaceae	*Melocantus peruvianus*	cactus	1	lomas (fog vegetation)/desert	wild	medicinal?
Bromeliaceae	*Tillandsia* sp.	achupalla	3959	lomas (fog vegetation)	wild	industrial
Amaryllidaceae	*Furcraea* sp.	cabuya	9	lomas (fog vegetation)	wild	industrial
Florideophyceae	Gigartinaceae	mococho	332	rocky shores	wild	Food

Preparing and Consuming Food in the Domestic Sector at Gramalote

The domestic residences are located around an open space or plaza, a pattern suggesting that family groups lived very close spatially (Prieto 2018a). Most domestic structures have a roofed area of seventy to 80 square meters, usually with three rooms: a resting area, a food consumption area, and a small storage facility. Food preparation was always outside the roofed area, located in the patio or open space surrounding the house (Fig. 2.3). Interestingly, boundary walls do not divide the houses. Instead, the patios and food preparation areas seem to have been the limits between domestic units, suggesting either that families shared food preparation areas and patios for common domestic activities or that each family had their own food preparation areas intentionally next to the other for social interaction. If this was the case, then food preparation was a major social activity among the settlers in fishing settlement (Prieto 2018a).

Economic and social roles were clearly determined by gender. For example, the distribution of adults buried at the site indicated a pattern: adult females and children are buried exclusively in the domestic sector, while adult males were primarily buried in the ceremonial part of the site. In terms of physical activity, muscular attachments on the bones on some male individuals suggest that they were engaged in fishing practices, and/or paddling reed boats. The characteristic external auditory meatus of multiple adult males also implies that diving was an important activity. In contrast, female adults showed activity marks on their skeletons related to long load-bearing walks, or performing manual activities: spinning or twisting fibers, weaving fabrics,

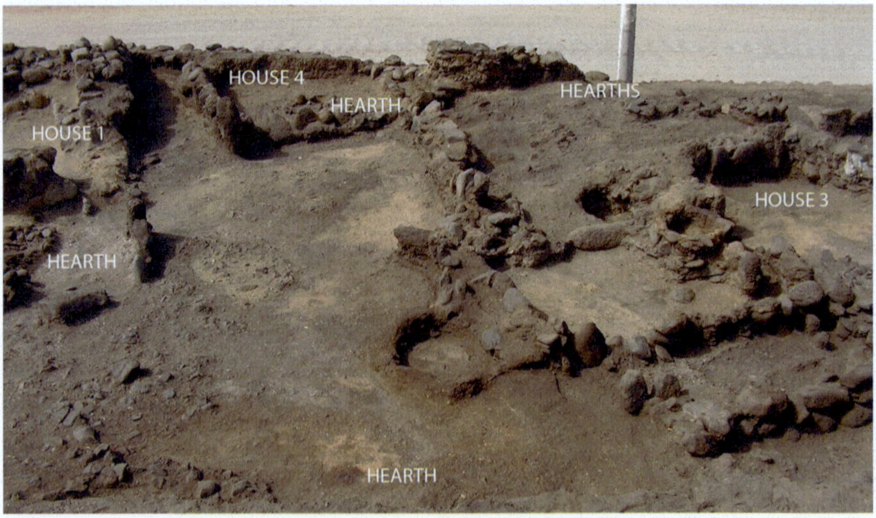

Fig. 2.3 Spatial distribution of hearths located in the patios of each residential unit

making mats and basketry. This is consistent with the evidence of male-fishing implements, female-weaving tools, and artifacts associated with food preparation. These areas or spaces were, however, not gender specific in the structured social realm of the Gramalote occupants. Remarkably, we uncovered fishing nets and net weights in the process of being manufactured along with weaving implements in the food preparation areas suggesting that these spaces could have served as meeting point for the family members. In short, food preparation and consumption areas were meeting places where families shared their lives and their day-to-day experiences. The archaeological evidence suggests food preparation and consumption activities served as "glue" and the food preparation areas represented a physical space where children, females and males left behind their gender roles to become a social unit.

The ethnographic evidence is consistent with the archaeological model; modern fishermen frequently congregate in their yards or open spaces between their houses with neighbors (relatives, other fishermen) and tell stories while singing, eating, and drinking in a festive environment. These feasting events are also perceived as informal meetings to share food, in the intimacy of a domestic space, where traditional families in the Moche valley traditionally congregate (Gillin 1947; Anhuaman 2008). These social gatherings called "*causeo*" from a dish called "*causa*," in Huanchaco consists of dried fish with manioc and sweet potato as side dishes. It is served in a single large and open gourd bowl called a "*lapa*." Community members usually sit around a gourd bowl with food and talk, sing, etc., together. During these informal festivities, large quantities of maize beer are also generally consumed. Among contemporary Huanchaco families, such informal social gatherings reinforce their community ties and help to establish new alliances and friendships among families. Presumably, similar processes of building social cohesion took place in pre-Columbian times, as suggested by the spatial layout at Gramalote.

Food preparation areas are indicated by two different cooking structures. One is a circular hearth delimited by stones, filled with ash, chunks of charcoal, and food remains. The average hearth sizes are forty centimeters in diameter. Some had two flat stones in the center placed in a "V" shape, which may have served to hold the rounded bottom of ceramic pots. This evidence suggests food processed in this hearth was cooked for long periods and the heaths and stones were designed to avoid direct contact of the food with the fire. Slow-cooking techniques are common for certain marine resources (Fig. 2.4a).

Another cooking structure is similarly circular in shape. Its center is filled with numerous small river cobbles covered in soot (Fig. 2.4b). The diameter of such cooking areas is slightly larger than hearths (50 cm). An absence of ash and charcoal contrasts with the presence of food remains, suggesting that they were stone grills. Possibly, the cobbles were first heated and then covered with other stones to increase the temperature. Once the cobbles reached a high temperature, seafood was grilled on its hot surface. Experimental archaeology recreating the stone grills showed mollusks (especially clams and mussels) are cooked instantaneously. Fish can be prepared in minutes as well, with cooking time varying depending on thickness of the portion.

Excavation at the domestic sector produced a large number of ceramic vessels, particularly neckless ollas that are big enough to hold large-sized portions of fish,

Fig. 2.4 Cooking structure to hold ceramic vessels, possibly designed for a low-cooking process (**4a**). Circular stone grill for processing fish and sea mammal meat. Alternatively, this structure may have been used for roasting tubers (**4b**)

sea lion, and seabird meat. Three ranges of mouth diameters for the neckless ollas show that local families had access to different vessel sizes, and therefore they may have alternated the uses of these ceramic vessels to prepare different volumes of food (Fig. 2.5). The analysis of residues on the ceramic fragments shows the presence of starch and phytoliths of different food plant species, suggesting plants were processed together with seafood in pots. Starch grains and phytoliths of cucurbits, beans, maize, and chili peppers were more common in domestic areas of the site than in the ceremonial/public sector.

Once prepared food was ready to be eaten, inhabitants of Gramalote used gourd plates or bowls, locally known as "*mates*" (medium size) and "*lapas*" (large size) instead of ceramic containers to serve their grilled, boiled or stewed food. Ceramic bowls and plates were absent in the domestic sector but some of them were found in and around the ceremonial public architectural compound located in the northeast sector of the site. A few ceramic bottle fragments were tested for microbotanical residues. Results showed most ceramic bottle fragments were used to hold maize juice or maize beer (*chicha*) (Huaman 2012). The consumption of fermented intoxicants such as *chicha* is common throughout the the ancient Andes and even today. There is extensive archaeological evidence suggesting that *chicha* or maize beer was common

Fig. 2.5 Two fragmented ceramic vessels found along the southern wall of House 3, residential sector of Gramalote. They were possibly used to cook marine products and then left behind in the southern patio of this residential unit

to rituals, rites, and often left as offerings at *huacas* or sacred places throughout the Andes.

The massive presence of grinding stones or *manos* of different sizes, along with flat stones or *batanes* indicates that these devices were used repetitively to grind food plants such as cucurbits, beans, and tubers in the domestic sector (Fig. 2.6a). In addition, wooden tools such as stirring spoons were also common in food preparation areas (Prieto 2017). These wooden artifacts may have been used for cooking stews, soups, and drinks as well as for beating and processing other food products like dried/salted meat before it was cooked or consumed (Fig. 2.6b).

From the various types of materials and spatial evidence we can surmise the pattern of daily food practices. A daily meal in the domestic sector was based on fresh fish (mainly shark) either grilled or stewed. It is possible that fish and sea lion meat were flavored with shellfish when boiled for stews. Similarly, fish and sea lion

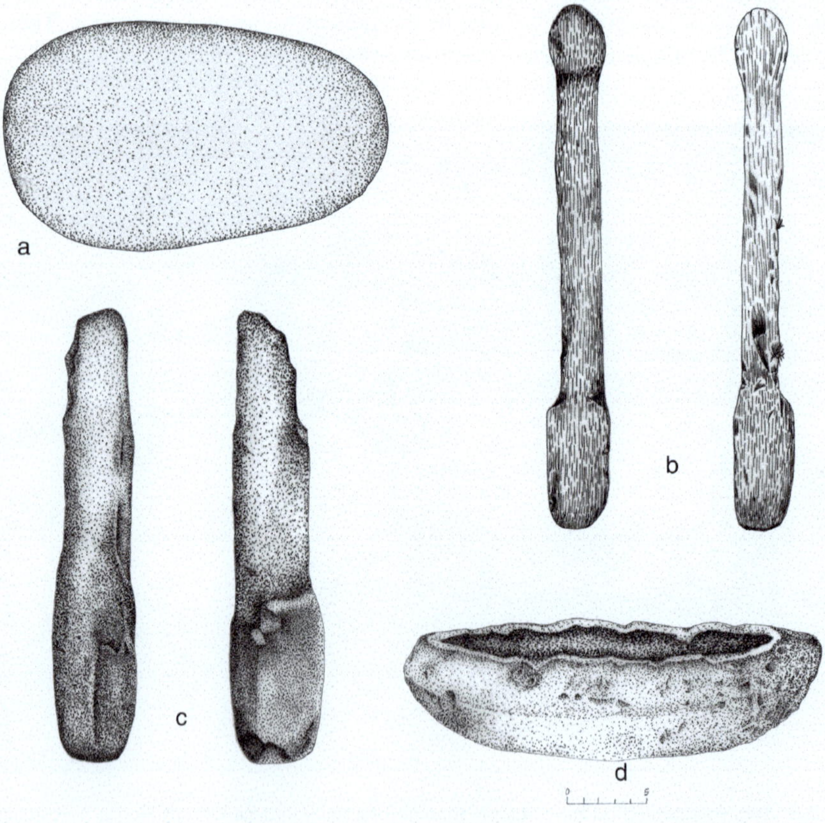

Fig. 2.6 Cooking implements used by the Gramalote inhabitants: griding stone (**a**); wooden stirring spoon (**b**); stone pestle (**c**) and bowl carved out of a block of colony of polychaetes. The latter was possibly used to hit tubers and seafood

meat were also flavored with sea salt (abundant in the area) and chili peppers as local fishermen do today. Smashed or roasted tubers such as manioc or sweet potatoes were consumed as side dishes.

Interestingly, sweet potato starch grains are present in many ceramic vessel fragments from the earliest to the last occupational phase of the domestic sector. Sweet potato starch grains are almost exclusively present on ceramic vessels indicating they were used almost exclusively for cooking. Today, the local fishermen say when they travel to the fishing grounds north of Huanchaco they collect certain types of cobbles to roast sweet potatoes with fish.

Manioc remains are concentrated on lithic artifacts. Indeed, most manioc starch grains were identified on elongated stone pestles in the domestic units (Fig. 2.6c, d). A few pot fragments had manioc starch grains. Interestingly, evidence of manioc represents the lowest presence of starch grain when compared with other species identified such as sweet potatoes, cucurbits, maize, and beans.

The high frequency of manioc residue on lithic artifacts and its low presence on ceramic vessels suggests that manioc was probably first roasted on the circular stone grills. Then it was peeled and then ground into a pulp using the stone pestles (see Fig. 2.6c). Manioc can also be consumed in portions or as a mash accompanying dried/salted fish, a practice that modern fishermen still do with sweet potato. Traditional Huanchaco fishermen call this dish "*causa.*" The dish is served in the bottom of a large gourd container with portions of sweet potato or manioc and topped with many pieces of dried/salted fish.

Mococho (*Gracilaria chamissoi*) a black-algae was probably served as a side dish with fish or sea lion meat. Traditional Huanchaco dishes mixed with *mococho* algae include clam stews or stews made with snails or gastropods. The presence of *mococho* algae in domestic contexts suggests it was important in the daily diet. A relative abundance of squash and beans in the archaeological deposits suggests these plant species were regularly consumed on a daily basis as well. They were possibly boiled in ceramic vessels. Large mussels and clams were grilled and largely consumed between the end of the austral winter and the beginning of the summer (September–November) when these species were most abundant along the coast. Fruits such as avocados (*Persea americana*) and *lúcuma* (*Pouteria lucuma*) were commonly eaten as complementary foods with such "marine and plant" resources.

Processing and Consuming Food in the plaza of the residential sector and around the Ceremonial Facility of Gramalote

Ethnohistoric documents highlighted the "festive" spirit of pre-Hispanic fishermen in the Andean region (Ramirez 1995; Marcus 1987). Similarly, ethnographic accounts demonstrate that feasts were generally organized by maritime communities after successful fishing expeditions (Gillin 1947; Sabella 1974; Anhuaman 2008). Huanchaco fishermen boat's crew state that after they land on the beach, they split the catch into equal parts for its various members (Gillin 1947). The largest fish were saved for consumption by the participants who shared their meat with the community. Fishermen believe large fish, for example, a large "robalo" (*Robaloscion wieneri*), should be reserved because it was considered an "old fish" and therefore, it constitutes a special treat. Today, when the entire community eats these "unusual" catches, a small celebration occurs on the beach. Such events are known as "*minga*" in Huanchaco. Such social congregations are sponsored by common members of the community rather than a high status or ranked individual of the same group. Similar community-directed or family-organized food sharing probably occurred in the Gramalote times.

Excavations in the plaza area and in the ceremonial facility (public architectural compound) uncovered numerous fish bones with hyperostosis, which are commonly known as "Tilly bones" (or "*hinchados*" in Spanish). Hyperostosis is described as a productive change in bone tissue characterized by an increase of the periosteal ossification combined with reabsorption of the bony tissue. Bony elements that are indicative of hyperostosis in teleost fishes are skulls, claviculae, pterygiophores, and haemal and neural spines (Jawad 2013: 1145). Hyperostosis develops mainly in large, mature individuals in most species (Smith-Vaniz et al. 1995: 579). This suggests large and old fish were primarily consumed on special occasions in the public or ceremonial spaces. This practice no doubt related to community-level events or celebratory feasting after successful fishing expeditions. If reserving large, older fish for sharing was a common practice, then it can be argued that community-level, food consumption was not only an economic act, but also a bond that held together the entire community through small and casual feasting and ritual celebrations.

A similar situation could be argued for the distribution pattern of fish otoliths. Most otoliths were found in the public plaza and around the ceremonial architecture. This suggests that fish-head soup (locally known as "*chilcano*") was commonly consumed on special occasions as part of community-level activities. Alternatively, high concentrations of fish otoliths in public sectors may be the result of fish processing.

The refuse from seabirds and shellfish also differ between houses and the public space. Bones from seabirds are mainly concentrated in the public plaza of the domestic sector suggesting that they were preferentially consumed in a public domain rather than in houses. However, the largest sample of shellfish remains was concentrated in and around the ceremonial sector.

The turban shell (*Prisogaster niger*), one of the most popular shellfish species, is believed to have been a condiment for food consumed at Gramalote. It is possible this small gastropod was consumed as a condiment to flavor the meat of shark, fish, and sea lion. Another gastropod, *Stramonita chocolata* commonly found in the ceremonial parts of the site was rarely encountered in the residential areas. Such gastropods were adapted to deeper waters along the beach rather than along the shoreline, which made collecting them difficult. The high percentage of this species in and around the public ceremonial architecture suggests it was a valuable food, commonly consumed in the context of ritual activities. Similarly, the unusual abundance of surf clams (e.g., *Donax peruvianus*, *Spisula adamsi* and *Lithophaga truncata*) and small gastropods (e.g. *Nassarius* spp., *Polinices uberinus, Mitra orientalis*) in the public domains, but uncommon in the diet of the residential sector, may indicate that the inhabitants consumed these taxa as part of ritualized consumption in and around the ceremonial architecture (Prieto 2018b).

Chili peppers (*Capsicum* sp.) were virtually absent in and around the ceremonial architecture. However, *ají* is a major condiment in pre-Hispanic cuisine by traditional fishermen on the north coast of Peru. Ethnohistoric accounts from the sixteenth century relate that during certain celebrations adult members of fishing communities were asked to fast, avoiding principally salt and chili peppers (Rostworowski 2004). The evidence suggests that *aji* was not used with foods consumed around the ceremonial facility. Conversely, fruits of the *Physalis* genus (probably "*aguaymanto*"), molle (*Schinus molle*), the elderberry locally known as "*sauco*" (*Sambucus* spp.) as well as the condiment "achiote" (*Bixa orellana* L.) and the highlands tuber "ajipa" (*Pachyrhizus ahipa*) were all exclusively found in and around the ceremonial architecture and not in the domestic sector of the site. Similarly, evidence of camelids are virtually absent with only five bones recovered during our excavations. Four were found in the ceremonial architecture, indicating that if camelids were consumed or their bones used as amulets or body adornments, camelid use was restricted to the sacred sector of Gramalote. The rarity of non-local fauna suggests that ancient occupants made a special effort to obtain products not necessarily "exotic," but unusual to the daily diet in order to celebrate their community-level ceremonies.

The ceramic assemblage in ceremonial contexts was similar to that of the domestic sector, except there were slightly more ceramic bowls in and around the ceremonial architecture. This contrasted with the dominant gourd bowls found in the domestic sector. A high concentration in and around the ceremonial sector of the site of stone pestles and stone bowls that were possibly used to process food in the context of community-level celebrations also distinguished the ceremonial areas (Fig. 2.7).

Discussion and Conclusions

The evidence presented in this chapter indicates an abundance of resources consumed in this early fishing village on the Peruvian north coast during the second millennium B.C. These data indicate that the paradigm of "staple food," such as the anchovy or maize generally theorized and discussed as critical food staples during the early stages of Andean prehistory does not apply to Gramalote. More importantly, by contextualizing the distribution of food remains we uncovered important differences in what was consumed in daily domestic activities compared to feasting, rituals, and public events.

These data emphasize the importance of contextual provenience in the identification of faunal and botanical remains obtained from early Andean maritime fishing communities. This bottom-up approach to different levels of economic and social interaction between early maritime communities and their agricultural counterparts in the mid-valleys reveals the complexity and variability of how and what was consumed

Fig. 2.7 Fancy stone pestles and fragments of stone bowls found in the Public Architectural Compound, Residential Sector, Gramalote. These implements were used to grind legumes, squash, fruits among other cultigens

among ancient Andean cultures (Burger 1985,1992; Quilter and Stocker 1983). As a result, we are able to articulate local economies of the second millennium B.C. and from there elaborate on a broader perspective regarding the social dynamics and economic interactions of early complex societies in the Andean region.

Our results at Gramalote suggest that instead of dried/salted anchovies, large quantities of dried/salted shark and sea lion meat were consumed on a daily basis, and possibly exchanged with inland communities. Indeed, the quantity of shark vertebrae and sea lion bones recovered far surpassed local needs, and are best understood as a result of food processing of resources to be exchanged for resources unavailable to this region of the coast, such as fruits, tubers, and other edible plants, which are present in considerable amounts at this site.

A variety of tubers, fruits, beans and cucurbits discussed here, certainly were the counterpart of a diet centered on marine proteins. Future research should determine the amount of marine proteins and plant carbohydrates in contemporary Initial Period fishing and farmers communities as has been done for other periods in the Andean region (e.g., Burger and Van der Merwe 1990).

The massive quantity of shellfish remains at Gramalote may lead us to an erroneous interpretation of food remains. These data are critical since it has been proposed that this site was a specialized shell-gatherer community (Pozorski 1976: 95, 106; Pozorski and Pozorski 1979: 424, 429; Velasquez 1987: 35). Furthermore, it has been argued that shellfish was the most important protein, more important than fish, sea mammals, or seabirds (Pozorski 1976: 98; Pozorski and Pozorski 1979: 424). Although more than two hundred thousand complete shells and pieces were recovered, they were probably consumed as a complementary food, filling gaps when major sources of marine resources such as sharks, large drums, and sea lions were unavailable. The evidence strongly suggests shellfish were a side dish, an aperitif for the main course (Prieto 2015). This is the way traditional fishermen today use shellfish, emphasizing that in certain periods when fish or other products are unavailable due to weather or other natural condition, shellfish become vital to the diet for short periods of time. The abundance of fish and marine mammal remains along with their greater meat yield in comparison to that of shellfish suggests that this culinary pattern also characterized the ancient past.

I am not claiming that current perspectives regarding food consumption should be applied into the second millennium B.C. diet, but rather that the abundant remains of sharks, large drums, and sea lion suggests these were the major "staple" resources in this coastal community. However, all these marine resources are subject to seasonal cycles and to a number of natural and climatic factors such as El Niño events, which with a limited fishing technology, are difficult to obtain year-round. How then did these inhabitants manage to get these and other products year-round for their diet?

The archaeological evidence presented here suggests that the basic necessities for survival and daily consumption relied upon five food resources: shark, large- and medium-size drums, sea lion, and perhaps the purple crab. Interestingly, of the five marine resources, four are seasonal, available only during a limited time of the annual cycle. Medium-sized drums are year-round fishes, but their occurrence in the faunal assemblages are surprisingly low compared to the other four resources. Sharks and large drums are only available during the summer season when large shoals of sharks reach the coastline for mating and for giving birth, making them accessible to human hunters.

Large drums frequent the coastline to feed on the surf clams and sandy crabs that are especially abundant during the summer, while sea lions mate between December and January, at the beginning of the austral summer. During this part of the annual cycle large colonies of sea lions inhabit the beaches, which are used as nurseries and for mating, making them easy prey for local fishermen. Beyond the presence of sea lions on the shore, fishermen could also have reached the offshore islands to hunt sea lions (Prieto 2015).

It is interesting that Gramalote fishermen relied on seasonal resources for the bulk of their diet. This suggests that exploitation of their major subsistence resources was carried out over a brief time in the annual cycle, specifically during a short period of the summer when they were able to procure enough food for the rest of the year. If this pattern of seasonal meat procurement was the case, it represents a strategic and an organized fishing and hunting activity well beyond the capability of the nuclear family. This pattern of supra-household communal behavior has been ethnographically documented in Huanchaco. As shown with modern fisherfolk, fishing practices are best understood as an activity extending beyond the household in which members beyond the nuclear family (cousins, grandsons, in-laws) all work together to guarantee optimal exploitation of the desired resources. During the Initial Period, this was likely the case when the abundant and desired resources were at their peak densities during the summer. Interestingly, modern traditional fishermen in northern Peru call this season (the austral summer) “abundance time” (Prieto 2013). As shown here the Gramalote fishermen were organized at an extended family level so that they could efficiently exploit the three main animals that provided a large quantity of meat (sharks, large drums, and sea lions) during three or four months of the annual cycle. This hunting and fishing pattern guaranteed the community enough food to sustain them for the rest of the year.

During the annual winter cycle, the absence of these major resources was partially compensated for by people capturing fresh purple crabs (*Platyxanthus orbignyi*). Purple crabs are well represented in the collections and evenly distributed among the different sectors of the site. Other resources such as seaweed (*Gracilaria chamissoi*), abundant during the winter, as well as the mussel, a mollusk that prefers cold waters, were possibly other alternative food options during the winter. During the transitional period between the end of winter and the beginning of summer, (September–October–November) large white clams (*Semele* sp.) and yellow clams (*Protothaca thaca*) were consistent and abundant resources, possibly exploited on a regular basis, as documented by their ubiquity in the shellfish collections. Finally, the pearly top shell (*Tegula atra*) and the dye shell (*Stramonita haemastoma*) are two varieties of gastropods present year-round and also abundant during the summer season. All these species were consumed fresh along with medium-size drums as well as the processed dried shark and sea lion meat that formed the central part of the daily diet.

Elsewhere, I have presented detailed descriptions of food procurement and food preferences during the three occupational phases and among the domestic units of Gramalote (Prieto 2015). These data demonstrate that there was a notable variability in the species of marine resources consumed among the different houses of Gramalote. This variability shows that food choices played a significant role within an apparent homogenous diet of an early fishing settlement. Previous to the excavations, it was assumed that there was a uniform pattern of marine products exploitation at the site. However, when marine resources were analyzed based on their provenience in domestic units, the results suggest a very heterogeneous scenario. In light of these findings, future research needs to be more inclusive when generalizing foodways patterns for early maritime communities on the Central Andean Region.

The diversity of seabirds, but infrequent quantity of seabird remains at Gramalote suggests that they were also a regular, but not an abundant, source of protein. Coastal inhabitants of the second millennium B.C. probably did not consider seabirds as an “additional source of protein” but probably as a delicacy that, when available, provided their diet with welcome variety in flavor and texture. Seabird meat may have served to break the routine of shark, fish, sea lion, and other more common foodstuffs. Elsewhere, I have detailed how among traditional populations of modern fishermen along the north coast seabirds are a preferred food for special events such as baptisms, birthdays, weddings, and funerals (Prieto 2015: 774). The phrase "*ceviche es de diario y pepián es para fiesta*" in which the word *ceviche* broadly means "seafood" and *pepián* refers to a special stew made with seabird or duck meat in the contemporary cuisine of Huanchaco; therefore, seabird meat is regarded as a food for special occasions in the local culture. Seabird remains at Gramalote may be the result of occasional meals but may also have been consumed during special celebrations by extended families or as part of communal events.

Seabirds also had non-culinary roles in both modern and ancient coastal cultures. Modern fishermen use the presence or absence of seabirds to determine if it is a good or bad day to fish. The recognition of seabirds as powerful beings is expressed on the designs of the Late Preceramic textiles from Huaca Prieta where varied seabird iconography is common (Bird et al. 1985). The discovery of a miniature ceramic vessel decorated with avian motives found in a domestic unit at Gramalote enhanced the idea of seabirds as ritualized creatures (Fig. 2.8). Therefore, early in the cultural trajectory of the maritime communities of the Peruvian North Coast, seabirds denoted an abundance of fish and good luck in fishing activities. It is, therefore, possible that seabirds were eaten with the intent of acquiring their skills and predictive power to identify and catch fish.

The Gramalote excavations provided data to distinguish food procurement and consumption patterns between domestic and ceremonial sectors at the site. This data demonstrates the importance of sampling both domestic and ceremonial contexts to infer foodways and its potential to explore paleonutrition of small groups as opposed to collective dietary analysis that then are generalized to entire regions and cultural periods. Thereby, the unusual high concentration of certain fish species, marine mammals, seabirds, gastropods, and clams in, for example, the ceremonial facility of the site, suggests that cultural selection was made among certain products to be consumed preferentially in ritualized occasions.

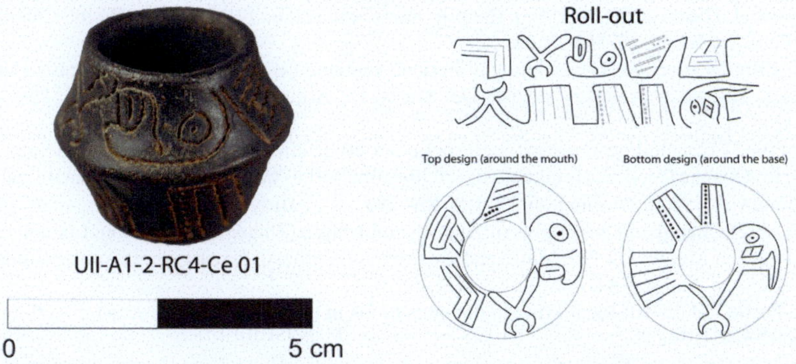

Fig. 2.8 Ceramic vessel miniature depicting two birds. This unique piece was found in one of the houses of the residential sector and it may have been used for domestic rituals. It suggest the importance of seabirds in Gramalote's mythology

In this light, the Gramalote investigations open the discussion if it may be valid to question whether the context in which seafood remains found at inland ceremonial centers of the coastal valleys are the result of daily food consumption or the leftovers of ritualized meals in which marine products are significant for feasting activities. Similarly, marine resources were not indiscriminately exploited only to satisfy nutritional demands at Gramalote. Their archaeological appearance was a result of cultural practices related to community well-being, preference, and flavoring that were beyond mere subsistence strategies.

In conclusion, the study of food consumption, and more importantly the food choices made by the early settlers at Gramalote during the second millennium B.C. provided a basis for understanding the adaptation process that was shaped by their cultural choices. Daily food and meal patterns were proposed based on the seasonal availability of marine resources as well as exchange patterns with valley populations. Food consumed in public and ceremonial areas of the site suggests that special choices were made for these events and that food acquired an intrinsic meaning determined by the context in which it was consumed. Evidence at Gramalote suggests food remains are powerful sets of evidence to not only uncover subsistence practices, but also how everyday people in the ancient Andes shaped the path toward traditional gastronomy still intact in modern coastal communities such as Huanchaco and many other towns in coastal Peru.

References

Anhuaman P (2008) Cultura Viva Muchik-Chimor de la Costa Norte del Peru. Historia, tradiciones, leyendas y personajes. Trujillo, Peru

Ben-Shlomo D, Hill AC, Garfinkel J (2009) Feasting between the revolutions: evidence from Chalcolithic Tel Tsaf, Israel. J Mediterr Archaeol 22(2):129–150

Bird J, Hyslop J, Dimitrijevic M (1985) The preceramic excavations at the Huaca Prieta, Chicama Valley, Peru, vol 62. Part I, New York

Briceño J, Billman B (2008) Gramalote y el Periodo Inicial en el valle de Moche. Nuevos Datos de un viejo sitio de pescadores. Revista Del Museo De Arqueologia, Antropologia E Historia 10:175–208

Burger R (1985) Concluding remarks: early peruvian civilization and its relation to the chavin horizon. In: Donnan C (ed) Early ceremonial architecture in the Andes. Dumbarton Oaks Research Library and Collection, Washington, DC, pp 269–289

Burger R (1992) Chavin and the origins of Andean civilization. Thames and Hudson, London

Burger R, Van Der Merwe J (1990) Maize and the origin of highland chavín civilization: an isotopic perspective. Am Anthropol (New Series) 92(1):85–95

Curet LA, Pestle WJ (2010) Identifying high-status foods in the archeological record. J Anthropol Archaeol 29(4):413–431

Diaz-Alvitez F (1999) La Pesqueria Artesanal en la Caleta de Huanchaco durante los meses de setiembre a diciembre de 1999. Licenciatura Thesis, Escuela Academico Profesional de Pesqueria, Facultad de Ciencias Biologicas. Universidad Nacional de Trujillo

Gillin JP (1947) Moche, a Peruvian coastal community. Smithsonian Institution Institute of Social Anthropology Publication, Washington DC

Goldstein D, Shimada I (2010) Food and craft production in the Middle Sican Period (AD 950–1050). In: Klarich E (ed) Inside ancient kitchens: new directions in the study of daily meals and feasts. University Press of Colorado, Boulder, pp 161–189

Gumerman G (2010) Big hearths and big pots. Moche feasting on the North Coast of Peru. In: Klarich E (ed) Inside ancient kitchens: new directions in the study of daily meals and feasts. University Press of Colorado, Boulder, pp 111–131

Hastorf CA (2008) Gender, space, and food in prehistory. In: Preucel R, Hodder I (eds) Contemporary archaeology in theory. Blackwell Publishing, pp 460–484

Henderson JS, Joyce RA, Hall GR, Jeffrey Hurst W, McGovern PE (2007) Chemical and archaeological evidence for the earliest cacao beverages. Proc Natl Acad Sci 104(48):18937–18940

Huaman L (2012) Analisis de Fitolitos y Granos de Almidon en Sedimentos, Liticos y Ceramicos del Proyecto Arqueologico Pampas Gramalote (Huanchaco, La Libertad). Laboratorio de Palinologia y Paleobotanica, Universidad Peruana Cayetano Heredia. Report in possesion of the author

Isbell W, Groleau A (2010) The wari brewer woman. Feasting, gender, offerings and memory. In: Klarich E (ed) Inside ancient kitchens: new directions in the study of daily meals and feasts. University Press of Colorado, pp 191–219

Jawad LA (2013) Hyperostosis in three fish species collected from the sea of Oman. Anat Rec 296(8):1145–1147

Joyce A (2010) Food preparation, feasting and the social negotiation of gender and power. In: Inside ancient kitchens: new directions in the study of daily meals and feasts. University Press of Colorado, pp 221–239

Joyce A, Henderson JS (2007) From feasting to cuisine: implications of archaeological research in an early Honduran Village. Am Anthropologist 109(4):642–653

Lavallee D, Julien M (2012) Prehistoria de la Costa Extremo-Sur del Peru. Los Pescadores Arcaicos de la Quebrada de los Burros (10000—7000 a.P.). Travaux de l'Institut Francais d'Etudes Andines 297. Instituto Frances de Estudios Andinos and Fondo Editorial PUCP, Lima, Peru

LaViolette A (2008) Swahili cosmopolitanism in Africa and the Indian Ocean World, A.D. 600–1500. Archaeologies 4(1):24–49

Lu H, XiaoyanYang MY, Liu K-B, Xia Z, Ren X, Cai L, Naiqin Wu, Liu T-S (2005) Culinary archaeology: millet noodles in late Neolithic China. Nature 437(7061):967–968

Marcus J (1987) Late intermediate occupation at Cerro Azul, Peru. A preliminary report. University of Michigan Museum of Anthropology Report. Technical Report 20, Ann Arbor, Michigan

McTavish R (2013) Faunal subsistence strategies among initial period coastal fishers at the Gramalote site in the Moche Valley of Peru. Department of Anthropology, University of Wisconsin-Milwaukee

Moseley ME (1975) The maritime foundations of Andean civilization. Cummings Pub Co, Menlo Park, California

Moseley ME, Feldman RA (1988) Fishing, farming, and the foundations of Andean civilisation. In: Bailey G, Parkington J (eds) The archaeology of prehistoric coastlines. Cambridge University Press, pp 125–134

Nash D (2010) Fine dining and fabulous atmosphere. Feasting facilities and political interaction in the Wari Realm. In: Klarich E (ed) Inside ancient kitchens: new directions in the study of daily meals and feasts. University Press of Colorado, pp 83–109

Norman NL (2010) Feasts in motion: archaeological views of parades, ancestral pageants, and socio-political process in the Hueda kingdom, 1650–1727 AD. J World Prehistory 23(4):239–254

Olsson M, Isaksson S (2008) Molecular and isotopic traces of cooking and consumption of fish at an Early Medieval manor site in eastern middle Sweden. J Archaeol Sci 35(3):773–780

Osborn A (1977) Strandloopers, mermaids and other fairy tales: ecological determinants of marine resource Utilizationv- the Peruvian Case. In: Binford LR (ed) For theory building in archaeology. Academic Press, New York, pp 157–205

Piperno DR, Dillehay TD (2008) Starch grains on human teeth reveal early broad crop diet in Northern Peru. Proc Natl Acad Sci USA 105(50):19622–19627

Pozorski S (1976) Prehistoric subsistence patterns and site economics in the Moche valley, Peru. Ph.D. dissertation, Department of Anthropology, University of Texas at Austin, Ann Arbor
Pozorski S, Pozorski T (1979) An early subsistence exchange system in the Moche Valley, Peru. J Field Archaeol 6(4):413–432
Prieto G (2009) Tres aspectos etnográficos del pueblo de Huanchaco. Revista Del Museo De Arqueologia, Antropologia E Historia 11:277–306
Prieto G (2013) Apuntes etnograficos de algunos pueblos de pescadores tradicionales del Norte Chico y la costa norte del Peru. Nosotros 6:139–191
Prieto G (2014) The early initial period fishing settlement of Gramalote, Moche Valley: a preliminary report. Peruvian Prehistory 1(1):1–46
Prieto G (2015) Gramalote: domestic life, economy and ritual practices of a prehispanic maritime community. Unpublished Ph.D. Dissertation, Department of Anthropology, Yale University, New Haven, CT
Prieto G (2017) Utensilios de subsistencia y produccion hechos de madera del sitio Gramalote. Periodo Inicial (1500–1200 a.C.). In: Ministerio de Cultura del Peru (ed) Actas del II Congreso Nacional de Arqueologia 2015, vol 1. Ministerio de Cultura del Peru, Lima, pp 67–83
Prieto G (2018a) The social dynamics and economic interactions of the households at Gramalote, a small-scale residential settlement during the Second Millennium BC on the North Coast of Peru. Latin Am Antiq 29(3):532–551
Prieto G (2018b) The temple of the fishermen: early ceremonial architecture at Gramalote, a residential settlement of the Second Millennium B.C., North Coast of Peru. J Field Archaeol 43(3):200–221
Prieto G (2020) The fisherman's garden: horticultural practices in a second millennium maritime community of the North Coast of Peru. In: Prieto G, Sandweiss DH (eds) Maritime communities of the ancient Andes. Universty Press of Florida, Gainesville, Florida, pp 218–246
Prieto G, Wright V, Burger RL, Cooke CA, Zeballos-Velasquez EL, Watanave A, Suchomel MR, Suescun L (2016) The source, processing and use of red pigment based on hematite and cinnabar at Gramalote, an early initial period (1500–1200 cal. B.C.) Maritime Community, North Coast of Peru. J Archaeol Sci: Rep 5:45–60
Quilter J, Stocker T (1983) Subsistence economies and the origins of Andean complex societies. Am Anthropologist 85(3):545–562
Ramirez S (1995) De pescadores y agricultores: Una historia local de la gente del valle de Chicama antes de 1565. Boletin Del Instituto Frances De Estudios Andinos 24(2):245–279
Raymond JS (1981) The maritime foundations of Andean civilization: a reconsideration of the evidence. Am Antiq 46(4):806–821
Rostworowski M (2004) Costa peruana prehispánica. Instituto Peruano de Estudios Andinos (IEP), Lima
Sabella J (1974) The fishermen of Caleta San Pablo. Latin American Studies Program, Dissertation Series. Cornell University, Ithaca
Sandweiss DH (2009) Early fishing and inland monuments: challenging the maritime foundations of Andean Civilization? In: Cooper JM, Williams PR (eds) Andean civilization. A tribute to Michael E. Moseley, vol Monograph 63. Cotsen Institute of Archaeology. University of California Los Angeles, Los Angeles, California, pp 39–54
Seeman M, Nilsson N, Summers G, LarryMorris PB, Dowd E, Newman M (2008) Evaluating protein residues on Gainey phase paleoindian stone tools. J Archaeol Sci 35(10):2742–2750
Smith-Vaniz WF, Kaufman LS, Glowacki J (1995) Species-specific patterns of hyperostosis in Marine Teleost fishes. Mar Biol 121(4):573–580
Smith ML (2006) The archaeology of food preference. Am Anthropologist 108(3):480–493
Sobolik K (1994) Paleonutrition: the diet and health of prehistoric Americans. Southern Illinois University, Center for Archaeological Investigations, Carbondale, Occasional Paper
Sutton MQ, Sobolik K, Gardner J (2010) Paleonutrition. The University of Arizona Press, Tucson

Velasquez P (1987) Recursos Marinos y Vegetales Durante el Formativo Temprano en el Sitio de Gramalote, Sector Huanchaquito: Valle de Moche. Facultad de Ciencias Sociales, Universidad Nacional de Trujillo, Trujillo, Peru, Escuela de Arqueologia

Weir G, Bonavia D (1985) Coprolitos y dietas del Preceramico Tardio de la costa Peruana. Bulletin De l'Institut Français D'Études Andines 14(1–2):85–140

Wilson D (1981) Of maize and men: a critique of the maritime hypothesis of state origins on the Coast of Peru. Am Anthropologist 83(1):93–120

Yang Y, Shevchenko A, Knaust A, Abuduresule I, Li W, Xingjun Hu, Wang C, Shevchenko A (2014) Proteomics evidence for kefir dairy in early bronze age China. J Archaeol Sci 45:178–186

Chapter 3
Camelids as Food and Wealth: Emerging Political and Moral Economies of the Recuay Culture

George Lau

Abstract Camelids have long played a key role in assessing the economies of Andean groups, not least in discussions of key anthropological themes such as domestication, diet, exchange, and verticality. Most discussions have focused on human intervention in animal use—most commonly how animals are incorporated and used by humans in their economic worlds. In this paper, I want to shift emphasis and examine how camelids came to be part of a mutually constitutive socio-cultural system. Two faunal assemblages (one from a small village and the other a major fortified center) and art imagery are assessed to describe the uptake of intensive camelid herding/consumption practices in the Recuay culture (c. AD 1-700) and the dramatic cultural transformations that ensued. At the same time that hunting diminished greatly, the faunal materials indicate increasing reliance on camelids in diverse residential food practices, corporate feasts, and fiber production. Just as important, camelids become the signal expression of the emerging wealth and status of nobles, which drive the production of effigy ceramics, figurines, and camelid fiber garments. Rather than being epiphenomenal to social change, camelids were therefore active elements, indeed co-participants, which shaped the look and reproduction of an ancient Andean society.

Introduction

Research on global warming and industrialized food production has highlighted the great impact that humans have on the environment, including the plants and animals that nourish our worlds. Much archaeological and anthropological research has also documented the broad impacts of human–environment interactions in the past, as seen in scholarship on the origins of agriculture, species extinctions, the globalization of foods, and spread of food production technologies. The emphasis has generally been

G. Lau (✉)
University of East Anglia, Norwich, UK
e-mail: george.lau@uea.ac.uk

J. E. Staller (ed.), *Andean Foodways*, The Latin American Studies Book Series,
https://doi.org/10.1007/978-3-030-51629-1_3

on human agency in these transformations. Archaeologists very often need to focus on plant and animal resources as sources of energy and nutrition.

Yet plants and animals also have profound effects on human groups. They are crucial, for example, in fashion, cuisine, recreation, and religious practices throughout the world. Put simply, plants and animals can also shape people and cultures just as much as the other way around. This may be due to their availability, their uses and effects, and their diverse cultural meanings. The intensification of plant and animal domestication has also transformed cultures and cultural adaptations in the past (e.g., Flannery et al. 1989: Chap. 12).

The following research examines the impact of camelids for ancient groups of the Central Andes, specifically of Ancash Department, Peru (Fig. 3.1). It focuses on the region's archaeological record of camelids during the Early Intermediate Period, ca. A.D. 1–700, after the decline of Chavín religious ideology and before the rise of Wari[1] culture, the first major Andean state (see also Chapter 4 by Tantaleán, and Rodríguez and Chapter 5 Toohey for the roles of food and domesticates to rank and status among Andean cultures).

During this period, dramatic transformations occurred in the look and composition of the cultures of the Central Andes. Many of these changes were related to the intensification of camelid production, consumption, and their use as pack animals in the transportation of resources between the coast and various regions of the highlands. Beyond their dietary significance, camelids came to structure various core features of Recuay culture and society, including social organization, settlement patterns, and their cosmology.

Four major innovations distinguish the Recuay culture from the preceding period, and they are pegged to different categories of evidence. The first concerns the emergence of architecture, settlement patterns, and land use practices centered on camelid production. The second sees increasing reliance on domestic camelid resources, including their meat and hair fiber. The third concerns their consumption within domestic and ceremonial contexts. The fourth is the development of artworks and imagery depicting camelids. The artworks visualize the role of camelids in mediating social relations, as items of wealth, display, and offering.

Different lines of evidence indicate that camelids and camelid production oriented major cultural transformations in highland Ancash during the first millennium A.D. The Recuay came to greatly rely on the raising and consumption of camelids. In addition to inspiring innovations in architecture and land-use, camelids were critical in new ritual practices, many requiring their representation on ceramic vessels and figurines. Camelids also became embodiments of wealth and elevated the value of high grazing lands. Recuay peoples integrated the camelid and its various products into an emergent ideology that saw political authority embedded in labor organization, mutual nourishment, and reciprocity. In short, with the help of camelids, Recuay groups ushered in a new form of Andean society characterized by both moral and political economies.

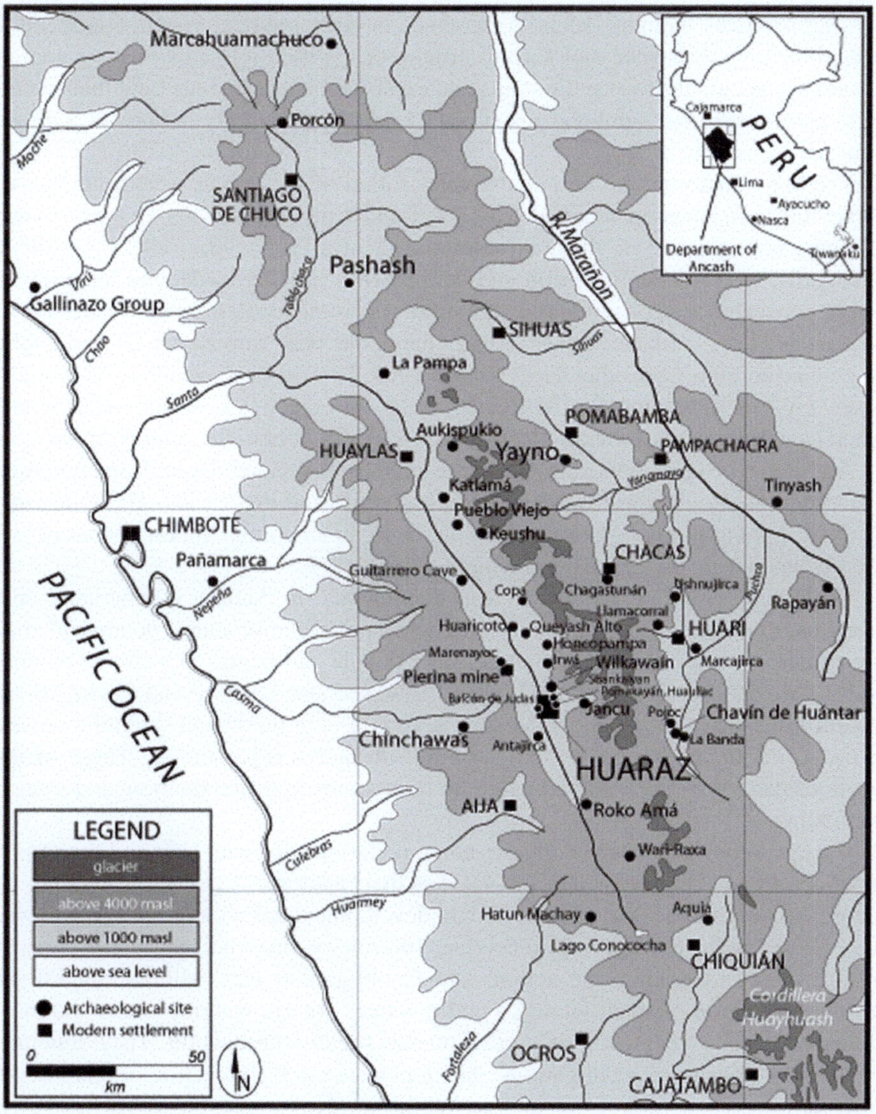

Fig. 3.1 Map of north-central Peru and places mentioned in the text. Drawing by author

Camelids in Andean Political and Moral Economies

Animals are crucial to the formation and maintenance of complex societies. The Central Andes constitutes one of the key case regions for examining the myriad roles of animals and food resources, in general, in early political economies (deFrance 2010; Murra 1965). Such research has generally centered on how camelids and herding practices maintain social differences in large political systems, especially in optimizing subsistence and surplus strategies, politico-religious display, and/or feasting. The emphasis is on the status and exploitation of camelids (and their products) as material and symbolic resources of people, especially nobles, in a given political arrangement.

The key transformation here is the valorization of animals as wealth items—as something to be amassed, retained, transacted, and displayed. Groups[2] of the Bolivian altiplano saw camelids as "the main resource" (Murra 1965: 190): leaders controlled vast herds, there was competition and conflict over grazing lands, and large herds were confiscated as spoils of war. In addition, it was the ideal of every kinsperson—peasant and lord—to have llamas; without them, one was "impoverished" and might be granted animals from other herds (Murra 1965: 193–194).[3] If wealth is "essential" to any civilization (Baines and Yoffee 2000: 15), not much is known about how and when camelids came to occupy the status as a wealth item in the central Andes.

Additional insights that complement the emphasis on camelid wealth and function concern how camelids were associated to social life and experience. The following analysis is focused upon the role of camelids in an emerging "moral economy" of the Recuay during the Early Intermediate Period.

Reciprocity and interaction are well documented in Andean ethnography and ethnohistory. They are components of a widespread native moral economy[4] that lay emphasis on the recognition and care of people and beings in a shared natural world (Allen 1988; Dransart 2002; Gose 1994; see also Overing and Passes 2000; Viveiros de Castro 1996). Since feeding is vital in both literal and symbolic senses to the world, the exchange of food and fluid substances is paramount. These social relations therefore operate under an idiom of mutualism, (re)production, and shared well-being.

For the Andes, and specifically the human relationships with land and economic practices, Penny Dransart (2006: 7) observes that "analogies between the tending of natural growth and of processes of human development are widespread." It follows that actions with food (giving, receiving, offering, eating) become highly meaningful: for in nourishing, the actions satisfy obligations that facilitate conditions for the restoration of spent vitality. Food practices are cyclical in process and may center on production/growth as part of overall social reproduction. They generate cultural relationships, or obligations, that anticipate the reciprocation or reciprocity of resources at some future point in time. While moral economies have been primarily characterized in terms of kin groups and village communities, their workings and principles are integral to how complex Andean political systems encompassed and managed wider communities and the extreme environments (e.g., Ramírez 2005; Salomon 1986). Camelid herds, even during the time of the Inca state, were the prerogatives of (and managed by) *ayllu* kin collectives of the provincial ethnic groups (Murra 1965: 191, 211). Animals were distributed as gifts and their products entered into complementary exchange networks (Murra 1965: 194, 201, 203; Bonavia 2008: 267–274).

Moral economies were not necessarily restricted only to humans. For many highland and lowland Andean cultures, the natural world and its biota are frequently heavily entangled in mutual coexistence and benefit, including, for example, plants, supernatural forces, ancestors, and their physical effigies seen as animate in many cosmologies (e.g., Allen 1988; Flores Ochoa 1977; Salomon 2018). It is the understanding that humans are not alone as actors in the world that incite people to acknowledge and treat other distinct beings in a manner befitting their status and potential to act in the world.

Periodic ritual offerings or *pagos* ("payments") of alcohol, foodstuffs, and coca leaves commonly were made by Andeans to supplicate the earth (or its divine agents, such as *Pachamama*, or other numina) and to ensure crop abundance and success (e.g., Allen 1988). The *pago* is commonplace today and are directly linked to ritual practices that extend back to pre-Columbian times. Ancestors, or named progenitors, were also propitiated in this way; offerings made to the esteemed deceased sought to nurture them for their continued favor and overall well-being (e.g., Isbell 1997).

The kind of non-human beings of particular interest is terrestrial mammals, specifically, camelids (Dransart 2002; Flores Ochoa 1977). Pre-Columbian Andeans were particularly dependent upon camelids, as a source of meat protein and for long-distance transport of goods and resources. Once herding in the Central Andes became firmly entrenched and widespread, camelids came to hold privileged status among various cultures and particularly to highland political economies for nearly two millennia. With the coming of Europeans and the introduction of Eurasian and European domesticated food crops and animals, camelids were largely replaced. The conquest of the New World transformed tastes and new demands powered the adoption of foreign animals, foodways, and economic practices (Flannery et al. 1989; Gade 1999). Old World domesticates (sheep, goat, cattle, and pig) effectively displaced camelids, often referred to colloquially as "native sheep" (*carneros de la tierra*) (see Bonavia 2008: 207–265), as sources of meat and for their fiber. New species had to be combined to provide what camelids did for many centuries.

Murra (1965: 192) asserted that, "In Lupaca…llamas were a major capital good, and most households expected to acquire rights in llamas as part of successful participation in *altiplano* economic life. The absence of camelids limited the opportunities to weave cloth and carry our long-distance trade, both important since camelids were essential to cloth, and the interaction of *chicha* or maize beer, coca leaf, and other hospitality goods needed in a society where, in the absence of forms of currency or markets, institutionalized generosity was the obligation and privilege of individuals of high status as well as commoners. Also, in an emergency, at times of drought and frost…llama products could always be bartered for food." In describing the adaptive role of camelids for groups of the Bolivian *altiplano*, Murra (1965) clearly implies their logic more broadly for herding groups of the Central Andes: camelids as strategic resources and wealth, and particularly their importance to exchange, as well as their role in social relations and political authority.[5] Murra's observations about the core principles of Inca herding may have implications for considering the rise of similar institutions and complexity in early Andean prehistory.

Early manifestations of an emergent moral and political economy involving camelids can be observed in Recuay culture, around A.D. 300. Their importance to subsistence economies was a pathway for political authority. Recuay cultures were among the first Andean peoples to widely perceive camelids as a form of wealth and as a key component to sacrificial rites and ritual offerings.

Recuay Culture History and Camelids

Recuay culture developed during the Early Intermediate Period, ca. A.D. 1–700, a time characterized by the rise of regional cultures and corporate art styles, associated with large ethnic polities. The period was preceded by the Early Horizon period, associated with the spread of Chavín religious ideology, and followed by the Middle Horizon, the time of Wari state expansion in the Central Andes. Recuay groups flourished in the Department of Ancash (see Fig. 3.1) in the north highlands of Peru (Lau 2011; Orsini 2007). Yet it had cultural and exchange relationships with different parts of the coast, highlands, and the eastern slopes of the Andes. The Recuay depended on intensive agriculture and herding especially in the rich *suni* and *puna* production zones, just below the icecaps of the Cordillera Blanca (Fig. 3.2). Major political centers arose in different parts of highland Ancash and appear to have been the political seats of independent complex societies.

Most Recuay groups shared a predilection for finely crafted kaolinite pottery, stone carving, and a range of shared iconographic images (Lau 2011: 15–17). In addition, Recuay mortuary practices and the construction of tomb buildings were similar throughout the Recuay region. All these elements bear important relation to, if not basis in, an emergent mode of religious ideology and social organization centered on ancestor veneration (Lau 2000). Many of the traditions established during Recuay times continued well into later prehistoric periods and into colonial times. This is particularly the case surrounding cultural traditions of "chiefly house" and ancestors as organizing principles, and elaborate cultural production of warfare (Lau 2011, 2013). Many Recuay cultural and social forms appear to have been taken up by

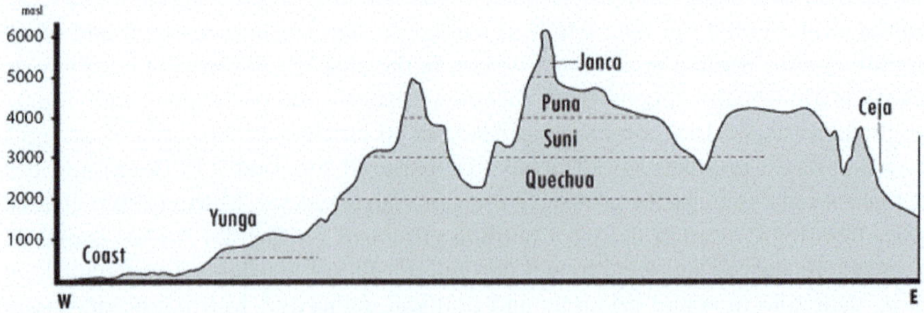

Fig. 3.2 Schematic profile of Central Andes, showing principal ecological zonation. Beneath the puna grasslands ideal for grazing animals are zones for intensive high agriculture, including the suni (potatoes, tubers, amaranth grains) and quechua (fruits, maize, beans). Drawing by author

later highland groups (e.g., Bazán del Campo and Wegner 2006; Gero 1990; Ibarra 2009; Mantha 2009).

There are four species of camelids, and this research is focused on the domesticated varieties, the llama (*Lama glama*) and alpaca (*Lama pacos*). While some use of the wild species such as guanaco (*Lama guanicoe*) and vicuña (*Lama vicugna*) cannot be ruled out, the current record indicates that most highland settlements were reliant in herding the domesticated animals by Recuay times (Lau 2007; Ponte 2008).

The beginnings of plant and animal domestication in the Andes are thought to have begun some six thousand years ago (Bonavia 2008: 205; Wheeler 1998). Highland groups initially focused on cultivation began intensive herding a millennium later. Dietary reliance on camelid meat, the uptake of herding practices, and displacement of existing strategies were highly variable across the Central Andes (Bonavia 2008; Flores Ochoa and McQuarrie 1994).

The first millennium B.C. was the crucial period of transition in the north highlands of Peru, associated with the florescence of the Chavín cult (Miller and Burger 1995: 450–451; see also Shimada 1982; Shimada 1985). Earlier periods (Initial Period and early Early Horizon) were characterized by a mixed diet represented by substantial proportions of cervid (esp. white-tailed deer, *Odocoileus virginianus*; taruca, *Hippocamelus antisensis*) and camelid meat. During the Chavín Horizon, faunal assemblages feature dramatic increases in the percentage of camelids in different parts of the north highlands. The predominance of camelids, as well as other indicators (e.g., widespread corrals, size variation, different organic remains), demonstrates a shift from intensive hunting to herding strategies by sedentary societies in Ancash (see Bria 2017; Ibarra 2003; Herrera 2005; Miller and Burger 1995; Ponte 1999; Rofes 1999; Sawyer 1985).

The focus on herded animals continued into Recuay, and domesticated animals have remained the principal meat source for highland populations up to the present (Bria 2017; Lau 2007; Rofes 1999; Sawyer 1985). Once the idea and technology of herding were established, camelids became a backbone and enduring feature of highland Andean life. In addition to their contribution to local diets, camelids undoubtedly served other functions. Unfortunately, evidence for the ancillary functions of camelids, for prehistoric contexts, are often less clear. Indigenous to the Andes, camelids flourish on montane vegetation and their general demeanor and surefootedness on narrow montane paths make them reliable as pack animals (Fig. 3.3). Adult llamas, the larger of the two domesticates, can carry some 25–50 kg, and perhaps as much as 60–80 kg (see Bonavia 2008: 414–424) (Fig. 3.4). Commonly, corrals and proximity to exchange routes and prehistoric roads have been cited as evidence for cargo-bearing animals (Wilson 1988) and of camelid caravans. Camelids may have been used intensively to transport exotic materials (ceramics, obsidian, maritime resources) desired by groups participating in the Chavín interaction sphere which linked coastal, highland, and tropical forest regions (Miller and Burger 1995: 451–453).

Technical analyses of camelid bones add to the idea of considerable Early Horizon mobilization and movement of camelids. In particular, recent isotopic research of bone collagen has documented that coastal camelids (in Nepeña) may have been

Fig. 3.3 Herd of camelids, adults with their young, grazing in the lush valley of the southern Cordillera Blanca. Streamlets meander slowly through flat bottomland and feed the boggy wetland bofedales, the favored grazing places of camelids at high altitude. Photo by author

reared locally but were part of coast-highland camelid transport networks (Szpak et al. 2015a).[6]

Ceramic imagery in different pottery styles, such as Moche, Wari, and Chimú, shows camelids with load bags on their backs. Recuay ceramics, as documented in this research, do not often depict camelids as transport animals (with load bags); rather, their imagery usually emphasized their importance as forms of wealth, property, gifts, and sacrificial animals.

Camelids were reared for their meat protein and fiber for the production of cloth, and were beneficial for many other kinds of necessities important to humans.[7] Skins and hides can be used for coverings and blankets, and sinew can be used for cordage, ties, and grips. Dung can be used as a source of fuel, to burn for cooking and warmth—as well as for fertilizer. Camelid bone was used to make various durable, handheld items, such as weaving implements, projectile points, awls, spatulas, pins, and adornments, as well as playthings such as dice, tokens, and whistles. Perhaps the most important raw material, however, was the fleece or the hair fiber, of the camelid. This was spun into yarn and woven into textiles, which were and are still today among the most important expressions of prestige, value, and cultural production in the high Andes (Dransart 2002; Lechtman 1993; Murra 1962).

Today, alpaca and vicuña fibers are especially valued for their fineness, weight, color, and particularly their warmth. Recuay people actively used camelid hair fiber in their most important weavings (Brito 2016; Lau 2014). Although only a few *bona*

Fig. 3.4 Llamas at the Chavín de Huántar site. The Chavín civilization used camelids widely, especially for the transport of long-distance preciosities and in the production of bone tools. It was also during Chavín that Ancash groups developed a reliance on camelid meat. Photo by author

fide Recuay style cloths are known, the activity of fiber processing was ubiquitous, as reflected in the widespread presence of spindle whorls in Recuay period settlements (Lau 2007, 2010). These are the discs attached to the ends of rods or "spindles", which, when rapidly turned, provide weight to draw and twist raw fiber into thread. The yarn wraps around the spindle, which when complete, is more suitable for weaving and transport.

Spindles are relatively rare in highland archaeological contexts. Given that most were of wood, their poor preservation is not surprising. But in Recuay culture sites whorls are relatively common in domestic contexts, and investigations have found them in a variety of sizes, materials, quality, and states of completeness. The most common whorls were made out of broken potsherds; edges were ground down, and a hole was drilled in the center to insert the spindle. Some sherds were clearly selected for having particular kinds of painted decoration on the flat faces. Many ceramic whorls are not found intact. Whorl fragments were broken while drilling the center hole; others many have broken during use or transport (Lau 2007). Making whorls out of sherds probably did not require great skill, and broken pottery, as raw material, would be plentiful in most villages and ancient sites. They were probably made by the same people who tended animals or as household work during free time.

Fig. 3.5 Spindle whorls (stone and fired clay), and polishing stone, found at Yayno site. Note the punctations and small holes to add visual spinning effect. Photo by author

Other whorls in Recuay culture were more elaborate and produced with greater effort and planning. These are of fired clay, ground stone, or of bone; these are less common at archaeological sites. Most are smaller and lighter in weight (Fig. 3.5). Some fired clay and ground stone examples are truncate in profile and show decoration on the base. The smaller whorls may have been for spinning cotton. These fancier whorls are occasionally found with burials, suggesting their status as prestige items. Ground sherd whorls, in contrast, were not used often as grave goods.

Settlements and Cultural Architecture

The importance of herding camelids in Recuay culture is indicated through new forms of settlements and architectural features in the landscape (Lau 2000, 2011: Chaps. 2–3). One of the critical developments in the first millennium involved an increase in the number and size of sites in higher altitudes—some above the upper limits of agriculture (Herrera 2005; Ibarra 2009; Lau 2011; Orsini 2006; Ponte 1999). This was probably due to a number of factors, such as population growth, defense, and environmental changes, but the significant variable was a reliance on herding camelids. Early Horizon settlement patterns do not show much evidence for dense occupation of these bleak and inhospitable high areas. From A.D. 400 to 500, however, large nucleated centers were built at high *puna* elevations above 4000 masl, such as Yayno, Tinyash, and Aukispukio, many with large enclosures for managing and protecting herds.

In addition, a distinctive kind of settlement pattern also developed in association with the Recuay culture, especially during mid to late times (Lau 2010). "Clusters" of neighboring villages and hamlets, usually on or near hilltop adjacent hilltops, emerged that surrounded a central place (Lau 2010). Located at high altitude and steep slopes, the cluster of settlements encompassed multiple ecological floors, including high tuber farming lands as well as those dedicated to camelid grazing, that occasionally reach down into riparian zones. It was a settlement adaptation that wedded collective defense with vertical integration to ensure access to complementary resources.[8]

Five or six sizeable farming hamlets (about 4–5 ha. in size) formed around Yayno, which was a major Recuay settlement in the northern Conchucos region. Each of the hamlets was situated near small pockets of arable land. As the seat of political power in the region, Yayno integrated the various hamlets, we assume through collective ritual and defense as the principal fortified and monumental settlement of the cluster. Corrals are located throughout the higher zones, and usually constructed adjacent to the hamlets.

During the Late Horizon (fifteenth–sixteenth centuries), purpose-built corrals proliferated across many parts of highland Ancash, particularly the rich grasslands of the Cordillera Negra and Cordillera Blanca below the highest, freezing environmental zones. Corrals were typically round or irregular in form, and their walls adapted to the contours and features of the landscape (Lane 2006). The walls were constructed using rough fieldstones (piled, without mortar), and occasionally abutted up against existing walls. In addition, residential communities, large and small, began to add corralling areas on terraces and near housing clusters—presumably to help pen animals. It is not always clear what the walled enclosures are for, but they may have also had ceremonial purposes (Herrera 2008; Ibarra 2003). Some Recuay ritual or shrine sites have their own corral features (Tello 1930). The pattern may be an early manifestation of later practices, best known from Inca and colonial accounts, which saw *huaca* (sacred places, instantiations of the divine) with their own herds that were managed by *ayllu* kin collectives (Hernández Príncipe 1923; Murra 1965: 201–203).

Most Recuay settlements had defensive constructions with nested walls, sometimes with four to five walls, providing the dual benefit of securing areas for penning camelids as well as protecting residential dwellings (Lau 2011). Excavations in these areas have uncovered deposits of camelid dung. Elsewhere, these have been called "concentric-ring sites," because of the multiple perimeter walls (Parsons et al. 2000).

Corrals are also identified archaeologically in the coastal valleys. In the Santa Valley walled enclosures were identified along trails leading from the highlands to the coast. They show evidence of Recuay-related kaolinite potsherds and desert geoglyphs of llamas. The enclosures were probably corrals which played a major role in coast-highland interaction caravans (Wilson 1988: 171–176, 189–193, 355).

Large-scale water management and irrigation developed in many regions of the Ancash highlands to optimize the terrain for the production of camelids (Lane 2006). Stepped revetment walls dammed streams in high-altitude valleys in the Cordillera Negra to help conserve water, fill reservoirs, and create tracts of wetland grazing with enriched soils and the succulent vegetation preferred by camelids. The best evidence for these "silt dams" is during the Late Intermediate Period, ca. A.D. 1000–1450 (Lane 2006), but such developments probably had local antecedents.

Iconographic Depictions of Camelids in Recuay Culture

Recuay culture showcased one of the first art styles in the ancient Andes to regularly depict the camelid (Hohmann 2010; Reichert 1977). The presence of camelid imagery represents a major difference from Chavín iconography. Around 500 B.C., camelids began to supplant deer as the primary source of meat protein (Miller and Burger 1995; Rosenfeld 2010; Sawyer 1985). Nevertheless, Chavín imagery (on ceramics, stone sculpture, or bone), rarely, if ever, depicts camelids. Chavin artisans primarily portrayed mythical beings, often creatures incorporating body parts from fearsome, predatory species, particularly raptors and jaguars. With the early spread of Recuay, significant changes occurred in the imagery and iconography, only a few centuries after the final decline of the Chavín civilization. Deer came to have a minor importance in Recuay art.[9] Meanwhile, camelids began to appear on the ceramic imagery with some regularity. There are two main forms of representation: stone and ceramic figurines of individual camelids and pottery multi-figure scenes, which show humans with camelids.[10]

Small Recuay camelid figurines are fairly common. They were made in stone and ceramics and in both solid and hollow forms. Their function probably varied, but most appear to have been small individual charms, ornaments, and/or playthings (see Fig. 3.6). Some are votive figurines, meant as offerings to be put in tombs or in other sacred places.

A number of camelid figurines or effigies have been found in contexts at Queyash Alto, Hualcayán, Yayno, and Chinchawas (Bria 2017; Burke 1990; Gero 1990; Lau 2010). Some figurines at Chinchawas have a hole to run string through, perhaps to be hung as pendant or dangled from clothing or the body. If so, such representations of camelids may have signaled social identity. Others, such as at Queyash Alto and Hualcayán, were made of pottery and may have been intentionally broken. These bear some resemblance to Inca and colonial era *conopas*, which had holes in the back of the figure, where camelid fat was placed. Such objects were magical items used by Andeans as offerings and in rituals that aimed to increase herds and their fertility. The camelid figurines in Recuay imply that beliefs and practices typically associated with Inca *conopas* were already in development long before the Late Horizon.

Most Recuay camelid figurines emphasize the fulsome torso and body, compared to the undersized head and limbs. Other examples from Yayno emphasize the four teats on the underbellies of the animals (Figs. 3.6 and 3.7). These elaborations on

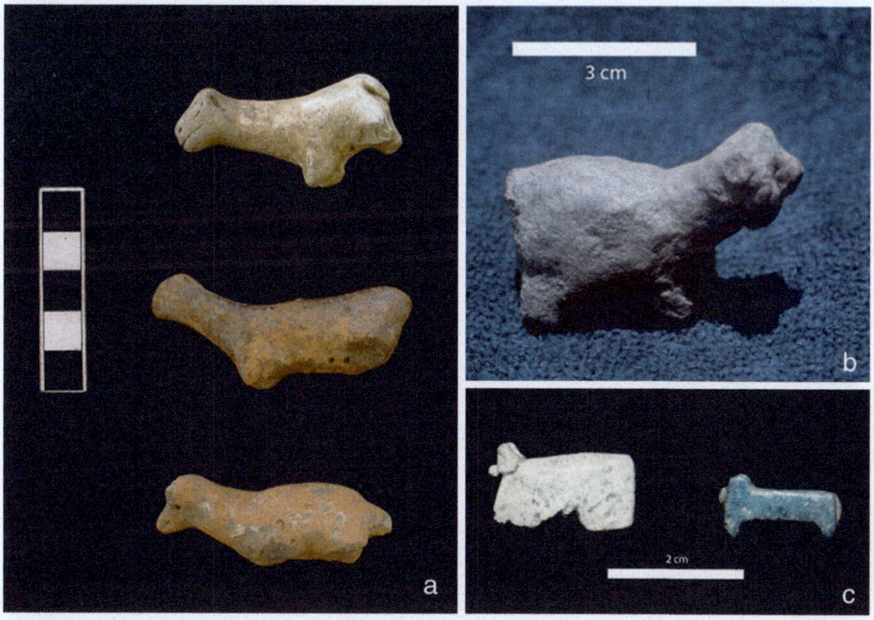

Fig. 3.6 Camelid figurines from Recuay sites: Yayno [**a** (fired clay)] and Chinchawas [**b** (fired clay); **c** (stone)]. Emphasis was given to the chunky torso and the neck fleece of the camelids (**b**), and deemphasized the limbs. Teats are shown in A (top, middle). The stone figurine (**c**, right) has a hole for use as a pendant. Photos by author

the chunky animal trunk—which mark key anatomical parts for the animals' meat, growth, and reproduction—are indicators of how they valorize the camelid as a source of wealth in Recuay culture.

The other main form of camelid imagery involves multi-figure scenes, and these are among the best known and characteristic of the Recuay culture. Such vessels depict a male next to a camelid, holding the animal with a rope (Fig. 3.8). The rope leads one to believe that the camelid to be a domestic animal, habituated to humans. The male figure is always bedecked in fancy attire and a headdress, as if for a ceremonial event rather than the more mundane activity of tending animals. Holding weapons or panpipes, the dignitary very likely is presenting the animal for sacrifice (Carrión Cachot 1955: 69) or as a gift/offering. The llamas look rather austere compared to their highly costumed male counterparts (Reichert 1977: 172–176). As is the case with most multi-figure scenes in Recuay pottery (see Lau 2011: Chap. 5), the animal is not necessarily the focus of the image; rather, the center of attention is the male lord, as was the chiefly act of presenting or receiving the animal.

Related practices were widely documented in the colonial accounts of Ancash and central Peruvian groups in general (Arriaga 1999: 51; Doyle 1988: 207–210; Duviols 1986; Hernández Príncipe 1923: 42; Polia 1999). Ritual sacrifice of camelids was one of the most important rites among highland cultures (Fig. 3.9); it was often obligatory in annual agro-pastoral ceremonies, burial rituals, house building events,

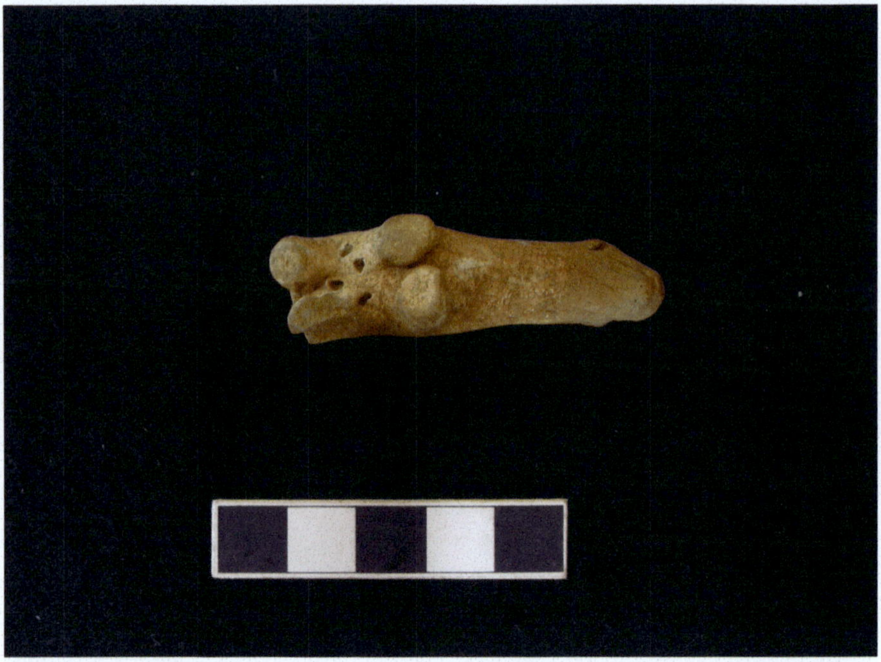

Fig. 3.7 Underside of camelid figure, showing expedient marks to show teats, from site of Yayno. Photo by author

and dedications. The hearts of camelids were removed and may have been eaten raw; other innards were often raised to the mouth and blown. Blood was splattered to the ground, or on cult effigies and people. During festivals, the host customarily slaughtered the camelid and provisioned the meat along with other food and drink. Attendees were expected to reciprocate with labor contribution for the lord and/or the host group (Gero 1990).

Camelids were a critical component of the "institutionalized generosity" and hospitality as reciprocity is characteristic of Andean food practices among both ancient and modern indigenous communities (Murra 1965: 192). Ceramic vessels depicting males with camelids are usually hollow, and the llama figure is invariably for holding and pouring liquids. Thus, the camelid is the "offering": its body, not the human's, is the physical receptacle, probably for a fermented intoxicant like *chicha*. The flow of nourishment is mirrored by the sharing of camelid meat itself, a gift of food also issued by the host lord. Use of camelid crania spoons and rib skewers—literally eating from the camelid—also reinforces the assertion that the camelid was more than mere food and calories; it was a vehicle for conveying vitality.

As foodstuffs and as ceremonial offerings, camelids grew increasingly significant in the political–economic practices of chiefly elites. Recuay imagery reveals that camelids served as a form of physical and symbolic currency in rituals of

Fig. 3.8 Recuay style effigy vessels showing man with camelid theme, which depicts a ritual presenting camelids for offering/sacrifice. Each shows a well-attired male, holding a shield (**a**, **b**) and mace (**a**), while leading a camelid with a rope. The camelid is the physical receptacle for holding and distributing liquids. Photos by author

display and nourishment that enhanced the prestige and labor recruitment of increasingly powerful community elites and their families. Overall, the crucial function of camelids was not simply as a source of meat protein; the Recuay record makes very clear that the camelid—in imagery and in ritual—became essential in mediating social relations in local moral economies during the Early Intermediate Period.

Camelids in Recuay Domestic and Ceremonial Contexts

Colonial accounts are for the most part quick to condemn Andean "pagan idolatry" in this case, how camelids were used in rituals and rites. But relatively few chronicles illuminate how they were processed and consumed. The principal evidence for the preparation and consumption by Recuay groups is shown by faunal remains and associated materials buried in archaeological deposits. Few Recuay investigations in Ancash report evidence of faunal remains; consequently, a complete picture must await further research.

Fig. 3.9 Sacrifice of black llama by male lord. Felipe Guaman Poma de Ayala, El primer nueva corónica y buen gobierno (1615), (Copenhagen, Royal Danish Library, GKS 2232 4°: facsimile at https://www.kb.dk/permalink/2006/poma/242/). 242v. Reproduced with permission of Royal Danish Library

Those Recuay faunal assemblages documented thus far appear to reflect fairly consistent patterns in diet and taphonomy (Sawyer 1985; Rofes 1999; Lau 2002, 2007; Ponte 2007; Bria 2017; Miller and Burger 1995; Rosenfeld 2010). In general, camelids dominate faunal remains, representing 75% or more of assemblages,[11] followed by small amounts of deer and guinea pigs, canines, rodents, and birds. Maritime resources also occur occasionally. The camelids were primarily domesticated llamas or alpacas. Guinea pigs were probably raised locally at the household level.

Investigations at Queyash Alto, a ceremonial center in the Callejón de Huaylas dating to the Early Intermediate Period, revealed walled enclosures[12] and areas associated with group-scale preparation and consumption, especially of camelid and cuy (guinea pig) meat (Gero 1990, 1991, 1992). Evidence of colanders and large jars indicates their use in the preparation and storage of *chicha* or maize beer. Food and drink were served in painted Huarás-style bowls and large spoons made from camelid crania. The walled enclosures served as special spaces for regular, seasonal feasts, presumably hosted by increasingly powerful families to create obligations.

Group-level feasting was also identified at Chinchawas, a small village and shrine site in the Cordillera Negra just west of Huaraz. Large quantities of animal bone, over 200 kg, were recovered during excavations (Lau 2002, 2007, 2010). Unlike at Queyash Alto, where feasting remains were found abundantly *within* walled enclosures, the refuse deposits of Chinchawas were found directly *outside* the enclosures. The bulk of the assemblage resulted from two excavation units sunk into ancient terraces used as middens during late Recuay times. In one pit, a sample contained a minimum of 97 camelid individuals; bone was frequent, as was broken cooking and serving pottery (Lau 2002, 2007). Most of the Chinchawas sample came from animals falling in the size range between modern llamas and alpaca, and it seems possible that the community raised large-sized alpacas, for provisioning of meat and desirable hair fiber (Lau 2007).

Objects manufactured from bone were used to serve and eat meals. These included pointed rib tools (perhaps skewers, some showing charring) and cranial spoons (Fig. 3.10). Just outside another side of the enclosure were several hearths and broken jars, probably the main food preparation area for the feasts. Recuay tombs and stone sculptures which depict ancestor themes link the public ceremonies to the veneration of ancestors (Lau 2002). Colonial sources make clear ancestor veneration was the principal reason for festive celebrations involving camelid sacrifice and widespread drinking in "*borracheras*" (Arriaga 1999; Hernández Príncipe 1923; Millones 1979). The veneration of ancestors was central to ethnic identity and status among Andean cultures.

Well-preserved camelid remains were also found at the summit of the Chinchawas site, at the foot of a circular ceremonial building—probably a shrine dedicated to a low natural outcropping of rock in its center (Lau 2007). Excavations inside and just outside its concentric walls produced fancy bowls and special objects made of rare stone, metal, and shell, which reflect high-status activities involving camelid offerings and consumption.

Fig. 3.10 Top and obverse views of three camelid cranial spoons from Chinchawas site. The curved crania were cut and the edges were ground down to use as eating and serving implements. Photos by author

The taphonomy of faunal remains is notable because they provide contextual evidence (Lau 2007). In refuse associated with domestic dwellings at Chinchawas, bones were more highly fragmented and in smaller pieces, as well as more sporadic in the refuse (also Rosenfeld 2010: 291). Some were burnt and found in very compacted strata, a depositional characteristic of everyday refuse, which tends to build up slowly. Domestic refuse often demonstrates greater human disturbances and animal activities as well as more variability in the fauna. Greater fragmentation also relates to getting at more difficult-to-access meat and marrow. This patterning of bone preservation was also apparent at Yayno, in various sample test pits in and around residential compounds.

In contrast, the feasting refuse at Chinchawas was discernibly different. Bones were abundant, not nearly as fragmented, and the deposits were thick and not particularly very well compacted. On occasion, soil did not build up between the bones, and the deposits were often loose with other associated artifacts (pottery, modified bone artifacts) suggesting that little time transpired between when bones were discarded and when they were covered. Also, ceramics were typically finer and the feasting deposits also had a great proportion of rare items, such as metal and lapidary items (Lau 2010). Finally, compared to the common everyday refuse, the camelid remains from the feasting contexts at Chinchawas were relatively well preserved. They do not show much evidence of having been burnt or for roasting meats, as much as they seem to have been butchered in large chunks. They were probably boiled for eating in stews.

Discussion: Camelids and Cultural Synergies

Intensive camelid transport on long-distance trade networks appears to have discontinued during the Early Intermediate Period. The abundances and variety of surviving long-distance trade materials (e.g., obsidian, marine shell, rare stones, fancy pottery) are relatively limited, especially when compared to the networks of the Early Horizon and later associated with spread of the Wari state. "Cosmopolitan" groups during the horizon periods were actively acquiring material from various different regions and ethnic groups (Burger et al. 2006; Contreras 2011; Lau 2014; Sayre 2010). The decrease of long-distance exchange is especially puzzling given that camelid production appears to intensify and become more widespread during the Early Intermediate Period. Chavín also set a precedent by opening interregional trade routes using camelids. If the Recuay used camelids as pack animals, it may have been more restricted in spatial extent.

The mobility of ancient camelids in Ancash is now being examined through characterization studies. Stock raising and their location of production can be informed through isotopic compositions of bone collagen and textile fiber (Szpak et al. 2015a, b). Analyses of thirteen Chinchawas bone samples revealed their composition to be relatively homogeneous, showing low carbon isotopic variability. The most parsimonious interpretation for the observed data is that the Chinchawas samples of camelids were raised in high-altitude grasslands.[13] The new data indicate that the camelids consumed at Chinchawas feasts were very likely coming from locally reared (highlands) animals.

Another point for fuller research concerns the impact of camelids on other cultural domains. For instance, how did camelids and their raw materials get integrated into rituals? The archaeological record suggests a rise of festive events associated with the generosity of emergent elites, while camelid production transformed rural agrarian economies during the Early Intermediate Period. Dramatic changes in Recuay religion also occurred in mortuary practices specifically involving ancestor veneration (Isbell 1997; Lau 2002, 2013).[14]

Among Andean cultures, ancestor veneration centered on physical effigies of the deceased. The practice of wrapping the mummified corpse was a vital part of making and refurbishing cult images. Wrapping was a ritual act that conferred ancestor-hood and potency to the object (Lau 2014). Periodic rewrapping with new high-quality textiles also helped to show continued reverence for the esteemed deceased; they were gifts that showed their continued importance for their respective communities. Camelid resources therefore helped to "feed" ancestors, to entreat them for continued favor and well-being. The few surviving Recuay textiles are of great sophistication and were the result of tremendous labor investment (Brito 2016). Burgeoning ancestor cults at the end of the Early Intermediate Period created a new demand and curation of finely woven cloths; more data are needed to study if there was a concomitant development and/or intensification of camelid varieties with fine fleece. It is impossible at present to verify whether these changing patterns are correlated; nonetheless, one can still aver that without intensive camelid production during the period, wrapping traditions basic to ancestor cults, especially using the finest and most labor-intensive cloths, would have been constituted very differently.

Another common thread links ancestor veneration and agro-pastoral production. Both were cultural domains favoring the growth and maintenance of territories and the surrounding landscape. Many factors encouraged the formation of Recuay territories during the first millennium AD. Among the most important were competition for prime production zones, language, endemic conflict, growth of chiefly polities, and demographic circumscription (Herrera 2006; Lau 2011; Mantha 2009).

The veneration of ancestors contributed to this synergy. Ancestor veneration in the Andes was often associated with tombs, cemeteries, and burial monuments, which marked the boundaries of communities (Lau 2000, 2002). Ancestors were also intimately tied to unique and prominent landscape features, such as springs, caves, outcrops, and mountains that were seen as places of origin, celebrated through oral traditions, or instantiations of the ancestor itself.

Intensive herding, of course, also requires regular access to lands with adequate resources to feed the animals (Lane 2006). These were not always present in the Central Andes during Recuay times, when a defensive orientation, hilltop locations, and fortifications characterized most settlements, large and small alike. As discussed earlier, Recuay communities also began to create clusters of settlements around central places. The extent to which Recuay conflict is owed to competition over herding lands remains unclear.

My final point returns to the notion of work, and specifically, the rise of gendered work practices in which camelids were deeply embedded. Because of the highly stylized and formulaic character of the figurative pottery (Lau 2011), we are uniquely able to observe some of the ways that Recuay groups saw and imaged themselves. Besides the collective nature of ritual, perhaps the most illuminating dimension of Recuay imagery is its emphasis on constructing female and male identity (Cromphout 2014; Gero 1999).

Recuay imagery most prominently distinguishes gender through the fancy costuming of the figures. However, patterns of actions and partnered elements are just as important. I have already mentioned vessels that show a well-attired male figure leading a camelid, and this is consistent with historical accounts. In the descriptions of herding practices of the Lupaca kingdom and Inca state, young men as well as older headmen were tasked with the job of managing community herds (Murra 1965: 197–199); the job satisfied annual labor obligations but could also be full-time and performed by lifelong retainers/servants.

Camelids and herding have male associations that are complemented by women and their associations with cultivation, maize and its product, maize beer (or *chicha*, or *aswa*) (Gero 1992). This too is marked in Recuay imagery, which shows women, carrying large jars of *chicha* and around male lords offering drink or receiving/raising drinking cups.

Recuay effigy vessels are ultimately expressions about the chiefly status of leaders and their kin (Lau 2011: Chap. 7). They depict group rituals including offering, pouring, presenting, and feeding that legitimate their special status. Because they are vessels, they become functionally integral to making those actions and meanings possible. Notably, the camelid meat from males produced from herding explicitly complements the maize beer resulting from cultivation and brewing (typically gendered female in highland Ancash).

This suite of gender symbolism manifested in the social relations and settlement organization of highland groups during mid-colonial Central Peru, including of Ancash (Arriaga 1999; Duviols 1973). Communities comprised peoples who identified as farmers (*huari*) or herders (*llacuaz*) resided and worked together, sharing resources, but their existence was framed through structural opposition with each other. Huari were seen as the aboriginals and farmers, associated with arable lands and irrigation. The llacuaz were newly arriving foreign "conquerors"; associated with herding and high (*puna*) grazing lands, they formed an "upper," and male component within the community. The llacuaz divinity was Lliviac (*Illapa*) who was associated to lightning and camelids; the Huari, meanwhile, venerated a telluric underworld divinity responsible for agriculture and maize.

The basic tenets of this complementarity—pairing upper:lower, herding:farming, male:female, foreigner:aboriginal, sky:earth—were already established over a millennium earlier in association with Recuay culture (Lau 2013). But it is not probably coincidental that the well-known martial imagery of Recuay (chiefly warriors, weapons, trophy heads) developed hand-in-hand with that of camelids (see Fig. 3.8). It is particularly tantalizing to relate Recuay patterns to these later developments, which linked camelids and warfare to foreign "llacuaz" conquerors—associated with *puna* herding lands and a vengeful lightning and war god. Notably, plant (*chicha* or maize beer) and animal (camelid, meat) resources play a crucial role in overall social complementarity. It is through their partnership and sharing, as comestibles, that Andean collectives in ancient Ancash were seen as complete, self-sufficient, and vital, an ideal first depicted systematically in Recuay times.

Conclusions

This chapter has described various lines of evidence to demonstrate that camelids were linked to major cultural changes to Recuay settlement, economy, and art. Compared to only a few decades ago, much more archaeological data are available to confirm that camelids were integral to the diets and food practices of high Andean peoples, long before the Inca. Butchered camelid remains occur in common residential spaces as well as communal and elite contexts, indicating a wide set of activities in which camelids and their meat featured. Many Recuay settlements exploited higher zones close to the *puna* regions, so as facilitate access to the grasslands and well-watered vegetation favored by camelids. The Recuay also intensively settled in ecotones adjacent to both high-altitude tuber farming lands and associated pastures. Corrals proliferated throughout the north highlands by the Early Intermediate Period.

Camelids became increasingly integrated into new modes of domestic production. Fiber processing, weaving, and bone tools were widespread, and achieved new technical levels. Spun fiber probably served as a key material for exchange. Regular access to resources such as sinew, dung, and leather must also have afforded new kinds of material engagements and technical choices.

Camelids also became integral components in public rituals intensively employed to provision public feasts, which were cultural and social exchanges involving food, sacrifices, and labor recruitment. Domestic camelids thus became key elements of group social identity and especially political authority of elites, who in northern Peru began to engage them directly as overt signs of wealth and prestige. This is particularly evident in pottery imagery, which illustrates camelids as vessels for vitality mediated by chiefly lords and venerated ancestors. Recuay leaders, it seems, added a new dimension to the moral economy, where they intervened as key actors regulating the flows of work, resources, and ritual obligations crucial for group or community well-being.

In short, camelids were fundamental in the evolution of ancient north highland societies. The introduction of camelids during the Early Horizon and their increased reliance during the late Early Horizon led to fundamental changes in Recuay society and culture. Within a few centuries of the demise of the Chavín culture and interaction sphere, north highland groups began to depict camelids. The growing importance of camelid herding translated into new attitudes and interactions with the physical world reflected in artworks. They are quickly integrated into both the moral and political economies of Recuay societies.

In summary, it is salutary to return to the introductory point concerning the impact of food resources on humans. The Recuay case furnishes a crucial example of great cultural transformations affecting groups once new animals and foodways are integrated. This is particularly evident because of Recuay imagery, which provides a window into new behavioral patterns (esp. ritual) and cultural values of the first millennium AD.

Scholars must continue to reconsider our Western orthodox notions about human–animal interactions. In Andean culture animals were not simply meat "protein on the hoof." Recuay peoples lived with and experienced animals, which are expressed in art and symbolic meanings. Scholars need to consider multiple uses, contexts, and meanings reconstructed from both the archaeological record and ethnohistorical sources, in which traditional Ancash practices are especially well represented. A more generous look at the role and meanings of the plants and animals in the Andean world will help to shed light on their contribution to overall social life and reproduction—in addition to how they can be and were often consumed as food.

End Notes

1 To help disambiguate, I follow the general current convention is to use "Huari" for the archaeological site in Ayacucho, and "Wari" for expansive archaeological culture and political network.

2 Of various kingdoms and regions comprising "Collasuyu," one of the four regions making up the Inca polity (Murra 1965: 190).

3 Yet Murra notes that these issues cannot be disentangled from changes and exigencies of the early colonial period, precisely the time when land and wealth were being appropriated, and indigenous systems were in flux. The comments also bear out his leftist politics and interest in the workings of the Inca state without capital.

4 Here I borrow the term mainly from economic anthropology: an economy that is based on co-existence, well-being and mutuality—especially smaller, closely knit food-producing communities whose members depend on each other.

5 The quote comes from a section where Murra is contrasting patterns of the Lupaca ethnic group (heavily invested in camelid herding) with the farming-centric patterns of Huánuco where "llamas were not a strategic resource" (Murra 1965: 192). Interestingly, Huánuco is located directly to the southeast of Ancash.

6 Faunal materials from Ancash have not been studied to research skeletal pathologies that may result from regular carrying of loads.

7 Little research has been done on camelids as pets or as long-lived actors with social, biographical capacities.

8 Similar clusters occurred elsewhere in the Recuay world: Pashash in Cabana, Aukispukio in Quebrada los Cedros and perhaps in Huaraz (Lau 2011).

9 On those occasions when deer are represented, there is generally emphasis on taruca deer—and its distinctive dual-tine antler.

10 A third principal form might be those representations of camelids on rock art—in caves, cliff walls and overhands. Unfortunately, due to lack of scholarship,

Chapter 4
Feast, Food, and Drink on a Paracas Platform, Chincha Valley, Southern Coastal Peru

Henry Tantaleán and Alexis Rodríguez

Abstract There has been considerable analytical and theoretical debate with regard to the role of feasting on the ancient world. In the Andes, the ancient existence of such activities and practices has been uncovered in various archaeological contexts, and such activities have greatly influenced anthropological theory with regard to the roles of ritual feasting and offerings to pre-Columbian cultures. Recent archaeological research has documented evidence for ritual feasting and the consumption of fermented intoxicants in various cultural and chronological contexts. The archaeological evidence for how such practices were carried out suggest an ever-increasing complexity and variability with regard to the kinds of compositions and their cultural particularities and characteristics. The following analysis presents archaeological evidence of ritual feasting associated with a sunken patio on a platform at the site of Cerro del Gentil in Peru. Pre-Historic occupation and archaeological evidence indicate that these ritual practices are associated with the Paracas culture (500 B.C.–200 B.C.) of the Chincha valley of the Southern Peruvian Coast. We use evidence from botanical and zoological remains, broken ceramic vessels, and other indications of ritual feasting to address the sociopolitical significance of feasting practices to the Paracas culture.

Introduction

For members of the Paracas culture in southern coastal Peru commensal politics was an important aspect of social life. Paracas is a complex culture dated to between 800 B.C.–200 B.C. and is the direct antecedent of the famous Nazca culture of the same geographic region. Initially documented archaeologically in the second decade of twentieth century by the father of Peruvian archaeology, Julio C. Tello, archaeologists have established that the Paracas culture extended between the Cañete and Nazca valleys (Fig. 4.1). The Chincha valley is located inside this influence area

H. Tantaleán (✉) · A. Rodríguez
Universidad Nacional Mayor de San Marcos, Lima, Peru
e-mail: henrytantalean@yahoo.es

J. E. Staller (ed.), *Andean Foodways*, The Latin American Studies Book Series,
https://doi.org/10.1007/978-3-030-51629-1_4

Fig. 4.1 Satellite view of the southern coast of Peru. Landsat image. NASA

and a number of scholars has noted its relationship with the eponymous site of this culture (Kroeber 1944; Wallace 1971, 1985; Canziani 1992, 2009; Proulx 2008; Lumbreras 2008).

Cerro del Gentil is one of the primary monumental Paracas centers in the Chincha mid-valley. The Chincha Archaeological Program (Programa Arqueológico Chincha-PACH) has conducted large-scale, intensive excavations at this site for several years. Our excavations uncovered architectural configurations of a building with a sunken court and its associated contents as well as different periods of building, rebuilding, decommission, and abandonment (Tantaleán et al. 2013, 2016). Here, we describe and interpret a part of site associated with ritual activities related to feasting that consisted of the consumption of large quantities of food and drink followed by the burial of the sunken court. The deposit is associated with Paracas elite material culture including ceramics, lithics, botanical, malacological, and zooarchaeological remains, and other items. We discuss the significance of the contexts related to the ancient sunken court as well as what transpired during the final feasting event before its abandonment.

Our analysis provides evidence the event documented at Cerro del Gentil was related to the development of Andean rituals and political feasting. Significantly, some sites documenting examples of political feasting recorded in the Andes have similar material correlates to those described at Cerro del Gentil (see Chicoine 2011; Lau 2002; Mesía 2014). However, despite these similarities with other contexts located elsewhere, we highlight the specific historical-social particularities involved in this event. Therefore, our interpretation focuses on the material evidence and food refuse as well as the unique social and historical factors that characterize Paracas culture. (see also Chap. 5 Toohey and Chap. 7 Bonzani for additional perspectives on feasting).

The Archaeological Site of Cerro Del Gentil

In the latter part of the 1950s, Dwight Wallace carried out a settlement survey in the Chincha valley, recording a number of Paracas sites in the mid-valley, of which Cerro del Gentil (PV.57–59[1]) was one (Wallace 1959, 1971). Later in 1980s, Luis Lumbreras and a team of archaeologists from the Instituto Andino de Estudios Arqueológicos (INDEA) carried out a research program in the Chincha valley. According to José Canziani (1992, 2009), the team identified numerous Late Paracas monumental centers characterized by stepped platforms with sunken courts built with conical adobe bricks across the valley (Fig. 4.2). This pattern of construction is found at both El Mono and Cerro del Gentil. Although Lumbreras and his team dug various mounds at El Mono (Lumbreras 2008), Cerro del Gentil was not investigated previously.

Although Cerro del Gentil is one of the main Paracas sites in the mid-valley, it is comparatively smaller than the monumental platforms of Huaca Limay, La Cumbe, Huaca Alvarado, Huaca Santa Rosa, Huaca Soto or Huaca Partida, located in the lower Chincha valley closer to the ocean (Canziani 1992, 2009). Cerro del Gentil is in the southern margin of the valley, and it was constructed on the edge of a terrace elevated from the surrounding cultivated valley at 192 masl (Fig. 4.3). Cerro del Gentil consists of two mound-platforms, Building A measuring 70 × 30 m and 6 m high and Building B measuring 30 × 20 m and 2 m high. Our excavations focused upon the larger structure, Building A (Tantaleán et al. 2013, 2016). We uncovered four major constructive phases (Tantaleán et al. 2013, 2016; Stanish and Tantaleán 2012, 2014, 2015). The first three related to the Paracas ceramic style, while the fourth is associated with the Topará ceramic tradition believed to be related to the Paracas Necrópolis culture initially identified by Julio Tello (Menzel et.al. 1964;

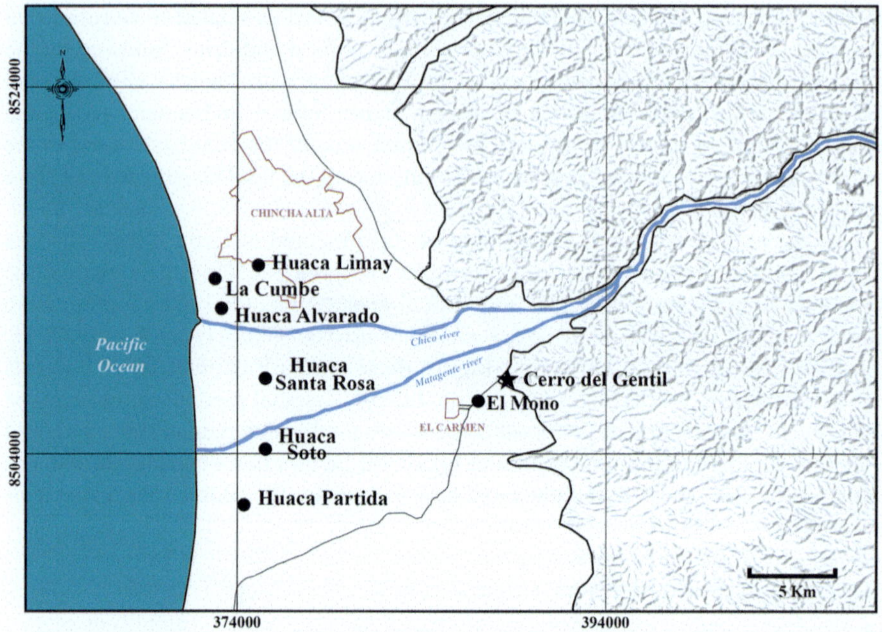

Fig. 4.2 Chincha Valley with the location of Cerro del Gentil and other monumental Paracas sites

Fig. 4.3 View of Cerro del Gentil from the Southwest

Wallace 1970a, b; Proulx 2008: 564). According to our ^{14}C dates, Paracas phases occurred between 500–200 B.C. (Tantaleán et al. 2013; Tantaleán et al. 2016).

Cerro del Gentil's sunken court is located in the central area of Building A. In the first phase of construction (yellow phase), builders leveled the natural surface and placed a clay floor above which were constructed adobe walls. The clay floor covered a 12 × 12 m sunken court at a depth of 2.50 m. The second construction (gray phase) indicated that the sunken court was reduced in size, resulting in another 7 × 7 m sunken court 2.30 m in depth. During the third construction (brown phase) the original configuration of the sunken court changed. We found a sunken space of 7 x 2 m (Enclosure FM-1) to the west and a small platform to the east (Figs. 4.4, 4.5, and 4.6).

In some moment in the third century B.C. activity in the court ended and the area was filled with layers of earth with great quantities of stones, plant and animal remains, fragments of ceramic vessels, and other debris. Multiple layers of fill document the use of the court through time (Strata B, C, D, E, and F) (see Tantaleán et al. 2016). The feasting and final filling episodes (Strata B, C, D, E, and F) of the ancient sunken court of Building A took place over a brief period of time. We interpret these layers as evidence of food consumption during a special feasting event that occurred in conjunction with the covering of the sunken court.

Stratum B is a roughly 1 m thick deposit composed of silty soil with rock inclusions and adobes fragments along with an abundance of cultural material. This stratum surface covers approximately the full extent of the ancient sunken court (Yellow

Fig. 4.4 View from the east of the excavations of the sunken patio of Building A of Cerro del Gentil

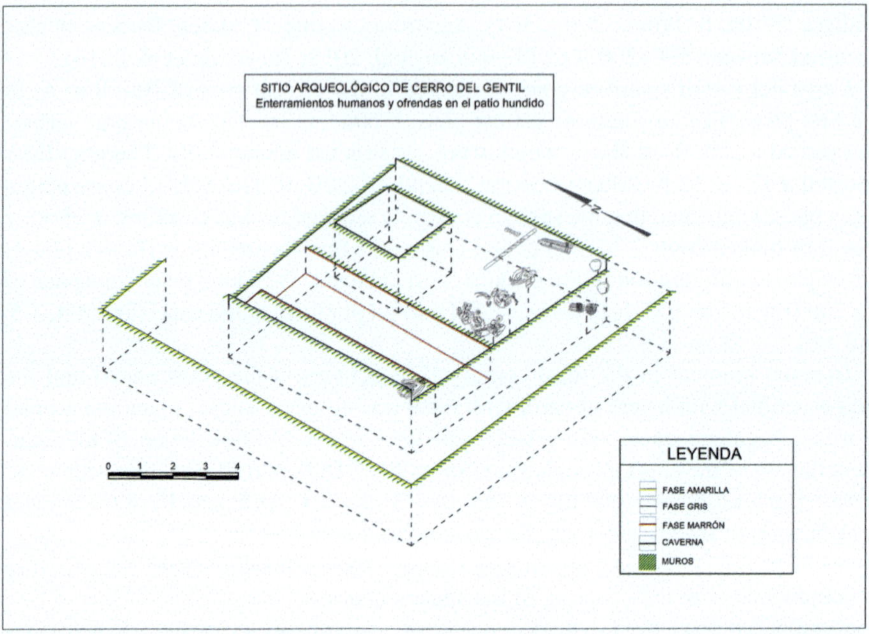

Fig. 4.5 Reconstruction of the sunken patios in the Building A of Cerro del Gentil. Note the location of the offering found during the excavations

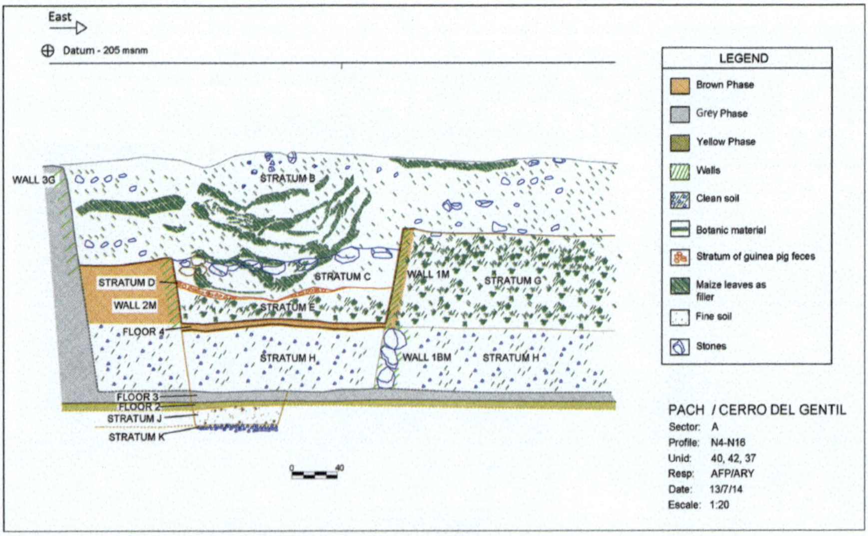

Fig. 4.6 Cutting drawing of the fill layers of the sunken patio

phase). During the excavation, we noted the intrusion of some materials from later culture periods associated with Topará and Carmen occupations in the most superficial part of this stratum. Below this disturbance, we recovered Paracas material in situ from Strata B to Strata E. These deposits, which are described and analyzed in this chapter, contain a variety of material including ceramics, lithics, textiles, botanical, and malacological remains, as well as offerings, and mortuary contexts.

Plant and Animal Remains and Human Consumption at Cerro Del Gentil

Archaeobotanical Remains

Archaeobotanical species identified at Cerro del Gentil indicate a variety of species. We identified 15 taxa from a sample of 5672 analyzed specimens (Table 4.1). Here, we describe taxa that were consumed or used as food as well as aspects of food storage and service. The diverse plant species identified in our excavations include maize (*Zea mays* L.), guayaba (*Psidium guajava*), pacay (*Inga feuilleei*), palillo (*Campomanesia lundiana*), string beans (*Phaseolus vulgaris* L.), "frejol" or lima beans (*Phaseolus lunatus* L.), "pallar" *Canavalia plagiosperma* ("pallar de los gentiles"), peanuts (*Arachis hipogaea* L.), calabash (*Lagenaria siceraria*) ("mate"), junco (*Schoenoplectus acutus*), and caña brava (*Gynerium sagittatum*) (Fig. 4.7).

Table 4.1 Botanical species documented at Cerro del Gentil

Class	Famiy	Species	Common name
Monocotolydoneae	Myrtaceae	*Psidium guajava*	Guayaba
		Campomanesia lineatifolia	Palillo
	Cucurbitaceae	*Lagenaria siceraria*	Mate
	Bromileaceae	*Tillandsia sp.*	Achupalla
	Poaceae	*Zea mays*	Maíz
		Zea mays (Kculli)	Maíz morado
		Gynerium sagittatum	Caña Brava
	Cyperaceae	*Schoenoplectus sp*	Junco
Dicotyledoneae	Malvaceae	*Gossypium barbadense*	Algodón
	Fabaceae	*Phaseolus vulgaris*	Frijol
		Phaseolus lunatus	Pallar
		Canavalia plagiosperma	Pallar de los Gentiles
		Inga feuillei	Pacae
		Arachis hypogaea	Maní
	Euphorbiaceae	*Manihot esculenta*	Yuca

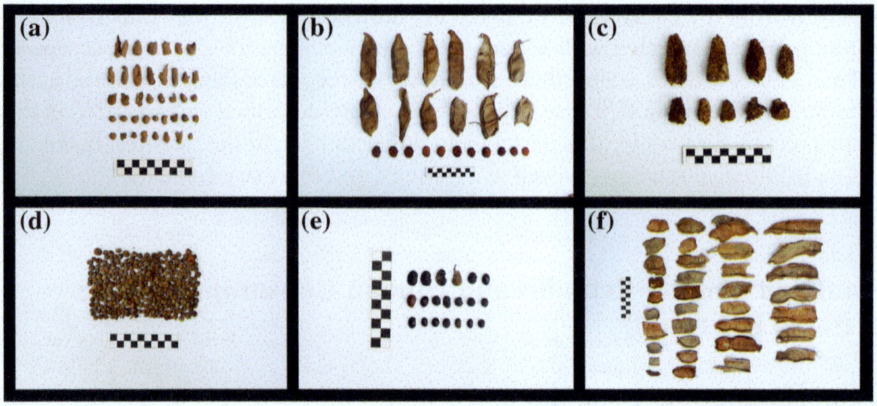

Fig. 4.7 Sample of botanical materials from the filling layers. **a** *Phaseolus vulgaris*, **b** *Canavalia plagiosperma*, **c** *Zea mays*, **d** *Campomanesia lineatifolia* **e** *Phaseolus lunatus*, **f** *Inga feuillei*

Ethnohistoric, ethnographic, historic, and archaeological sources have information or infer methods of food preparation and how the residents served maize. Maize appears to have been prepared as food in different ways. Toasted maize was consumed as "*cancha*" then grounded and boiled inside cornhusks as "*humitas*" (Bonavia 2008b; León 2013: 177). However, one of the most important ways to process maize was boiling and fermenting it to make maize beer or "*chicha*" (Bray 2003b: 99–100; Bonavia 2008a, b; Hayashida 2008; León 2013: 179). With regard to processing food, ethnohistoric and archaeological accounts which referred to Inca expansion, they suggest beans ("frejoles") and butter beans ("pallares") appear to have been consumed raw (in oils) or toasted and boiled as stews (León 2013: 191, 203, 205). The great quantity of fruits in our sample, guayaba, pacae, and palillo, indicates that these were consumed either raw, or boiled and made into beverages (León 2013: 209, 262, 272). Peanuts may have been consumed either toasted or boiled, but also made into beverages such as *chicha de maní* (León 2013: 187). The bottle gourd ("mate") (*Lagenaria siceraria*) is not only important as food but also for use as containers (León 2013: 239). Finally, junco grass (*Schoenoplectus* sp.) and caña brava (*Gynerium sagittatum*) were also used to create baskets (León 2013: 128).

Zooarchaeological Remains

We recovered remains of both mollusks and vertebrate in the site strata. These were found mostly interspersed with other materials within the burial layers that cover the sunken patio of Cerro del Gentil.

Malacological Remains

We found a significative variety of malacological species that were used for human consumption. The fill of the sunken court at Cerro del Gentil had at least 21 taxa from a sample of 1429 specimens (Table 4.2). All these species were associated with and adapted to the littoral of the Chincha valley and surrounding areas (Alamo and Valdivieso 1997; Zúñiga 2002; León 2013). The majority of these taxa generally live on intertidal rocks or submerged in the sand covered by seawater.

The identified several species that probably were consumed as food (Bonavia 1982: 185, 2003; Vásquez and Rosales 1998: 184; cited by León 2013: 462) including *Argopecten purpuratus* (concha de abanico o señorita), *Aulacomya atra* (choro, cholgo, or mejillón), *Brachidontes* sp. (chorito), *Brachidontes variabilis* (chorito), *Choromytilus chorus* (choro zapato, cholga, or choro), *Semimytilus algosus* (chorito, choro negro, chorito negro, or chorito lustroso), *Donax obesulus* (palabrita or almeja), *Mulinia edulis* (almeja), *Mulinia* sp. (almeja), *Calyptraea trochiformis* (picacho or trochita), *Crepipatella lingulata* (pique or señorita), and *Fissurella* sp. (lapa) (Fig. 4.8).

Table 4.2 Maritime Species documented at Cerro del Gentil

Taxonomy	Class	Family	Species	Common name
Mollusca	Bivalvia	Pectinidae	*Argopecten purpuratus*	Concha de abanico
		Mytilidae	*Aulacomya atra*	Choro
			Brachidontes sp.	Chorito
			Brachidontes variabilis	Chorito
			Choromytilus chorus	Choro zapato
			Semimytilus algosus	Choro negro
		Donaciae	*Donax obesulus*	Palabrita
		Mactridae	*Mulinia edulis*	Almeja
			Mulinia sp.	Almeja
	Gastropoda	Calyptraeidae	*Calyptraea trochiformis*	Picacho
			Crepipatella lingulata	Pique
		Fissurellidae	*Fissurella sp.*	Lapa
		Turbinidae	*Prisogaster sp.*	Caracolito negro
		Bulimidae	*Scutalus sp.*	Caracol de loma
		Tegulidae	*Tegula atra*	Caracol turbante
			Tegula corteziana	Caracol negro
			Tegula tridentata	Caracol turbante
		Muricidae	*Thais sp.*	Caracol
		Balanidae	*Perforatus perforatus*	Botella de mar
Arthropoda	Maxillopoda	Crustáceo N/I	Crustáceo N/I	

Fig. 4.8 Sample of malacological materials from the filling layers. **a** *Calyptraea trochiformis*, **b** *Mulinia edulis*, **c** *Fissurella peruviana*, **d** *Argopecten purpuratus*, **e** *Concholepas concholepas*, **f** *Semimytilus algosus*, **g** *Aulacomya atra*, **h** *Tegula Atra*, **i** *Donax obesulus*

Additionally, there are remains of several marine gastropods (caracoles de mar): *Tegula atra* (caracol turbante or caracol negro), *Tegula corteziana* (caracol negro), *Tegula tridentata* (caracol turbante or caracol negro), *Perforatus perforatus* (gran bellota de mar), and *Thais* sp. (caracol). Archaeological and ethnografic evidence indicates that mollusks generally were consumed after being either boiled or roasted (León 2013: 448).

Vertebrate Faunal Remains

The animals identified in our archaeological sample include a range of vertebrates. We recognized three classes (Osteichtyes, Aves, and Mammalia), seven families, nine species, and a total of identifiable specimens of 240 from a total sample of 270 zooarchaeological remains (Table 4.3) (Tantaleán et al. 2016). However, the main species consumed used were *Cavia porcellus* (guinea pig/cuy), *Lama* sp. (llama) and, occasionaly, *Otaria flavescens* (sea lions). Interestingly, the remains of *Canis lupus familiaris* (dogs) were largely complete and without any traces of butchering marks related to food processing suggesting their exclusion from consumption. The

Table 4.3 Zoological species documented at Cerro del Gentil

Class	Family	Species	Common name
Fishes	Fish N/I	Fish N/I	pez
Birds	Psittacidae	*Amazona sp.*	Loro amazónico
		Aratinga sp.	Loro
		Psilopsiagon aurifrons	Perico cordillrano
		Psittacidae N/I	Loro
	Accipritidae	*Accipiter sp.*	(Ave rapaz)
	Ave N/I	Ave N/I	Rata
Mammal	Muridae	*Rattus sp.*	Roedor
		Muridae N/I	
	Caviidae	*Cavia Porcellus*	Cuy
	Otariidae	*Otaria flavescens*	Lobo marino
	Canidae	*Canis lupus familiaris*	Perro doméstico
	Camelidae	*Lama sp.*	Camélido andino
	Mamífero N/I	Mamífero N/I	

presence of nearly completed dog remains as one of the first deposits on the floor of the brown phase before burial of the court may indicate the use of these animals in termination rituals similar to practices identified elsewhere in the Andes [Gamboa 2015: 89]) (Fig. 4.9).

Guinea pig or *cuy* (*Cavia porcellus*) is present in archaeological contexts in the Central Andes from the Late Archaic period (3000–1800 B.C.) either raised for meat or in an offering context (Rofes and Wheeler 2003; León 2003: 319). Evidence at Cerro del Gentil indicates that guinea pigs were an important animal for both feasts and as offerings to the dead during the Paracas occupation. There is a great quantity of cuy coprolites in Stratum D indicating that the animals were being raised in the court area. We also recovered an offering of a guinea pig inside a ceramic vessel in a human mortuary context located in the southwestern corner of the sunken court (Locus 246 in Stratum B).

Archaeological evidence suggests that another important meat source, llama (*Lama glama*)[2] was originally domesticated in the Central Andes at approximately 4000–3500 B.C. (Bonavia 2008a: 195). Llama is a primary source of meat protein in the region (León 2013: 300). Although the Puna in the high Andes is the principle habitat of llamas, they can also be reared in lower coastal elevations (see Shimada 1985; Bonavia 2008a). The specimens in our archaeological sample may have been brought in from highland ecological zones; however, we cannot reject the possibility that flocks of camelids existed on the coast during the Paracas period. Camelid bone

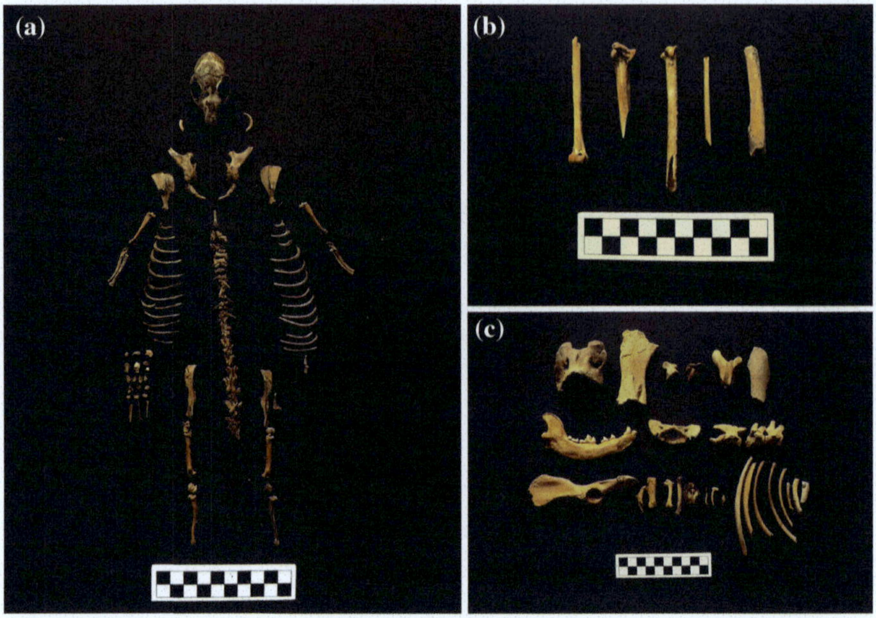

Fig. 4.9 Sample of zoological materials from the filling layers. **a** *Cavia porcellus* skeleton, **b** *Canis lupus familiaris* skull bones, **c** In the top row we can see the atlas and other unidentified bones of *lama sp.*; and in the next two rows we can see *canis lupus familiaris* bones (jaw, atlas, axis, vertebrae, scapula, phalanx, and ribs)

remains are found associated with Strata D and E. These remains have cut marks indicating processing for human consumption. Ethnohistorical and archaeological data indicate that people consumed camelid meat stewed, dehydrated (*charqui*), and roasted (León 2013: 308). Also, cooking camelid meat in an earthen pit dug in the ground (*pachamanca*) was also a common way that this meat was prepared (Bray 2003b: 101).

Finally, *Otaria flavescens* ("sea lion"), a species hunted in the littoral and transported to this part of the valley, was consumed at Cerro del Gentil. The remains of this species were found in small quantities only associated with stratum E and interspersed with different zooarchaeological, botanical, lithic, and ceramic remains. This species was also consumed in other areas of the pre-Hispanic Andes during the Pre-Ceramic period (León 2013: 342).

Ceramic Vessels and Their Use at Cerro Del Gentil

The ceramic vessels and sherds uncovered in excavation correspond to the Paracas Cavernas tradition (Tello and Mejía 1979), específically to the Pinta ceramic phase defined for Chincha valley by Wallace (1972, 1985). Many sherds originated from single vessels, although complete vessels were not discarded in single contexts. This could be because vessels were broken in a contained space and then only some sherds or fragments were interred in the strata covering the ancient sunken court. This practice is also known from other "ritual" contexts (Delgado 2013; Ikehara and Shibata 2005; Isbell and Groleau 2010; Nash and deFrance 2019). In these examples, the act of breaking vessels can be understood as a ritual act symbolizing the beginning of regeneration (Kaulicke 2005: 399).

We recovered 877 ceramic vessel fragments within the strata corresponding to uppermost layers of the sunken court and atrium. This quantity of fragments represents at least 500 vessels used during rituals related to the covering of this architectural context. Figure 4.10 presents the different vessel shapes found in the fill layers of Cerro del Gentil's Sunken Court. We defined a series of ceramic types with different functional implications, including ollas, necked jars, hemispherical bowls, hemispherical shallow bowls, plates, tazones, oversize bowls, vases, and bottles (see Fig. 4.10).

Regardless of vessel form and variants, we can make distinctions between cooking pots and service vessels (Ikehara and Shibata 2005). Inside the first group, we find *ollas* and necked jars as well as vessels such as bowls with slightly vertical wall (*tazones*), hemispherical deep bowls (*cuencos*), hemispherical shallow bowl (*escudillas*), plates (*platos*), bottles (*botellas*), glasses (*vasos*), and oversized bowls (*fuentes*) different variations. The great quantity of hemispherical bowls, shallow bowls, and

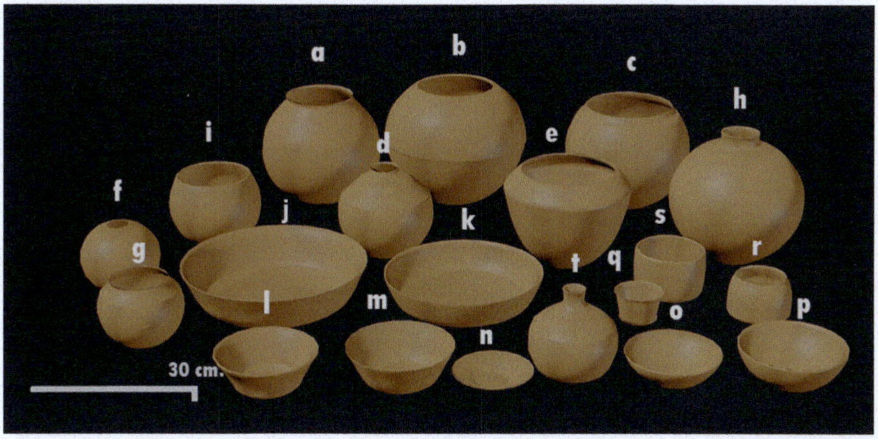

Fig. 4.10 Reconstruction of ceramic types of Cerro del Gentil: **a** Neckless olla, rim with brace A; **b** convex neckless olla; **c** neckless olla, rim with brace B; **d** small neckless olla with angular shoulder; **e** neckless olla with angular shoulder; **f** convex neckless olla; **g** curved "S" rim, neckless olla; **h** necked jar; **i** hemispherical deep bowl; **j** oversized tazón with slightly vertical wall; **k** oversized hemispherical shallow bowl; **l** tazón with slightly concave wall; **m** tazón with slightly vertical wall; **n** plate with slightly vertical wall; **o** hemispherical shallow bowl; **p** Hemispherical slightly deep bowl; **q** small tazón with slightly concave wall; **r** convex glass; **s** small tazón with right angle wall

tazones is indicative of the preparation of large quantities of food and beverages. These vessels all fall within the service group and are characterized as unrestricted vessels because one can easily see and access the contents of the vessel (Skibo 2013; Mesía 2014).

On the other hand, neckless *ollas* were used to cook or store food. *Olla* sherds included both those with and without traces of burning, thus providing clear evidence they functioned to cook or boil food and beverages. Other *ollas* were probably used to store and ferment beverages (*chicha* or *aqha*). Therefore, *ollas* with traces of burning are evidence of boiled food or food preparation resulting in stews or soups (Bray 2003b). Significantly, the finest *ollas* with post-fired paint and negative technique decorations have no traces of direct exposure to fire.

The sizes of *ollas* were important because this not only reflects volume of food that could be cooked or stored, but also what kinds of food these vessels contained (Mesía 2014: 323). Necked jars were vessels which served as containers for liquids. They have short mouth diameters and therefore are considered restricted vessels because the contents are difficult to pour out. While carrying out ritual feasting, it is highly probable liquids passed from *ollas* to *cántaros* (necked jars) and finally into smaller bottles and *tazones* (small *tazón* with slightly vertical wall) or glasses.

Another important factor regarding ceramic wares concerned oversized bowls (*fuentes*), with a diameter between 26–50 cm. These serving wares are comparable to other unrestricted vessels in that they hold a greater quantity of food. Additional vessels for personal use such as tazones, hemispherical bowls, hemispherical shallow bowls of lower diameter are also important for collective, shared use (DeBoer 2001).

Feasting at Cerro Del Gentil

All of the archaeobotanical and zooarchaeological remains, ceramic vessels, and lithic tools including granite and andesite ground stones, basalt and andesite knives, polishers, and spindle whorls indicate that the ancient sunken court of Cerro del Gentil was involved in the preparation, distribution, consumption, and discard of great quantities of food and beverages. It also appears that such feasting was associated with special events carried out in a relatively short period of time.

The great density, particular characteristics, and the high quality of the food and artifact remains in this context suggest they were associated with extraordinary activities of supra-domestic food consumption known to the anthropological and archaeological literature as feasting (Dietler and Hayden, eds. 2001; Hayden 2014). The assemblage of elements clearly follows the contexts of preparation, consumption, and discard of food. This hypothesis is significant if we take into account the great quantity of archaeobotanical, zooarchaeological, and malacological remains along with ceramic vessels that are the result of preparation, distribution, and consumption of a vast quantity of food. Also, this context takes on significance because of the presence of maize. As we have previously stated, production and consumption of alcoholic beverages play a transcendental role in feasts (also see Morris and Covey

2003; Dillehay 2003; Bray 2003a, 2003b, Bray 2009). This latter proposal is based on our identification of lithic artifacts including *manos* or grinding stones, which were probably used specifically for the grinding of maize grains for *chicha* (Camino 2008; Montibeller 2008).

A feasting model has been broadly applied to similar archaeological data and sites, especially those related to complex societies in Andean archaeology (Mesía 2014; Vega-Centeno 2005, 2007; Lau 2002; Chicoine 2011; Ikehara and Shibata 2005; Kaulicke 2005; Klarich 2005). Importantly, many of the referenced archaeological studies are similar to the findings at Cerro del Gentil.

Discussion

As we see in the archaeological and anthropological literature, the consumption of food is both a basic need for human subsistence and a cultural act. The preparation and consumption of food include an array of decisions and social processes such as selecting ingredients, preparation techniques, relationships between participants, variations of service, aesthetic values, types of consumption. Therefore, analysis of this kind of social practice interweaves different variables (e.g., history, geographical location, environment, exchange, production relations, etc.); however, once social practice related to feasting has been defined in a specific cultural milieu, it is very difficult to change (Dietler 2010). It is evident from the ceramic and archaeological data that ritual feasting represents the embodiment of identity of the participating social group (Dietler 2010: 16).

Feasts are considered extraordinary events conducted within special physical spaces. They are also characterized by having a proper liturgy of the ritual activity (performance), great quantity of food, a mechanism of dramatization, special service activities, and consumption specific to this activity. However, feasting also establishes a relationship with the commonplace consumption of food. This relationship is manifested principally in local ingredients, preparation techniques, the means of service, and consumption behavior. It is possible to detect archaeologically these discrete variables through the food remains, artifacts associated with food processing, cooking vessels, and evidence for the biogeography and consumption of certain foods.

Moreover, the creation of symbolic capital (sensu Bourdieu [1978]2013) on the part of the hosts creates a mechanism that permits the reproduction of feasting and, therefore, strengthens social roles. In the realm of the food, the aggregated symbolic value of the special vessels and common and exotic ingredients along with sophisticated food preparation techniques adds significant cultural importance to the feast. Therefore, with regard to Cerro del Gentil, it is likely that culinary expertise was developed and accumulated in the sunken court as this was where most feasting, rituals, and rites were carried out.

Dietler (2010) found that foods and consumption validated in feasting events enter into a broad and cyclic diachronic process in the annual ritual calendar (see Dillehay 2003: 358–359). Therefore, we can observe that food, and in general, consumption, accomplish an important role in the power relationships generated by these events. Additionally, feasting group membership, and differentiation with regard to "others", are set up and validated, thereby constructing Identity (Dietler 2001).

In the instance of Cerro del Gentil, the material remains of a feast, represented a closing event or ritual burial of an ancient sunken court (Tantaleán et al. 2016). Archaeological literature from the Central Andes indicates similar events were taking place in the earliest periods of the Late Archaic and continued throughout the greater part of pre-History (Vega-Centeno 2005; Ikehara and Shibata 2005; Chicoine 2011; Swenson 2011; Gamboa 2015: 88–89). The main goal of such feasting was the reproduction of the elite practices and discourses along with the communication and assimilation of these practices by guests. With this, elites ensured the mobilization of labor, establishment of their elite status, and renovation of social roles.

In Cerro del Gentil, the architectural characteristics and quality of the materials uncovered suggests that those in charge of feasting had some control over these material elements and over the attendants. The building was linked directly to Paracas elites. At the same time, this event was internalized to interested commoners in the population. These commoners, according to the archaeological evidence found at the site (Tantaleán et al. 2013; Tantaleán 2016), shared and were familiar with practices that constitute part of the local Paracas cultural traditions in the valley and beyond. Similar evidence of feasting was identified in Cerro Colorado on the Paracas peninsula (Tello 1959, 1979; Engel 1966; Pezo-Lanfranco et al. 2015; Knudson et al. 2015) and in Cerrillos in the Ica valley (Splitstoser 2009). The feasting events documented at Cerro del Gentil allow us to address the questions surrounding the preparation, distribution, and consumption of foods very important to reference with regard to elite status and rank, i.e., power relationships. These relationships are only possible when the guests recognize themselves as part of the social dynamic.

Finally, we can propose that the orchestration of the feast identified at Cerro del Gentil had important economic, political, and ideological components that were framed within the process of the burial of an ancient sunken court. One of our main goals was to elucidate the material aspects of the political and economical power that these Paracas elites maintained in the last moments of their leadership.

Conclusions

In our analysis, we presented the archaeological evidence of a political feasting event recovered during the excavation of multiple layers that were used to bury a discrete architectural feature at the Cerro del Gentil site: a sunken court. There are many examples of similar practices in pre-Hispanic Andean archaeology beginning in the Late Pre-Ceramic period (3000–1800 BC). In our case, the stratum or layers which buried this architectural space contained a huge quantity of archaeobotanical and zooarchaeological remains along with a diverse range of ceramics and stone tools associated with food processing.

We interpret the refuse as the consequences of celebrations and feasting related to the closure of the main building at Cerro del Gentil. Anthropological and ethnographical literature indicates that such activities were very common around and inside of public buildings, especially when they occurred in ritual or religious spaces. Also, the main goal of such feasting events served to include people from different communities and social status in order to create cultural cohesion through social and political bonds within the culture or community. The archaeological evidence analyzed at Cerro del Gentil, indicates a meeting of high-status leaders with the society took place in such a sacred or special place, accompanied with extraordinary food and drinks, and uncommon artifacts that this reaffirmed ethnic identities, as well as status and rank within the society (Dietler 2001).

Besides the economic and political activity clearly developed there, one of the most important motivations seems to be the offering of all of these remains and even intact artifacts to the *huaca* of Cerro del Gentil. *Huaca* is a central element in the Andean cosmovision and in the general perception and belief in a sacred landscape. Its sacredness posits its importance and such places were venerated in the ancient socially constructed landscape. Paracas cultures used such sacred places for their ritual activities and feasting based on and mediated by food. The case presented here documents a particular event that occurred at Cerro del Gentil, however, it is also an example across the world and through time where a great meal was a good excuse to share and meet with others.

Notes

1. In the system developed by John Rowe for the coastal Peruvian valleys, Chincha valley was enlisted as PV.57.
2. It is probable that bone remains excavated corresponds to *Lama glama*, because important differences in scapulae and ulna bones.

Acknowledgments The authors thank John Staller for inviting us to publish our research to this book and their help in translating this text. Also, we thank Charles Stanish, Co-Director of Chincha Archaeological Program, for allows us to present all of the data presented in this chapter. Also, we thank the Cotsen Institute of Archaeology at UCLA, the Cotsen Endowments, and Institute for Field Research (IFR), the National Geographic Society, and the National Science Foundation. We gratefully acknowledge the financial support of Harris Bass, Bruce Hector, and Charles Steinmetz and thank the Ministry of Culture for permits and supervision of our work. We especially acknowledge the assistance of Rubén García in his capacity as the representative of the Dirección Desconcentrada del Ministerio de Cultura of Ica. We particularly thank Kelita Pérez for her assistance in the field. Mr. Luka Baraka and Ing. Luis García, from the company Agroexportadora Virgen del Rosario, for generously supporting our research adjacent to their land, and facilitated our work. We finally acknowledge the work of Paolo Zorogastúa and Mary Ávila for the archaeobotanical and zooarchaeological analyses.

References

Alamo V, Valdivieso V (1997) Lista sistemática de moluscos marinos del Perú (Segunda Edición, revisada y actualizada). Instituto del Mar del Perú, Callao

Bray T (2003a) Inka pottery as culinary equipment: food, feasting, and gender in imperial state design. Latin Am Anti 14(1):3–28

Bray T (2003b) To dine splendidly: imperial pottery, commensal politics, and the Inca state. In: Bray T (ed) The archeology and politics of foods and feasting in early states and empires. Plenum Publishers, New York, pp 93–142

Bray T (2009) The role of Chicha in Inca state expansion: a distributional study of Inca Arîbalos. In: Jennings J, Bowser B (eds.), Drink, power and society in Andes State, pp 108–132. University Press of Florida, Gainesville

Bonavia D (2008a) The south American camelids. Cotsen Institute of Archaeology, UCLA. Los Angeles, An Expanded and Corrected edition

Bonavia D (2008b) El maíz: Su origen, su domesticación y el rol que ha cumplido en el desarrollo de la cultura. Fondo editorial de la Universidad San Martín de Porres, Lima

Bourdieu P ([1978] 2013) Symbolic capital and social classes. J Class Soc 13(2):292–302

Camino L (2008) Norte, chicha y sol. In: León R (ed.), Chicha peruana, una bebida, una cultura, pp 34–71. Fondo editorial de la Universidad San Martín de Porres, Lima

Canziani J (1992) Arquitectura y urbanismo del período Paracas en el valle de Chincha. Gaceta Arqueológica Andina 6:87–117

Canziani J (2009) Ciudad y territorio en los Andes: Contribuciones a la historia del urbanismo prehispánico. Fondo Editorial Pontificia Universidad Católica del Perú, Lima

Chicoine D (2011) Feasting landscapes and political economy at the Early Horizon center of Huambacho, Nepeña valley, Peru. J Anthropol Archaeol 30(3):432–453

DeBoer W (2001) The big drink: feast and forum in the upper amazon. In: Dietler MD, Brian H (eds.), Feasts, archaeological and ethnographic perspectives on food, politics and power, pp 215–239. Smithsonian Institution Press, Washington DC

Delgado C (2013) Feast and offering in Arcopata, Cusco. Andean Past 11:85–110

Dietler M (2001) Theorizing the feast: rituals of consumption, commensal politics, and power in African contexts. In: Dietler M, Brian H (eds.), Feasts: archaeological and ethnographic perspectives on food, politics, and power, pp 65–114. Smithsonian Institution Press, Washington DC

Dietler M (2010) Cocina y colonialismo. Encuentros culinarios en la Francia mediterránea protohistórica. In: Mata C, Pérez G, Vives-Ferrándiz J (eds.), De la Cuina a la Taula: IV Reunió d'Economia en el Primer Milenio, pp 11–26. Universitat de Valencia, Saguntum, Valencia

Dietler M, Hayden B (eds.) (2001) Feasts: archaeological and ethnographic perspectives on food, politics, and power. Smithsonian Institution Press, Washington DC

Dillehay T (2003) El colonialismo Inka, el consumo de chicha y los festines desde una perspectiva de banquetes políticos. Boletín de Arqueología PUCP 7:355–363

Engel F (1966) Paracas, cien siglos de cultura peruana. Editorial Juan Mejía Baca, Lima

Gamboa J (2015) Dedication and termination rituals in Southern Moche public architecture. Latin Am Anti 26(1):87–105

Hayashida F (2008) Chicha histories: pre-hispanic brewing in the Andes and the use of the ethnographic and historical analogues. In: Jennings J, Bowser B (eds) Drink, power and society in Andes State. University Press of Florida, Gainesville, pp 232–256

Hayden B (2014) The power of feast: from prehistory to the present. Cambridge University Press, New York

Ikehara H, Shibata K (2005) Festines e integración social en el período Formativo: Nuevas evidencias de Cerro Blanco, valle bajo de Nepeña. Boletín de Arqueología PUCP 9:123–159

Ikehara H, Paipay F, Shibata K (2013) Feasting with Zea mays in the Middle and Late Formative north coast of Peru. Latin Am Anti 24(2):217–231

Kaulicke P (2005) Las fiestas y sus residuos: Algunas reflexiones finales. Boletín de Arqueología PUCP 9:387–402

Klarich E (2005) ¿Quiénes eran los invitados? Cambios temporales y funcionales de los espacios públicos de Pukará como reflejo del cambio de las estrategias de liderazgo durante el periodo Formativo Tardío. Boletín de Arqueología PUCP 9:185–206

Knudson K, Peters A, Tomasto E (2015) Paleodiet in Paracas Necrópolis of Wari Kayan: carbon and nitrogen isotope analysis of keratin samples from the south coast of Peru. J Archaeol Sci 55:231–243

Kroeber A (1944) Peruvian archeology in 1942. Viking Fund Public Anthropol, New York

Lau G (2002) Feasting and ancestor veneration at Chinchawas, north highlands of Ancash, Peru. Latin Am Anti 13(3):279–304

León E (2013) 14,000 años de alimentación en el Perú. Universidad de San Martín de Porres, Lima

Lumbreras L (2008) La présence de Paracas a Chincha. In: Lavallée D (ed.), Paracas. Trésors Inédits du Perou Ancient, pp 34–39. Musée du Quai Branly, París

Menzel D, Rowe J, Dawson L (1964) The Paracas Pottery of Ica: A Study in Style and Time. University of California Press, Berkeley and Los Angeles

Mesía C (2014) Festines y poder en Chavín de Huántar durante el periodo Formativo Tardío en los Andes centrales. Chungará 46(3):313–343

Montibeller M (2008) Chicha: Vitalidad en los Andes. In: León R (ed.), Chicha peruana, Una bebida, una Cultura, pp 74–119. Fondo editorial de la Universidad San Martín de Porres, Lima

Morris C, Covey A (2003) La plaza central de Huánuco Pampa: Espacio y transformación. Boletín de Arqueología PUCP 7:133–149

Nash DJ, deFrance SD (2019) Plotting abandonment: excavating a ritual deposit at the wari site of Cerro Baul. J Anthropol Archaeol 53:112–132

Pezo-Lanfranco L, Aponte D, Eggers S (2015) Aproximación a la dieta de las sociedades formativas tardías del litoral de Paracas (Costa sur del Perú): evidencias bioarqueológicas e isótopicas. Ñawpa Pacha 35(1):23–55

Proulx D (2008) Paracas and Nazca. Regional cultures of the south coast of Peru. In: Silverman H, Isbell W (eds) Handbook of South American archaeology, pp 563–585. Springer, New York

Rofes J, Wheeler J (2003) Sacrificio de cuyes en los Andes: el caso de El Yaral y una revisión biológica, etnográfica y arqueológica de la especie Cavia porcellus. Archaeofauna 12:29–45

Skibo J (2013) Understanding pottery function. Springer, New York

Shimada M (1985) Continuities and changes in patterns of faunal resource utilization: Formative through Cajamarca periods. In: Terada K, Onuki Y (eds.), The Formative Period in the Cajamarca basin, Perú: excavations at Huacaloma and Layzón, 1982: report 3 of the Japanese scientific expedition to nuclear America, pp 289–305. University of Tokyo Press, Tokyo

Splitstoser J (2009) Weaving the structure of the Cosmos: cloth, agency, and worldview at Cerrillos, an early Paracas site in the Ica valley, Peru. PhD. Dissertation, Department of Anthropology, School of Arts and Sciences of the Catholic University of America, Washington DC

Stanish C, Tantaleán H (2012) Informe de campo e informe final: Proyecto de investigación arqueológica Cerro del Gentil. Chincha, Presentado al Ministerio de Cultura del Perú

Stanish C, Tantaleán H (2014) Informe de campo y final del proyecto de investigación arqueológica "Excavaciones en el sitio Cerro del Gentil y prospección del valle medio de Chincha." Entregado al Ministerio de Cultura del Perú, Lima

Stanish C. Tantaleán H (2015) Informe de campo y final del proyecto de investigación arqueológica "Excavaciones arqueológicas en Cerro del Gentil y el Complejo Soto, valle de Chincha". Entregado al Ministerio de Cultura del Perú, Lima

Swenson E (2011) Architectural renovation as ritual process in Late Intermediate period Jequetepeque. In: Zori C, Johnson C (eds.), From state to empire in the prehistoric Jequetepeque valley, Peru, pp 105–128. British Archaeological Reports, International Series 2310, Oxford

Tantaleán H, Stanish C, Zegarra M, Pérez K, Nigra B (2013) Paracas en el valle de Chincha: Nuevos datos y explicaciones. Boletín de Arqueología PUCP 17:31–57

Tantaleán H, Stanish C, Rodríguez A, Perez K (2016) The final days of paracas in Cerro del Gentil, Chincha Valley, Peru. PLoS ONE 11(5):e0153465. doi:10.1371/journal.pone.0153465
Tello, JC, Mejía T (1959) Paracas. Primera parte. Empresa Gráfica T. Scheuch S.A. Lima
Tello JC, Mejía T (1979) Paracas. Segunda parte: Cavernas y Necrópolis. Universidad Mayor de San Marcos/Institute of Andean Research de Nueva York, Lima
Vega-Centeno R (2005) Consumo y ritual en la construcción de los espacios públicos para el periodo Arcaico Tardío: El caso de Cerro Lampay. Boletín de Arqueología PUCP 9:91–121
Vega-Centeno R (2007) Construction, Labor Organization, and Feasting During the Late Archaic Period in the Central Andes. J Anthropol Archaeol 26(2):150–171
Wallace D (1959) Informe del reconocimiento del valle de Chincha. Revista del Museo Regional de Ica X(11):31–40
Wallace D (1970a) Informe de reconocimiento del valle de Chincha. Arqueología y Sociedad 2:13–17
Wallace D (1970b) Trabajo de campo en la costa sur del Perú. Arqueología y Sociedad 2:19–24
Wallace D (1971) Sitios arqueológicos del Perú (Segunda Entrega): Valles de Chincha y de Pisco. Arqueológicas 13:1–131
Wallace D (1985) Paracas in Chincha and Pisco: a reappraisal of the Ocucaje sequence. In: Kvietok P, Sandweiss D (eds.), Recent studies in Andean prehistory and protohistory, pp 67–94. Latin American Studies Program, Cornell University, Ithaca
Wallace D (1986) The Topará tradition: an overview. In: Sandweiss D, Kvietok P (eds.), Perspectives on Andean Prehistory and Protohistory, pp 35–47. Latin American Studies Program, Cornell University, Ithaca
Zúñiga O (2002) Macrofauna y algas marinas N°1 Moluscos [Guía de biodiversidad vol. 1]. Universidad de Antofagasta, Antofagasta

Chapter 5
Cuisine and Social Differentiation in Late Pre-hispanic Cajamarca Highlands of Northern Peru

Jason L. Toohey

Abstract Cuisine, the shared patterns of food preparation techniques and contexts of consumption, is a critical factor in addressing questions of community, social, and political organization, identity, and status differentiation. In the Cajamarca region of the northern Andean highlands, spatial patterning in both food remains and the ceramics involved in their cooking and serving are beginning to shed light on these aspects of social life at the Late Intermediate Period (AD 1000–AD 1465) fortified community of Yanaorco. Patterning in excavated food resources and ceramics indicate that, in addition to spatial and architectural differentiation, household social status was also marked by variation in the daily activities and practices of domestic cuisine. These patterns are most evident in differential access to camelid resources, variability in the state of faunal elements after cooking, and inter-household differences in the ratios of particular storage, cooking, and serving vessel forms. The utility of the organizing concept of cuisine is that it provides a holistic, multidimensional, and comparative means of evaluating social and political variation within households and communities, where its daily practice would have worked to both produce and reproduce social patterns that in some cases integrated the community, while in others internally divided it.

Introduction

Peruvian cuisine has received international acclaim in the last decade with high profile *novo andina* restaurants opening in many cosmopolitan cities of North and South America. Recently published cookbooks celebrate the "art of Peruvian cuisine," homogenizing its many influences under a national cuisine. Peru even has its own celebrity chef, Gastón Acurio, who has over 30 restaurants in more than 12 countries, and Lima has emerged as a culinary capital of South America and consequently a favorite destination for traveling foodies. The elevation of Peruvian cuisines to

J. L. Toohey (✉)
Department of Anthropology, University of Wyoming, Laramie, WI, USA
e-mail: jtoohey2@uwyo.edu

J. E. Staller (ed.), *Andean Foodways*, The Latin American Studies Book Series,
https://doi.org/10.1007/978-3-030-51629-1_5

haute cuisine has made traditional dishes at many high-end restaurants unaffordable to the average Peruvian. Traditional pre-Hispanic highland dietary staples, such as llama, alpaca, and guinea pig (*cuy*), are now served as the most expensive specialties in Peruvian restaurants. The ingredients, seasoning, and preparation techniques of today's Peruvian cuisine symbolize and communicate identity, particularly class identity. While these symbols are particularly marked in the early twenty-first century, food and cuisine also acted as symbols of identity in pre-Hispanic domestic and public contexts, marking class and status differences and similarities.

My research in the northern Andes addresses social and economic differentiation in terms of foodways and daily culinary practice. I compare elite and lower status domestic contexts at the Late Intermediate Period (~AD 1000–AD 1465) community of Yanaorco in the northern Peruvian highlands. The class and intra-site social separation is identified not only in the range of resources available to different domestic units, but also in the nature of food preparation techniques, serving practices, and how these are patterned across the community.

Due to its multidimensional scope, the cuisine concept is ideal for investigating issues of status and identity in the archaeological past. Cuisine, as a concept, is now widely utilized in comparisons of societies, cultures, and ethnic collectives. Cuisine encompasses a variable and patterned set of socially produced and reproduced "norms" of practice in the realm of food choice, preparation, consumption, and discard. It encompasses basic ingredients, cooking techniques, "etiquettes" of consumption, and flavor profiles. In the following study, I emphasize the aspects of cuisine most closely associated with differences in primary ingredients, food preparation techniques, and consumption (Farb and Armelagos 1980; Rozin 1982). Cuisine is often thought to define or distinguish a culture, to provide unity to an otherwise heterogeneous community and to cross class lines (Barthes 1979 [1961]; deFrance 2006). Caution should be taken to document subtle variability within a cuisine as this can indicate patterned class differentiation within a community.

Feasting and the Cuisine Concept

The preparation and consumption of food is both quotidian experience and expressive of identity, class, and social place within a community (Atalay and Hastorf 2006; Belasco 2008; Crown 2000; Douglas 1999; Farb and Armelagos 1980; Goody 1982; Hastorf 2012; Rozin (1982); Welch and Scarry 1995). This dynamic connection between foodways and identity is perhaps best thought of in terms of the concept of cuisine, which has been described as encompassing four primary points (Farb and Armelagos 1980). A cuisine encompasses a basic set of primary foods, or key ingredients. They are characterized by distinctive methods and sequences of food preparation (Rozin 1982). The cooking of this food is more than an economic dietary transformation; it also has cultural significance (Smith 2012). Cuisines are characterized by distinguishing flavors, brought about through the employment of particular combinations of seasoning elements. They involve more than the dishes themselves but are also integral to culinary practices, rules, and accepted manners which revolve around actions of preparation, serving, and consumption, either of everyday meals or large feasts. Each of these aspects is culturally produced and reproduced, forming what Barthes calls "a system of differences in signification" (1979 [1961]: 168). The cuisine is not static but, like other expressions of identity, changes according to the social context of a particular meal. Finally, cuisine is conservative and its rules of preparation and consumption change only very slowly over time (Crown 2000; Farb and Armelagos 1980; Goody 1982). These are understandings passed generationally within the walls of kitchens and between women and offspring, often daughters.

It may be tempting to characterize a society or culture on the basis of an overarching and integrated cuisine, but meaningful and signaling variation does occur within cuisines (Goody 1982). Rozin (1982) speaks to the potential for variation within a cuisine based on variability in the subtype of a particular basic ingredient, the use of different cooking fuels, or different classes of a particular seasoning agent. Such variability in the preparation of cuisine may be a marker of status and class difference (Goody 1982; Grant 2002; Gumerman 1991; Dietler 2001; Welsh and Scarry 1995).

Here, I focus on variability in food preparation methods as a marker of status difference. The dynamics of cuisine are practiced along a continuum of scale ranging from simple everyday household meals to small-scale commemorative household feasts, to larger scale commensal feasting events at the community and state scales. Feasting, the practice of extra-household, commensal food preparation, cooking, and consumption, is carried out within most societies in the past and present throughout the world (Dietler 2001; Dietler and Hayden 2001; Klarich 2010; Mills 2002). Feasting events are both sponsored and attended for a wide range of reasons, from work-party feasting (Dietler and Herbich 2001), to events meant to mark or commemorate stages in a life-cycle (Gumerman 2010; Lau 2002), to large-scale state-level political feasting meant to establish and periodically reiterate relations of power and asymmetrical reciprocity as well as certain points in the annual solar or lunar cycle (Bray 2003b; Staller 2008).

In trying to expand the breadth and scope of food research at broader level, this research is consciously focused upon foodways at both the household and community levels. I explore variation within the local cuisine with reference to social and economic differentiation. I subsequently revert to the concept of political or commensal feasting in terms of the differential capacities of domestic units within the community who sponsored such events.

One of the strengths of the cuisine concept is its breadth and inclusive scope, involving everything from the staple ingredients of a dish to the rituals and habits of a meal. These can all be observed in the ethnographic present, but reconstruction of a culinary tradition is much more difficult when it is generated from the archaeological record. It is the material record of cuisine, the kinds of ingredients encountered, and the objects involved in preparation, cooking, and consumption, as well as places or contexts within which such activities occur. While cooking features, ceramics, and faunal remains may be present and easily identified at archaeological sites, botanical remains are often absent, or only trace quantities are preserved unless one works in an arid environment. Many of the nuanced aspects of seasoning, symbol, and behavior must be carefully inferred from this material. Therefore, this investigation focuses on the archaeologically preserved materials including architecture, ceramics vessels, and faunal remains.

Food Studies in the Andes

Food production and consumption in the Andes has become increasingly important to archaeological investigations over the past 25 years. Data related to ancient foodways and cuisine come from ethnohistoric accounts, ethnographic information, and archaeological sources (Cuéllar 2013).

A number of ethnohistoric accounts have helped to establish not only the primary ingredients prepared and consumed in general through the Central Andes, but also some of the common culinary techniques employed. We know that the primary domesticated fauna consumed included camelids (both alpaca (*Vicugna pacos* L.) and llama (*Lama glama* L.), guinea pig (*Cavia porcellus* L.), muscovy duck (*Cairina moschata* L.), and, on occasion, dog. Important highland plant staples included various tubers, potatoes, maize, beans, squash, achira (*Canna indica* L.), and quinoa. In terms of cuisine, a critical flavoring element throughout the Andes was the ají chili pepper (*Capsicum baccatum* L.) Archaeologists have also documented something of the relatively limited set of cooking techniques employed, including grilling, roasting, stewing, boiling, and grinding (Cieza de Leon 1984 [1553]; Cobo 1990 [1653]). The drying of camelid meat, the creation of jerky (*charki)*, was also critical in the production of storable meat.

The importance of food and cuisine to the maintenance of ethnic identity, the marking of social status within communities, and in particular the political and social power of women in Andean cultures is clearly documented from ethnographic evidence over the past five decades (Isbell 1978; Meyerson 1990; Weismantel 1988). The power wielded by women not only within the kitchen and in the realm of food preparation, but also publicly in terms of the rules and practices of serving order both during small daily meals and larger community feasting events which underscores the importance of cuisine as a source of power for women in Andean societies (Isbell 1978; Meyerson 1990; Weismantel 1988). Women are also the source for the transmission of social and traditional knowledge of cuisine through teaching or by passing on foodways in the kitchen.

Weismantel (1988) notes inter-household variation in the realm of diet, with variation in wealth or affluence showing ethnographically in the degrees of freshness. Additionally, aspects of cuisine, particularly differential patterns of cooking and food preparation between households, are related to social and status differentiation. Just as variability in diet can reflect wealth and status variation, so too can variations in food preparation.

Food is not only an important indication of identity and social status, but it is also a nexus for women's political and social power (see e.g., Gumerman 1997).

Archaeologists have long focused upon food production and consumption patterns as reflecting to varying degrees of cultural and political integration of communities and households, as well as gender and power in early states (Bray 2003a; Dietler and Hayden 2001; Gero 1992). Much recent work has focused on the political and economic importance of feasting at the community and early state levels, as well as the reciprocal dynamics of feasting in creating and maintaining social and political organization (Cook and Glowacki 2003; Gero 1992; Hastorf 1991; Janusek 2004; Nash 2010). Critically, investigators are also focusing on the nature of everyday cuisine at the household and community scales of domestic economy (Cutright 2010; Hastorf 1990). Although the sponsoring of feasts certainly functioned to communicate, the everyday practice of food production and consumption was also critical in the creation and recreation of identity and status (Gero 1992; Gumerman 1991; Hastorf 1991; Sandefur 2001). Sandefur, for example, looks at both the elements present and estimates meat amounts in comparing elite to non-elite households in the Late Intermediate Period of the Mantaro Valley and concludes that although all households had access to camelid, elites had greater access to protein, i.e., better elements, "meatier" cuts (Sandefur 2001). On the north coast of Peru, Cutright (2010) has integrated botanical, faunal, and ceramic data at several north coast LIP sites to speak to shifting Lambayeque and Chimú cuisines. Working with well-preserved food remains and secure contexts, she analyzes food offerings in their ceramic vessels in order to shed light on the social significance of everyday meals.

Cajamarca Geography and a Brief Culture History

The Cajamarca Tradition developed over approximately 1500 years within the northern Peruvian highlands east of the westward flowing Chicama, Jequetepeque, Zaña, and Lambayeque Valleys, and west of the Marañón River. Topography here ranges from jagged and mountainous zones to rolling hills and basins. The Cajamarca Valley is the largest of several highland basins in the region and was as agriculturally rich in the past as it is today (Fig. 5.1).

Pre-Hispanic Cajamarca communities practiced an agro-pastoral economy. It is unclear whether communities in the Quechua zone (~2300–3500 masl) close to basin bottomlands (~2700 m) maintained access to high elevation grazing resources in the Suni/Paramo zone (above 3500 masl), or if people of these two zones exchanged resources (Murra 1972; Pulgar Vidal 1981). Cultivation was present in the region since the Initial Period, and maize appeared by at least 1000 BC (Grobman and Ravines 1974; Terada and Onuki 1985; Tykot et al. 2006). Food production intensified during the Early Cajamarca period (~AD 200–AD 600) as evidenced by the construction of formal agricultural terracing on the edges of the valley for the first time (Julien 1988).

Western travelers have visited this highland region since the nineteenth century, describing and illustrating some of its ruins, but systematic archaeology began in the mid-twentieth century (Squier 1877; Tello 2004; Wiener 1880). Henry and Paule

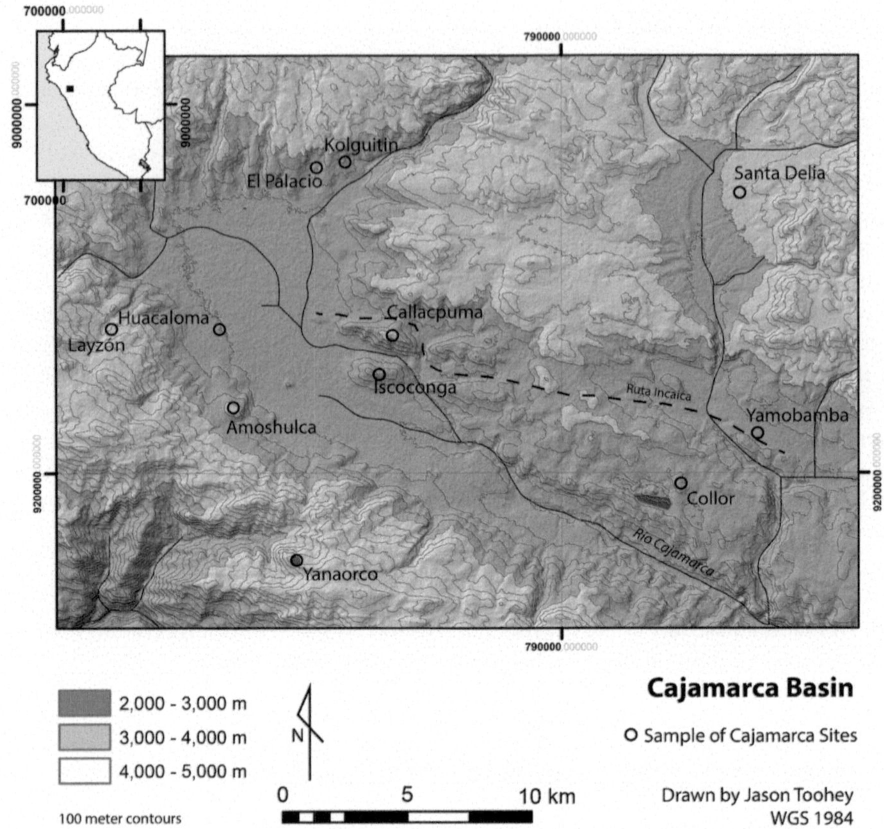

Fig. 5.1 Map of the Cajamarca basin indicating the location of Yanaorco (Illustrated by Jason Toohey)

Reichlen tested several archaeological sites in Cajamarca and created the tradition's first ceramic chronology, characterized by five periods (Reichlen 1970; Reichlen and Reichlen 1949, 1985). The excavation of important Formative Period sites in the 1970s and 1980s by the Japanese Expedition resulted in a new ceramic chronology including the Formative periods (Matsumoto 1982; Terada and Matsumoto 1985; Terada and Onuki 1982, 1985, 1988). This new five-stage chronology, later refined by Julien (1988, 1993) and Watanabe (2009), is the system most often utilized today (Cardich 1994; Ravines and Rogger 1968, 1976, 1985; Sachún 1986; Silva Santisteban 2000).

The Yanaorco Community

The Late Intermediate Period (AD 1000–AD 1465) began with the collapse of Wari Imperial influence through the central Andes. This was associated with increased competition and, in some areas of the highlands, conflict among small-scale chiefly societies as aspiring elites fought to fill local power vacuums (Conlee et al. 2004). This seems to have been the case in the Cajamarca region where by AD 1000, a settlement shift was underway with many people moving into higher elevation defensible communities (Julien 1988; Toohey 2009). It was within this social and political flux that the village of Yanaorco emerged (Toohey 2009, 2011, 2012).

Yanaorco is a fortified 14 ha. village located along a finger ridge overlooking the strategically important Gavilán Pass, connecting the Jequetepeque Valley to the intermountain Cajamarca Basin (see Fig. 5.1). This settlement was characterized by sectors of elite and non-elite domestic terraces, public plazas, three platform mounds, and a series of open corrals (Figs. 5.2 and 5.3). Unlike Huacaloma and other communities closer to the basin floor (see Fig. 5.1), Yanaorco was much nearer to high elevation (3550 masl) pasturage with grasslands within a 30-min walk to the east. There is also evidence of camelid husbandry in the form a large probable corrals at the west end and camelid dung in Architectural Unit (AU) 20X where two well-preserved spherical droppings were recorded on an activity surface.

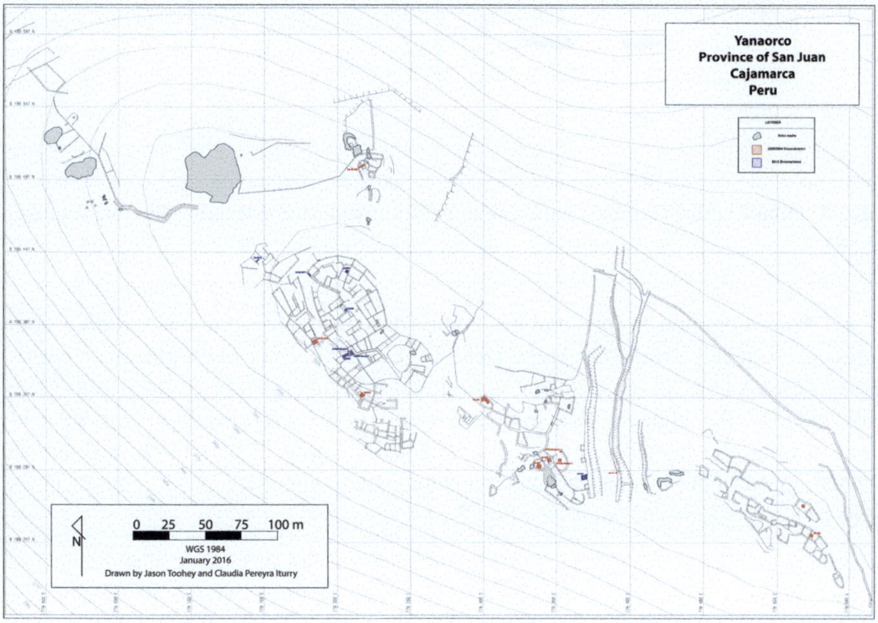

Fig. 5.2 Site map of Yanaorco (Illustrated by Jason Toohey)

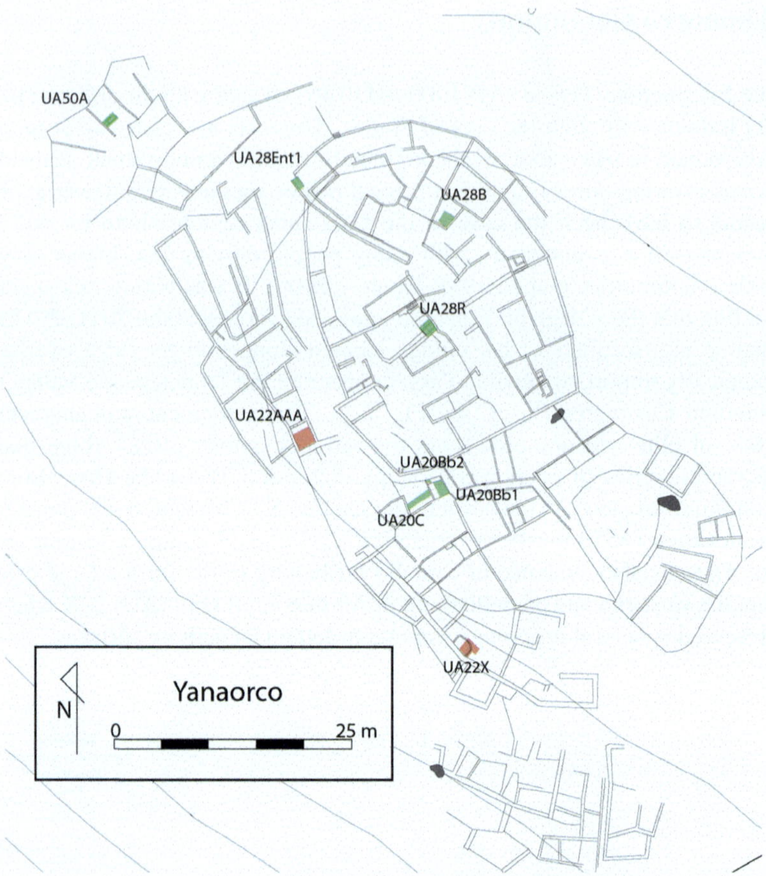

Fig. 5.3 Detailed plan of contexts and excavation blocks mentioned in the text (Illustrated by Jason Toohey)

All data integrated in this analysis come from excavated contexts in either open patios or domestic contexts. Areas 18A, 18D, and 22X are open plazas or patios. Areas 22AAA and 20I are elite and non-elite domestic contexts, respectively. Areas 20B and 20C are adjacent elite domestic contexts that are closely associated with a set of elite and possibly public spaces. Finally, room 28B is likely a domestic context, and 28ENT1 is an entrance corridor which was filled at some point in the past, creating a dense midden (see Fig. 5.3).

Treatment of Food in the Cajamarca Literature

Despite nearly seventy years of relatively consistent archaeological research in the Cajamarca region, reporting of food remains is rare, and discussion of the importance of food to understanding social and political organization is virtually absent. Most of what we know of food in pre-Hispanic Cajamarca comes from a long-term Japanese expedition that carefully described the palynology and faunal remains encountered from the early sites of Huacaloma and Layzón (Shimada 1982, 1985; Terada and Onuki 1982, 1985, 1988; Toohey 2009).

Resources available to Cajamarca communities were also ubiquitous in other highland regions and included potato, maize, beans, quinoa, and ají, as well as domesticated llamas and alpacas, guinea pig, and wild mammals including deer, rabbit, and others (Bray 2003b). One important finding of this work has been the identification of a clear shift from a hunting and gathering adaptation to that of camelid domestication (Shimada 1982, 1985). Throughout the Formative Period (~2000 BC–~200 BC), deer made up approximately 95% of the faunal assemblage. This focus on wild game shifted through the Early Layzón Period, and by the beginning of the Layzón Period, camelids made up about 95% of the assemblage (Shimada 1982, 1985). Deer remained present in local contexts, but they were rare. Evidence from Layzón period levels at Huacaloma also hints at an economy of exchange between valley bottom sites, i.e., Huacaloma and higher elevation communities. A lack of cranial and distal limb elements at Huacaloma has lead Melody Shimada (1982) to suggest that camelid meat was probably traded into the community as *charki* (dried meat) from higher elevation communities.

Elite and Commoner Spaces at Yanaorco

Any archaeological treatment of foodways within a community and its households should integrate multiple data sets (Atalay and Hastorf 2006; Twiss 2015; Van Derwarker and Peres 2010). Here, I analyze the ceramics, faunal remains, and to a lesser extent architectural patterns in order to gain a comparative understanding of elite and non-elite residential or high- and low-status architectural spaces at the site. The research takes into account a number of factors, ranging from the size of the space, the number of rooms, wall thickness and quality of construction, the elaboration of interior spaces, and the differential presence of high-quality artifacts, and other materials in domestic contexts (Hirth 1993).

The direct comparison of elite contexts like AU22AAA or AU20B with a non-elite domestic context is difficult, given the stark contrasts in the architecture. Elite domestic contexts at Yanaorco have well-defined thick walls faced with cut stone, and interior space is divided and elaborated by features like benches, dividing walls, and wall niches. Non-elite domestic contexts are characterized by open areas on domestic terraces with few well-preserved walls. Architecture in these lower status contexts may have been more ephemeral, perhaps produced of *quincha* (wattle and daub) walls. Domestic function was interpolated based on the presence of house-related materials, such as those involved in food production, preparation, or storage or those that were related with domestic activities, such as spindle whorls, bone tools, and lithics.

Yanaorco Ceramics

The frequencies of particular vessel types from space to space are presented in Table 5.1. Vessel forms included plates with walls angled less than 30°, four different kinds of bowls, both narrow and wide mouth jars, and broad, open, colanders. The original functions of these vessels were largely based on their formal qualities, with serving vessels such as plates and bowls, and jars used for in long- and short-term storage (Toohey 2011). Plates, convex bowls, and straight-sided, everted bowls are probably associated with serving dry or relatively dry foods, while both vertical and incurving sided bowls are more likely used to serve liquids (Hally 1986; Hendrickson and McDonald 1983). Jars are divided into narrow-necked jars, probably used in the storing of liquids, and wide-necked vessels more likely associated with storage of dry goods.

There are similarities in vessel forms present across all excavated contexts, but the relative frequencies of different forms, particularly at elite and non-elite domestic spaces, and also between different high-status zones, indicate different regimes of storage, cooking styles, and serving priorities. These differences also likely speak to greater facilities to store food and potentially to sponsor commensal consumption events by elites. Both elite and non-elite domestic zones utilized storage vessels, but in very different frequencies.

As would be expected, serving food frequently occurred in both elite and non-elite contexts, but had higher frequencies of bowls and plates in elite domestic contexts. Interestingly, incurving bowls are present in greater frequencies with regard to the serving repertoire in lower status contexts, possibly indicating a greater focus on the consumption of liquids.

Although a few colander fragments were found in non-elite spaces, these were much more common in elite and more public spaces indicating the processing or cooking of particular items in these spaces and not in lower status residential zones. Complete colanders were not discovered at Yanaorco, but the rims indicate they were shallow, broad (averaging approximately 50 cm across) basins having had rectangular punctuation removed while the clay was leather-hard before firing (Fig. 5.4). Complete colanders in a local private collection suggest that they may have been involved with open-fire grilling of larger elements of camelid or deer, but further research is required for such a confirmation. It is also possible that these broad colanders were associated with toasting or drying maize, possibly associated with the process of *chicha* production.

Table 5.1 Summary statistics for ceramics from contexts discussed in text

Excavation context	20I			20AAA			18D			18A			22XXX		
Cubic meters excavated	3.64			4.782			1.444			1.636			1.708		
	N	%	$\#/m^3$	N	%	$\#/m^3$	N	%	$\#/m^3$	N	%	$\#/m^3$	N	%	$\#/m^3$
Bowl straight	20	17.2	5.49	47	16.3	9.83	5	18.5	3.46	4	36.4	2.44	48	23.2	28.10
Bowl converse	38	32.8	10.44	89	30.8	18.61	3	11.1	2.08	4	36.4	2.44	55	26.6	32.20
Bowl vertical	10	8.6	2.75	12	4.2	2.51	1	3.7	0.69	0	0	0	11	5.3	6.44
Bowl incurving	16	13.8	4.40	13	4.5	2.72	4	14.8	2.77	2	18.2	1.22	7	3.4	4.10
Plate	10	8.6	2.75	13	4.5	2.72	3	11.1	2.08	0	0	0	7	3.4	4.10
Jar narrow neck	10	8.6	2.75	47	16.3	9.83	6	22.2	4.16	1	9.1	0.61	35	16.9	20.49
Jar wide neck	9	7.8	2.47	50	17.3	10.46	3	11.1	2.08	0	0	0	40	19.3	23.42
Colander	3	2.6	0.82	18	6.2	3.55	2	7.4	1.39	0	0	0	4	1.9	2.93
Total (diagnostic)	**116**			**289**			**27**			**11**			**207**		

(continued)

Table 5.1 (continued)

Excavation context	20I			20AAA			18D			18A			22XXX		
Cubic meters excavated	3.64			4.782			1.444			1.636			1.708		
	N	%	#/m³	N	%	#/m³	N	%	#/m³	N	%	#/m³	N	%	#/m³
Excavation context	**20B1**			**20C**			**28B**			**28ENT1**			**Total**		
Cubic meters excavated	**2.482**			**1.931**			**1.724**			**0.328**					
	N	**%**	**#/m³**	**N**	**%**	**#/m³**	**N**	**%**	**#/m³**	**N**	**%**	**#/m³**			
Bowl straight	58	26.6	23.37	93	37.3	48.16	52	39.1	30.60	72	35.1	219.51	**399**		
Bowl converse	42	19.3	16.92	13	5.2	6.73	15	11.3	8.70	24	11.7	73.17	**283**		
Bowl vertical	10	4.6	4.03	26	10.4	13.46	9	6.8	5.22	4	2.0	12.20	**83**		
Bowl incurving	20	9.2	8.06	21	8.4	10.88	13	9.8	7.54	15	7.3	45.73	**111**		
Plate	3	1.4	1.21	5	2.0	2.59	5	3.8	2.90	2	1.0	6.10	**48**		
Jar narrow neck	16	7.3	6.45	39	15.7	20.20	9	6.8	5.22	25	12.2	76.22	**188**		
Jar wide neck	37	17.0	14.91	39	15.7	20.20	28	21.1	16.24	57	27.8	173.78	**263**		
Colander	32	14.7	12.89	13	5.2	6.73	2	1.5	1.16	6	2.9	18.28	**80**		
Total (diagnostic)	**218**			**249**			**133**			**205**			**1455**		

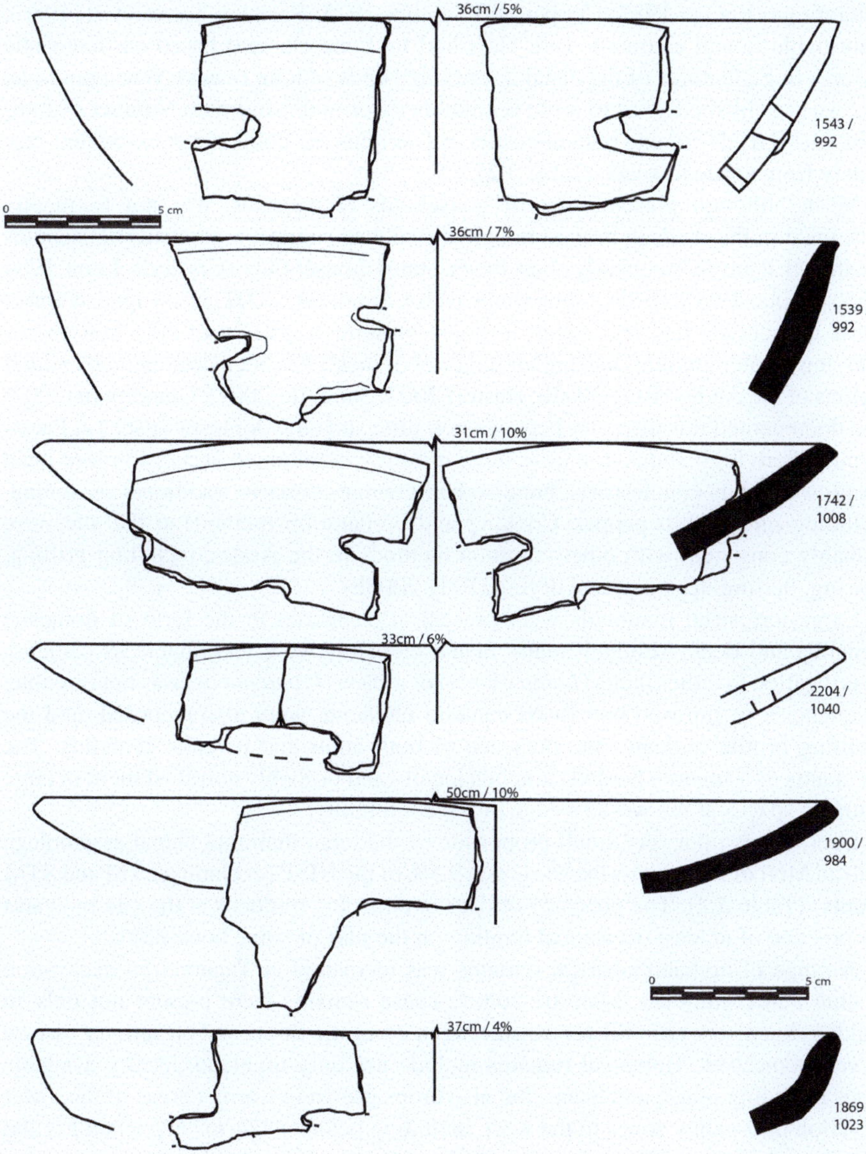

Fig. 5.4 Examples of colander vessels (Illustrated by Jason Toohey)

Faunal and Plant Remains at Yanaorco

All excavated bones were sorted into a series of broad categories including camelid, deer, Large Terrestrial Mammal (indet.) (LTM), guinea pig, Small Terrestrial Mammal (indet.) (STM), rabbit, dog, bird (indet.), fish, and terrestrial mollusk

(Altamirano Enciso 1983; Cooper and Schiller 1975; Pacheco Torres et al. 1986). Identifiable faunal elements were classified by bone element based on diagnostic features (e.g., features of the distal or proximal ends of long bones). When possible, side was established so as to more accurately determine Minimum Number of Individuals (MNI). NISP and both elements and weights per cubic meter excavated vary widely from space to space (Table 5.2).

Meat, although present in most households' economies, was not frequently consumed in the Andean past, although chroniclers suggest high-status households consumed it more frequently than lower status households (Cieza de León 1984 [1553]; Cobo 1990 [1653]; Gumerman 1991). Sandefur (2001) also suggests lower status households had less access to meat protein. A pattern of elite households enjoying greater amounts and higher qualities of meats has been documented in other regions of the Andes (Bray 2003b; Hastorf 2003; Sandefur 2001). Gumerman (2002) also documented this disparity between non-elite and elite domestic spaces at Pacatnamu, where high-status domestic units not only maintained access to more meat than lower status counterparts, but also had greater access to traditional seasoning, including *ají* or chili pepper. Cooking and preparation methods at the site were probably consistent with other common methods in the Andes, including grilling, stewing, boiling, toasting, and drying (Bray 2003b).

Large terrestrial mammals were present at Yanaorco in the form of domestic camelids and deer. Although many faunal elements were identifiable as camelid, identification to either llama (*Lama glama*) or alpaca (*Lama pacos*) was not possible. Frequencies of unfused and fused camelid elements were also recorded, and the resulting profile indicates the presence of both adult and juvenile camelids. The occupants of Yanaorco herded camelids locally and probably corralled them in large bounded spaces at the western edge of the community.

Deer makes up a very small proportion of the large mammal faunal assemblage with an MNI of four across the site and at 9.5% of the NISP (excluding LTM and STM counts) (Table 5.2). The presence of deer in domestic middens at the site indicates the practice of at least occasional hunting on the part of some households.

A range of archaeobotanical remains was recovered at Yanaorco in excavation of elite and commoner domestic zones. These remains were present not only in middens but also within fill deposits between domestic floors and on activity surface layers themselves. Botanical remains include not only highland, locally available crops including maize and beans, but also crops grown in warmer areas of the lower elevation mid-valley zones to the west, including peanut (*Arachis hypogaea* L.) and *ají*. The only botanical remains identified or recovered were carbonized.

Cuisine at Yanaorco

Evidence for diet and the domestic economy at Yanaorco is very similar to that recorded at many Andean communities. A great deal of overlap exists between domestic areas or "households," whether non-elite or elite status. All households

Table 5.2 Summary faunal data

	Excavation block	20I	22AAA	18D	18A1&2	22XXX	20B1	20C	28B	28ENT1	Total
	Vol. excavated (m^3)	3.64	4.782	1.444	1.636	1.708	2.482	1.931	1.724	0.328	19.675
Camelid *Lama sp.*	NISP	0	88	5	0	15	35	5	10	46	204
	Wt	0	960	63.5	0	103.25	190.3	25.9	157.5	279	1779.45
	Avg Wt	0	10.91	12.7	0	6.88	5.44	5.18	15.75	6.07	
	#/Vol	0	18.4	3.46	0	8.78	14.1	2.59	5.8	140.24	
	Wt/Vol	0	200.75	43.98	0	60.45	76.67	13.41	91.36	850.61	
	MNI	0	3	1	0	1	3	1	1	2	12
Deer Cervidae	NISP	0	28	0	0	1	1	0	0	8	38
	Wt	0	117.7	0	0	4	1.7	0	0	112.4	235.8
	Avg Wt	0	4.2	0	0	4	1.7	0	0	14.05	
	#/Vol	0	5.86	0	0	0.59	0.4	0	0	24.39	
	Wt/Vol	0	24.61	0	0	2.34	0.68	0	0	342.68	
	MNI	0	1	0	0	1	1	0	0	1	4
LTM indet	# elements	403	1609	179	0	107	179	23	26	195	2721
	Wt	100.6	1227.35	27	0	152.45	195.45	76.9	40.7	231.2	2051.65
	Avg Wt	0.25	0.76	0.15	0	1.42	1.09	3.34	1.57	1.19	
	#/Vol	110.71	336.47	123.96	0	62.65	72.12	11.91	15.08	594.51	
	Wt/Vol	27.64	256.66	18.7	0	89.26	78.75	39.82	23.61	704.88	
Guinea pig *Cavia porcellus*	NISP	0	71	7	0	1	31	1	2	14	127
	Wt	0	18.75	4.9	0	0.25	13.2	0.9	1.5	5.85	45.35
	Avg Wt	0	0.26	0.7	0	0.25	0.43	0.9	0.75	0.42	

(continued)

Table 5.2 (continued)

	Excavation block	20I	22AAA	18D	18A1&2	22XXX	20B1	20C	28B	28ENT1	Total
	Vol. excavated (m^3)	3.64	4.782	1.444	1.636	1.708	2.482	1.931	1.724	0.328	19.675
	#/Vol	0	14.85	4.85	0	0.59	12.49	0.52	1.16	42.68	
	Wt/Vol	0	3.92	3.39	0	0.15	5.32	0.47	0.87	17.84	
	MNI	0	2	1	0	1	2	1	1	2	10
Rabbit *Sylvilagus sp.*	NISP	0	5	0	0	1	0	0	0	0	6
	Wt	0	4.5	0	0	3	0	0	0	0	7.5
	Avg Wt	0	0.9	0	0	3	0	0	0	0	
	#/Vol	0	1.05	0	0	0.59	0	0	0	0	
	Wt/Vol	0	0.94	0	0	1.76	0	0	0	0	
	MNI	0	1	0	0	1	0	0	0	0	2
STM indet	# elements	0	91	46	0	0	16	0	0	9	162
	Wt	0	33.75	4.7	0	0	5.8	0	0	1.85	46.1
	Avg Wt	0	0.37	0.1	0	0	0.36	0	0	0.21	
	#/Vol	0	19.03	31.86	0	0	6.45	0	0	27.44	
	Wt/Vol	0	7.06	3.25	0	0	2.34	0	0	5.64	
Dog *Canis familiaris*	NISP	0	0	0	0	0	0	0	2	0	2
	Wt	0	0	0	0	0	0	0	7.4	0	7.4
	Avg Wt	0	0	0	0	0	0	0	3.7	0	
	#/Vol	0	0	0	0	0	0	0	1.16	0	
	Wt/Vol	0	0	0	0	0	0	0	4.29	0	

(continued)

Table 5.2 (continued)

	Excavation block	20I	22AAA	18D	18A1&2	22XXX	20B1	20C	28B	28ENT1	Total
	Vol. excavated (m^3)	3.64	4.782	1.444	1.636	1.708	2.482	1.931	1.724	0.328	19.675
	MNI	0	0	0	0	0	0	0	1	0	1
Bird	NISP	0	1	0	0	0	1	0	0	0	2
	Wt	0	0.25	0	0	0	0.4	0	0	0	0.65
	Avg Wt	0	0.25	0	0	0	0.4	0	0	0	
	#/Vol	0	0.21	0	0	0	0.4	0	0	0	
	Wt/Vol	0	0.05	0	0	0	0.16	0	0	0	
	MNI	0	1	0	0	0	1	0	0	0	2
Mollusk	NISP	2	11	0	0	0	9	0	0	0	22
	Wt	0.05	2.25	0	0	0	4.1	0	0	0	6.4
	Avg Wt	0.03	0.2	0	0	0	0.46	0	0	0	
	#/Vol	0.55	2.3	0	0	0	3.63	0	0	0	
	Wt/Vol	0.01	0.47	0	0	0	1.65	0	0	0	
	MNI	1	*	0	0	0	*	0	0	0	1*
		20I	**22AAA**	**18D**	**18A1&2**	**22XXX**	**20B1**	**20C**	**28B**	**28ENT1**	

*indeterminate

seem to be self-sufficient and have had access to many of the same food items, whether fauna or flora (Fig. 5.5). At Yanaorco, dog, fish, bird, and terrestrial gastropods are present in very small quantities. Although the analysis of guinea pig is ongoing, based on the current samples, identifiable remains are only present in elite domestic contexts. That said, the friable and very fragmentary state of faunal materials in the non-elite space AU 20I implies that if guinea pig remains had been present, they are unrecognizable.

It is likely that all domestic units at Yanaorco had access to foods typical of an Andean cuisine, including camelid, guinea pig, potato, bean, and quinoa. It is less clear, given the poor organic preservation, whether both high- and low-status households had access to exotic seasoning such as aji, or lower elevation crops such as peanuts. The presence or absence of these ingredients may have been patterned across the community, an indicator of class difference.

Domestic contexts in spaces 20I and 22AAA at Yanaorco exhibit the strongest variability in the elements of cuisine signaling status difference. Space 22AAA is an elite room connected to a broader open patio on a domestic terrace and surrounded by high-quality walls faced with cut stone. Excavation there also revealed a low stone wall dividing the room internally. No hearths or cooking features were observed here. Architectural Unit 20I, on the other hand, is a non-elite domestic space, most likely once roofed, on a terrace on the site's north slope. Walls, probably once made of *quincha*, have not been preserved. Excavation in AU20I revealed a small rectangular stone lined hearth at its center, clearly associated within a domestic activity surface. This space saw no elaboration in terms of the organization of space other than the small hearth.

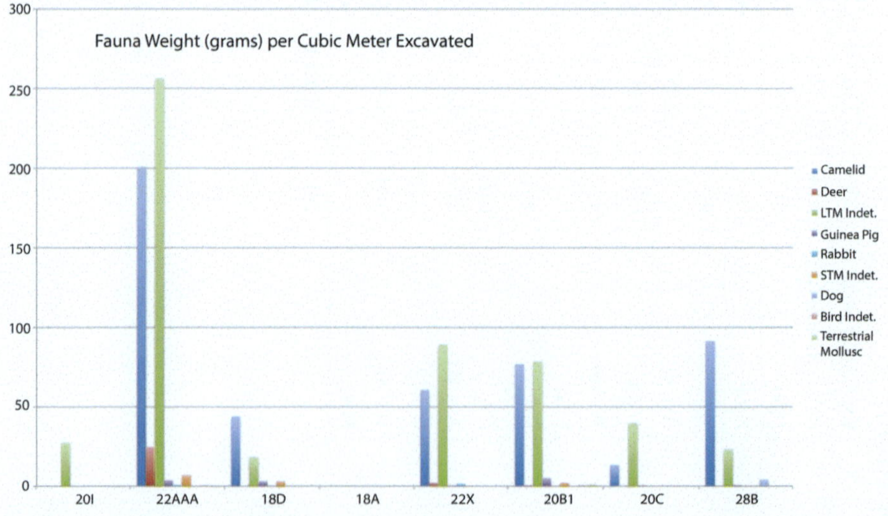

Fig. 5.5 Weight (grams) of fauna recovered per cubic meter excavated by context (Illustrated by Jason Toohey)

Space 22AAA contains a much richer suite of faunal remains than non-elite room 20I (see Figs. 5.2 and 5.7). Bone counts and weights of all the bones reveal substantial variation between these two contexts. Evidence for large mammals is present in both of these spaces, but fragments are much smaller and more friable in the non-elite space, so much so that no bone fragments are identifiable beyond classifications of large terrestrial mammals (see Table 5.2 for LTM counts and weights). There are also unidentifiably small remains in the elite context (AU22AAA), but they are relatively uncommon, and the great majority of bone fragments are larger and not friable. The average element weight for fauna in the lower status AU20I is 0.25 g (LTM) compared to 0.76 g (LTM) and 10.91 g for identified camelid fragments in higher status AU22AAA contexts (see Fig. 5.5). Moreover, bone remains are larger, and many are identifiable to element (Fig. 5.6). Many of these bones are also burned or partially burned indicating possible open-fire grilling. This grilling or roasting of

Fig. 5.6 Example of large identifiable faunal material (Photo by Jason Toohey)

large elements is in stark contrast to the friable bone remains common in the non-elite domestic contexts (AU20I), which probably indicates a focus on the extraction or rendering of bone grease through bone boiling.

The smaller size of faunal elements as well as the fact that they are exceedingly friable in the assemblage of the non-elite domestic context is interpreted here as representing intensive processing of bones to render bone grease or fat through boiling (Morin and Soulier 2017; Munro 2015) (see Fig. 5.7). Occupants of these spaces would most likely have fractured any available bone fragments into small fragments and then boiled these for several hours; skimming rendered grease in the process. The extraction of bone fat, whether through marrow extraction or grease rendering, is well documented in the archaeological record for both hunter-gatherers and food-producing societies (Morin and Soulier 2017; Munro 2015). However, rendering of bone fat is significantly more costly in terms of time, labor, and the necessary fuel than marrow extraction. It thus indicates a social and economic need to engage in the most intensive and least direct processes of food extraction. This is an example of what I would call a very conservative culinary practice in which nutrients are extracted exhaustively. Alternatively, it must be acknowledged that other post-depositional taphonomic processes might account for the high frequencies of very small and friable bone fragments in one context and not in another. If the lower status

Fig. 5.7 Example of friable bone fragments (Photo by Jason Toohey)

domestic space had also seen higher levels of rainfall or increased use for agriculture (resulting in increased trampling and mechanical disturbance), then we might expect to see the same kinds of differential patterning in the physical properties of faunal remains. Observation of these two contexts, though, indicated that both had witnessed low levels of agricultural use, probably leading to disturbance only to approximately 15–20 cm below the modern surface. Exposure to rainwater would similarly have been roughly the same in all areas of the site. Finally, both high- and low-status spaces were on terraces sided by stone retaining walls which would have resulted in roughly the same degree of drainage and pooling. This suggests that the patterning in faunal elements and processing is likely the result of cultural processes, differences in cooking technique.

Ceramics present in these two contexts also indicate important differences in storage capacities, food preparation, and cooking methodologies. While occupants of both spaces used storage jars, the number of both narrow and wide-necked jars was much greater in the elite domestic space (20.3/m^3) than in the lower status space (5.2/m^3) (Table 5.1). This pattern also holds for serving vessels, that is plates and bowls, which, although present on both areas, are much more common in elite space (see Table 5.1 and Fig. 5.8). An interesting observation here was that the frequencies of vertical-sided and incurving bowls, interpreted here as associated with the consumption of liquids, were relatively more common in the non-elite space, possibly indicating a consumption focus on liquids. These vertical or incurving bowls make up 40% of the bowl assemblage in AU20I versus 15.5% of the bowl assemblage in AU22AAA (see Table 5.1 and Fig. 5.8).

The presence of cooking vessels such as colanders indicates that there was considerable variability in the preparation of foods in the community. Rim and body fragments of shallow, broad colanders (ranging 35 to 75 cm, rim diameters) were recovered in many areas of the site (see Table 5.1 and Fig. 5.8). The high frequency of sherds

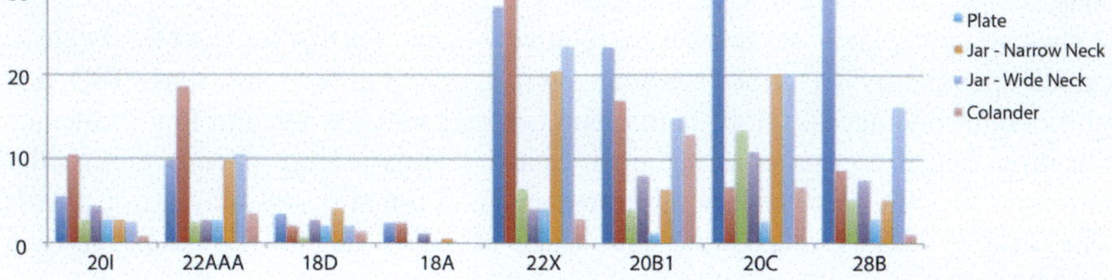

Fig. 5.8 Ceramic frequencies per cubic meter excavated by context (Illustrated by Jason Toohey)

with charred exteriors implies that meats were often grilled over fire. In terms of inter-household variation, the frequencies of colanders at Yanaorco are much higher in elite spaces like 22AAA, 20B, and 20C than non-elite domestic contexts, indicating that large higher value fragments of meat were grilled in these spaces, which would have resulted in the creation of the large, often partially charred elements (Sandefur 2001).

Greater frequencies of large and incompletely cooked or charred elements in room 22AAA suggest that larger quantities of meat were cooked, some incompletely grilled. This contrasts significantly with space 20I where only very small, friable remains were identified. These represent a culinary circumstance in which residents may have routinely practiced the extraction of as much nutrient value as possible from food elements, a more conservative cuisine variant than that in room 22AAA.

The ceramic assemblages at elite domestic spaces 20B and 20C are interesting when compared with those of AU22AAA. All of these spaces are located on elite terraces and each contains domestic artifacts. Architecturally, rooms 20B and 20C are larger and more elaborate than elite room 22AAA (see Fig. 5.3). Rooms 20B and 20C include a large bench along the NW wall of 20B with a low dividing wall through 20C that also contains a small central hearth. These rooms are associated with several large open plazas and a curious terrace/patio containing several benches creating the impression of terracing. All of these architectural elements appear to be connected by a long corridor. The presence of the large plazas near 20B and C indicates that they may have functioned for periodic gatherings by groups of people.

Rooms 20B and 20C also contained many more diagnostic sherds than elite space (22AAA), including higher frequencies of both serving vessels and storage vessels (see Table 5.1). Interestingly, the number of colander sherds in 20B and 20C is more than three times the frequency per cubic meter excavated than in AU22AAA. These high frequencies of ceramics indicate at least the capacity in these spaces to process, store, cook, and serve large quantities of food, compared to that of AU22AAA and AU20I.

Space 22X is characterized by an open, probably unroofed, patio on the terrace located south and southeast of architectural units 22AAA and 20B/20C. This space also contained the remains of a low D-shaped walled area, which resembles a pen of some kind. Very little faunal evidence was present in this space compared to the domestic areas discussed above. Large numbers of both serving and storage-related ceramic vessels indicate it may have functioned for the storage and preparation of foods.

The two open plazas or patios were relatively clear of artifactual and food debris. A small open patio on a terrace associated with a pair of elite rooms, space 18D, had tightly controlled access. Both faunal and ceramic remains were in low frequency indicating casual consumption at most. Space 18A is a large open semi-circular plaza near to 18D. Excavations recovered no food remains and only a few sherds from serving vessels, suggesting this space probably did not function for food consumption.

Context 28ENT1 was a narrow corridor allowing access to the western edge of one of three platforms. Our excavations there were focused upon better understanding access patterns to the community; however, we encountered a dense trash midden. This corridor had at some point been closed and used as a midden. It contained a dense accumulation of faunal remains, ceramics, and tools of stone, bone, and antler, with secondary, mixed deposits probably containing materials collected from a number of domestic spaces on the platform above.

The varied capacities for storage, cooking, and serving of food between domestic contexts at Yanaorco point to the differential ability among households to sponsor extra-household consumption events, or feasts of various scales. The faunal data from non-elite space 20I indicates that its occupants probably did not have the means to sponsor such events. Higher numbers of storage, cooking, and faunal remains in AU22AAA, a high-status area, may indicate storage capacity for substantial amounts of food for distribution during extra-household rites, rituals, and feasts. Finally, 20B/C, the elite domestic context, contained even higher frequencies of storage, cooking, and serving vessels. Its spatial association with larger open plaza spaces may also indicate that it functioned in some way with the production and serving of large amounts of food to groups of people.

Possible feasting at Yanaorco may have served a series of functions, from establishing and reinforcing status differences and community hierarchical organization to reciprocating for smaller scale labor projects (Isbell 1978; Meyerson 1990). The scale of the event would have depended on the extent of the labor necessary to carry it out by the sponsor. Small-scale private reciprocal events, *minka*, suggest the family received labor for house construction or roof building, and the clearing, sowing, or harvesting of fields. Such labor from support populations would have been reciprocated by providing sustenance to the labor. Larger scale labor involving the clearing of community canals outside of Yanaorco, or maintenance of its fortifications, was no doubt organized and sponsored by community leaders. Isbell (1978) describes these *faenas* (as does Meyerson 1990) as requiring a male laborer from every household. They also describe the reciprocal provision of food and or *chicha*, *coca*, and *trago* to laborers. Such consumption would have most likely taken place off-site, while labor inside the community would have occurred inside the walls of Yanaorco, perhaps nearby AU 22AAA or AU20B/C.

Conclusions

Social and economic status differences at Yanaorco were clearly represented and communicated during the Late Intermediate Period through both architecture and cuisine. Although cuisine involves a number of aspects including ingredients, preparation methods, flavor principals, and rules for consumption, archaeologically, we can best speak to the ingredients utilized, practices of preparation, cooking, and consumption. All households at Yanaorco had general access to similar ingredients and flavoring elements, but this is where close similarities end. Social differentiation is particularly apparent in the architecture, and also in patterned variation in certain aspects of cuisine; particularly food preparation techniques reflecting differences in the general level of what I am calling conservatism in food preparation. Households in Yanaorco had the same kinds of ingredients but were using them in very different ways within lower and higher status contexts, indicating substantial intra-community variation in status and access to certain resources, particularly meat protein. Also, as might be expected, both faunal and ceramic data indicate that some groups had much greater capacity to store foods in greater volumes, therefore a greater capacity to host or sponsor extra-household commensal feasting events. This evidence couches political activity at Yanaorco within a much broader regional pattern of dynamic connections among leadership, class, and cuisine in the Andean past.

The employment of a holistic concept of food and foodways, and in particular the multifaceted concept of cuisine, has indicated the presence of class distinction within the community of Yanaorco. This level of social segregation may have been common to many late pre-Hispanic Cajamarca communities, with social difference marked consciously or unconsciously by sometimes subtle variability in what was otherwise a shared northern Andean cuisine.

Acknowledgements The fieldwork upon which this paper is based took place over three seasons (2003, 2004, and 2012) under permit of the Peruvian Ministry of Culture in Lima and Cajamarca. These investigations were supported generously by Fulbright Hays, the University of Wyoming, and the University of California Santa Barbara. The project was co-directed by Jimmy Bouroncle, Claudia Iturry, and the author, and was assisted in the field by Peruvian and North American students as well as the people of Tamiacocha. Finally, I am indebted to several anonymous reviewers for their thoughtful and insightful comments and suggestions.

References

Altamiranco Enciso AJ (1983) Guía osteológico de cérvidos andinos. Universidad Nacional Mayor San Marcos, Lima, Peru

Atalay S, Hastorf CA (2006) Food, meals, and daily activities: food habitus at neolithic catalhoyuk. Am Antiq 71(2):283–319

Barthes R (1979 [1961]) Toward a psychosociology of food consumption. In: Forster R, Ranum O (eds) Food and drink in history, Johns Hopkins University Press, Baltimore, MD pp 166–173

Belasco W (2008) Food: the key concepts. Berg, Oxford

Bray TL (ed) (2003a) The archaeology and politics of food and feasting in early states and empires. Kluwer Academic/Plenum, New York

Bray TL (ed) (2003b) To dine splendidly: imperial pottery, commensal politics, and the Inca state. In: The archaeology and politics of food and feasting in early states and empires, Kluwer Academic/Plenum, New York, pp 65–92

Cardich A (1994) Descubrimientos de un complejo Preceramico en Cajamarca, Perú. Revista del Museo de la Plata (Nueva Serie) 9:225–237

Cieza de León P (1984 [1553]) Crónica del Peru, primera parte. Pontífica Universidad Católica del Peru, Lima, Peru

Cobo B (1990 [1653]) Inca religion and customs. Translated by R. Hamilton, University of Texas Press, Austin, TX

Conlee CA, Dulanto J, Mackey CJ, Stanish C (2004) Late pre-hispanic sociopolitical complexity. In: Silverman H (ed) Andean archaeology. Blackwell Publishing, Malden, MA, pp 209–236

Cook A, Glowacki M (2003) Pots, politics, and power: Huari ceramic assemblages and imperial administration. In: Bray TL (ed) The archaeology and politics of food and feasting in early states and empires. Kluwer Academic/Plenum, New York, pp 173–202

Cooper G, Schiller AL (1975) Anatomy of a guinea pig. Harvard University Press, Cambridge

Crown, PL (ed) (2000) Women's role in changing cuisine. In: Women and men in the pre-hispanic Southwest: labor, power, and prestige, SAR Press, School of American Research Advanced Seminar Series, Santa Fe, NM, pp 221–266

Cuéllar AM (2013) The archaeology of food and social inequality in the Andes. J Archaeol Res 21:123–174

Cutright RE (2010) Food, family, and empire: relating political and domestic change in the Jequetepeque Hinterland. In: Cutright RE, López-Hurtado E, Martín AJ (eds) Comparative perspectives on the archaeology of coastal South America. University of Pittsburgh, Pittsburgh, PA, pp 27–44

deFrance SD (2006) The sixth toe: the modern culinary role of the guinea pig in Southern Peru. Food Foodways 14:3–34

Dietler M (2001) Theorizing the feast: rituals of consumption, commensal politics, and power in African contexts. In: Dietler M, Hayden B (eds) Feasts: archaeological and ethnographic perspectives on food, politics, and power. Smithsonian Institution Press, Washington DC, pp 65–114

Dietler M, Hayden B (eds) (2001) Feasts: archaeological and ethnographic perspectives on food, politics, and power. Smithsonian Institution Press, Washington DC

Dietler M, Herbich I (2001) Feasts and labor mobilization: dissecting a fundamental economic practice. In: Dietler M, Hayden B (eds) Feasts: archaeological and ethnographic perspectives on food, politics, and power. Smithsonian Institution Press, Washington DC, pp 240–266

Douglas M (1999 [1975]) Implicit meanings: essays in anthropology, 2nd edn. Routledge, New York

Farb P, Armelagos G (1980) Consuming passions: the anthropology of eating. Houghton Mifflin, Boston

Gero J (1992) Feasts and females: gender ideology and political meals in the Andes. Norwegian Archaeol Rev 25(1):15–30

Goody J (1982) Cooking, cuisine, and class: a study in comparative sociology. Cambridge University Press, Cambridge

Grant A (2002) Food, status and social hierarchy. In: Miracle P, Milner N (eds) Consuming passions and patterns of consumption. McDonald Institute for Archaeological Research, Cambridge, pp 17–23

Grobman A, Ravines R (1974) Maíz prehispánico del valle de Cajamarca, Perú. Revista del Museo Nacional, Lima, vol 40, pp 135–138

Gumerman G IV (1991) Subsistence and complex societies: diet between diverse socio-economic groups at Pacatnamu, Peru. Ph.D. dissertation, Department of Anthropology, University of California, Los Angeles

Gumerman GIV (1997) Food in complex societies. J Archaeol Method Theory 4:105–139

Gumerman GIV (2002) Llama power and empowered fishermen: food and power and Pacatnamu, Peru. In: O'Donovan J (ed) The dynamics of power. Southern Illinois University Press, Carbondale, IL, pp 238–256

Gumerman GIV (2010) Big hearths and big pots: moche feasting on the North coast of Peru. In: Klarich E (ed) Inside ancient kitchens: new directions in the study of daily meals and feasts. University Press of Colorado, Boulder, CO, pp 111–131

Hally DJ (1986) The identification of vessel function: a case study from Northwest Georgia. Am Antiq 51:267–295

Hastorf CA (1990) The effect of the Inka state on Sausa agricultural production and crop consumption. Am Antiq 55:262–290

Hastorf CA (1991) Gender, space, and food in prehistory. In: Gero J, Conkey M (eds) Engendering archaeology. Blackwell Press, Cambridge, MA, pp 132–159

Hastorf CA (2003) Andean luxury foods: special food for the ancestors, deities, and the elite. Antiquity 77(297):545–554
Hastorf CA (2012) The habitus of cooking practices at neolithic Çatalhöyük. In: Graff SR, Rodríguez-Alegría E (eds) The menial art of cooking: archaeological studies of cooking and food preparation. University Press of Colorado, Boulder, CO, pp 65–86
Henrickson EF, McDonald MMA (1983) Ceramic form and function: an ethnographic search and an archaeological application. Am Anthropol 85(3):630–643
Hirth KG (1993) Identifying rank and socioeconomic status in domestic contexts: an example from central Mexico. In: Santley RS, Hirth KG (eds) Pre-hispanic domestic units in Western Mesoamerica: studies of the household, compound, and residence. CRC Press, Boca Raton, FL, pp 121–146
Isbell BJ (1978) To defend ourselves: ecology and ritual in an Andean village. University of Texas Press, Austin, TX
Janusek J (2004) Identity and power in the ancient Andes. Routledge, New York
Julien DG (1988) Ancient Cuismancu: settlement and cultural dynamics in the Cajamarca region of the North highlands of Peru, 200 B.C.-A.D. 1532. Ph.D. dissertation, Department of Anthropology, University of Texas
Julien DG (1993) Late Pre-Inkaic ethnic groups in highland peru: an archaeological-Ethnohisorical model of the political geography of the Cajamarca region. Latin Am Antiq 4(3):246–273
Klarich E (ed) (2010) Inside ancient kitchens: new directions in the study of daily meals and feasts. University Press of Colorado, Boulder, CO
Lau GF (2002) Feasting and Ancestor veneration at Chinchawas, North highlands of Ancash, Peru. Latin Am Antiq 13(3):279–304
Matsumoto R (1982) Seriated sequence of pottery types. In: Terada K, Onuki Y (eds) Excavations at Huacaloma in the Cajamarca valley, Peru, 1979. Report 2 of the Japanese scientific expedition to nuclear America, University of Tokyo Press, Tokyo, pp 119–179
Meyerson J (1990) Tambo: life in an Andean village. University of Texas Press, Austin, TX
Mills BJ (ed) (2002) Identity, feasting, and the archaeology of the greater Southwest. University Press of Colorado, Boulder CO
Morin E, Soulier M-C (2017) New criteria for the archaeological identification of bone grease processing. Am Antiq 82(1):96–122
Munro ND (2015) Bone fat extraction. In: Beaudry MC, Metheny KB (eds) Archaeology of food: an Encyclopedia, Rowman & Littlefield Publishers, Lanham, MD, pp 67–69
Murra, JV (1972) El 'control vertical' de un máximo de pisos ecológicos en la economía de las sociedades andinas. In: Visita de la provincia de León de Huanuco en 1562. Documentos por la historia y etnología de Huanuco y la Selva Central, vol 2, Universidad Nacional Hermilio Valdizán, Huánuco, Peru, pp 427–476
Nash DJ (2010) Fine dining and fabulous atmosphere. In: Klarich EA (ed) Inside ancient kitchens: new directions in the study of daily meals and feasts. University Press of Colorado, Boulder, CO, pp 83–109
Torres P, Victor R, Enciso AA, Porras EG (1986) *The Osteology of South American* Camelids. Translated by Elsie Sandefur Institute of Archaeology UCLA, Los Angeles
Pulgar Vidal J (1981) Geografía del Perú: las ocho regiones naturales del Perú, 8th Edición. Editorial Universo S.A, Lima, Peru
Ravines S, Rogger H (1968) Arqueología del valle de Cajamarca. Universidad Nacional Tecnica de Cajamarca: Facultad de Educacion (Publicaciones de la Facultad de Educacion), pp 1–37
Ravines S, Rogger H (1976) Sobre la historia y arqueologia del valle de Cajamarca. Historia de Cajamarca. Instituto Nacional de Cultura-Cajamarca, Cajamarca, Peru, pp 69–76
Ravines S, Rogger H (1985) Cajamarca pre-hispanica: inventario de monumentos arqueologicos. Instituto Nacional de Cultura-Cajamarca/Corporacion de Desarrollo de Cajamarca, Cajamarca, Peru

Reichlen H (1970) La sierra: reconocimientos arqueologicos en los Andes de Cajamarca. In: Ravines R (ed) 100 años de arqueología en el Peru: fuentes e investigaciones para la historia del Peru 3. Instituto de Estudios Peruanos/Petrolees del Peru, Lima, Peru, pp 463–501

Reichlen H, Reichlen P (1949) Recherches archeologiques dans les Andes de Cajamarca: premier rapport de la mission ethnologique Francaise au Perou septentrional. Journal de la Societe des Americanistes 38:137–174

Reichlen H, Reichlen P (1985) Reconocimientos arqueológicos en los Andes de Cajamarca. In: Santisteban FS, Soriano WE, Ravines R (eds) Historia de Cajamarca: 1. Arqueología, Instituto Nacional de Cultura-Cajamarca y Corporacion de Desorrollo de Cajamarca, Cajamarca, Peru, pp 29–54

Rozin E (1982) The structure of cuisine. In: Barker LM (ed) The Psychobiology of human food selection. AVI Publishing, Westport, CT, pp 189–203

Sachún J (1986) Patrones de asentamiento en el proceso cultural pre-hispanico del valle de Cajamarca (primera aproximacion). Materiales para la arqueologia de Cajamarca 3. Editorial Sudamerica, Trujillo, Peru

Sandefur EC (2001) Animal husbandry and meat consumption. In: D'Altroy TN, Hastorf CA (eds) Empire and domestic economy. Kluwer Academic/Plenum Publishers, New York, pp 179–202

Shimada M (1982) Appendix IV: zooarchaeology of Huacaloma: behavioral and cultural implications. In: Terada K, Onuki Y (eds) Excavations at Huacaloma in the Cajamarca valley, Peru, 1979. University of Tokyo Press, Tokyo, pp 303–336

Shimada M (1985) Appendix III: Continuities and changes in patterns of faunal resource utilization formative through Cajamarca periods. In: Terada K, Onuki Y (eds), The formative period in the Cajamarca Basin, Peru: excavations at Huacaloma and Layzón, 1982. Report 3 of the Japanese scientific expedition to nuclear America, University of Tokyo Press, Tokyo, pp 289–310

Silva Santisteban F (2000) Cajamarca: historia y paisaje/history and landscape. Minera Yanacocha/Antares Artes y Letras, Lima, Peru

Smith AK (2012) National cuisines. The Oxford handbook of food history. Oxford University Press, Oxford, pp 444–460

Squier E (1877) Peru: Incidents of Travel and Exploration in the Land of the Incas. Harper and Brothers, New York

Staller JE (2008) Dimensions of place: the significance of centers to the development of Andean civilization: an exploration of the Ushnu concept. In: Staller JE (ed) Pre-columbian landscapes of creation and origin. Springer, New York, pp 269–313

Tello JC (2004) Obras completas volumen I: arqueología de Cajamarca: expedicion al Marañon—1937. Universidad Nacional Mayor de San Marcos, Lima, Peru

Terada K, Matsumoto R (1985) Sobre la cronología de la tradición Cajamarca. In: Santisteban FS, Soriano WE, Ravines R (eds) Historia de Cajamarca: 1. Arqueología, Instituto Nacional de Cultura-Cajamarca y Corporacion de Desarrollo de Cajamarca, Cajamarca, Peru, pp 67–89

Terada K, Onuki Y (eds) (1982) Excavations at Huacaloma in the Cajamarca valley, Peru, 1979. University of Tokyo Press, Tokyo

Terada K, Onuki Y (eds) (1985) The formative period in the Cajamarca Basin, Peru: excavations at Huacaloma and Layzón, 1982: report 3 of the Japanese scientific expedition to nuclear America. University of Tokyo Press, Tokyo

Terada K, Onuki Y (eds) (1988) Las excavaciones en Cerro Blanco y Huacaloma, Cajamarca, Peru, 1985. University of Tokyo Press, Tokyo

Toohey JL (2009) Community organization, militarism, and ethnogenesis in the late prehistoric Northern Highlands of Peru. Ph.D. dissertation, Department of Anthropology, University of California, Santa Barbara

Toohey JL (2011) Formal and stylistic variation in the ceramic assemblage at the late intermediate period site of Yanaorco in the Cajamarca highlands of Northern Peru. Ñawpa Pacha 31(2):171–200

Toohey JL (2012) Recent excavations at the late intermediate period village of Yanaorco in the Cajamarca highlands. Andean Past 10:263–267

Twiss KC (2015) Methodological and definitional issues in the archaeology of food. In: Kerner S, Chou C, Warmind M (eds) Commensality: from everyday food to feast. Bloomsbury, London, pp 89–98

Tykot RH, Burger RL, Van Der Merwe NJ (2006) The Importance of maize in the initial period and early Horizon Peru. In: Staller JE, Tykot RH, Benz BF (eds) Histories of maize: multidisciplinary approaches to the prehistory, linguistics, biogeography, domestication, and evolution of maize. Academic Press, New York, pp 187–197

Van Derwarker AM, Peres TM (eds) (2010) Integrating zooarchaeology and paleoethnobotony: a consideration of issues, methods, and cases. Springer, New York

Watanabe S (2009) La cerámica caolín en la cultura Cajamarca (sierra norte del Perú): El caso de la fase Cajamarca Medio. Bulletin de l'Institut Français d'Études Andines 38(2):205–236

Wiener C (1880) Perou et Bolivie. Hatchette Edtts, Paris

Weismantel M (1988) Food, gender, and poverty in the Ecuadorian Andes. University of Pennsylvania Press, Philadelphia

Welch PD, Scarry CM (1995) Status-related variation in foodways in the Moundville chiefdom. Am Antiq 60(3):397–419

Chapter 6
Ancient Paria, Bolivia: Macrobotanical Remains Recovered from an Administrative Site on the Royal Inca Highway

Renée M. Bonzani

Abstract Tawantinsuyu was the great Inca Empire estimated to have encompassed 3,000 miles of western South America, including parts of present-day Ecuador, Peru, Bolivia, Argentina, and Chile. Many studies have focused on various aspects of Inca culture, on the major sites of the Empire, and on mapping the extensive Inca road system. Fewer studies address the varying use of plants at these major administrative centers. This present chapter presents the paleoethnobotanical investigations of the redistribution/administrative center, Khota Chullpa or the ancient site of Paria (Ce-01), and associated *qollqas* or storage structures (Ce-51) along the Inca highway in the Paria Basin, Department of Oruro, Bolivia. In particular, plant remains that indicate the subsistence strategies, as well as the use of possible hallucinogenic and medicinal plants. These plant remains provide a perspective on the functions of this administrative center as more than just a place to store and account for everyday economic goods like maize and potatoes.

Keywords Andes · Bolivia · Inca empire

Introduction

Empires face the daunting task of managing the production and distribution of foodstuffs to feed their subjects. The Inca Empire, which was called Tawantinsuyu or land of the four corners, encompassed an estimated 3,000 miles of the Andes cordillera including parts of present-day Ecuador, Peru, Bolivia, and Chile. Numerous archaeologists have focused on aspects of the Inca Empire, particularly its major sites, and the extensive road system referred to as the Inca highway or royal road (*Camino Real*) (Bauer 2002; Bauer and Alan Covey 2002; Covey 2006; D'Altroy and Hastorf 2001; Murra 1972, 1980; Pärssinen 2002, 2005; Rice 2012; Rowe 1946). Previous

R. M. Bonzani (✉)
University of Kentucky, Lexington, USA
e-mail: Renee.Bonzani@uky.edu

J. E. Staller (ed.), *Andean Foodways*, The Latin American Studies Book Series,
https://doi.org/10.1007/978-3-030-51629-1_6

research addressed subsistence adaptation and the political economy under the Inca rule (Bray 2002; Hastorf 1993; Jennings and Bowser 2009; Rossen et al. 2010).

The Inca Empire actively engaged in the storage and redistribution of foodstuffs as part of its political economy. Excavations carried out at the Inca administrative center called Ancient Paria, also referred to as Khota Chullpa, (Site Ce 01) and a possible mortuary site (Site Ce 51), located in the Paria Basin of the Department of Oruro, Bolivia (Gyarmati and Castellón 2014) included the excavation of *qollqas* or storehouses, where Inca administrators accumulated a variety of plant foods. Macrobotanical remains recovered from these excavations provide insights into Inca management of foods and the role of foodstuffs in the political economy and redistribution networks of the Inca state (see Lau, Chapter 2 and Tantaleán et al., Chapter 3 for additional insights into political management of food).

Although Ancient Paria was mentioned in ethnohistoric documents as an important Inca administrative center (Cieza de León 1973 [1553], 1985 [1553]; Vaca de Castro Cavellero 1908 [1543]), its exact location had remained unknown until recently. Archaeologists Trimborn (1967), and later Hyslop (1984), initially carried out some research in the area. During his survey of the Inca highway, Hyslop (1984: 145) identified Khota Chullpa, a site along the Jacha Uma River near the indigenous community of Balneario de Obrajes and east of the village of Paria. Excavations at the site by Carola Condarco Castellón (Condarco et al. 2002; Condarco 2009), fieldwork by Janós Gyarmati (2005), and Carola Condarco Castellón (Gyarmati and Castellón 2006) documented its exact location, and an extensive volume on the Paria Archaeological Project was recently published (Gyarmati and Castellón 2014).

Paria is situated at 3,810–3,815 masl, on the juncture of three rivers, the Jacha Uma, the Iruma, and the Obrajes (also called Chiaraq'i or Umutiri), which combine to form the Paria River (Fig. 5.1). The Inca highway follows along with the Rio Jacha Uma as a passageway over the Cordillera de Azanaques into the lower fertile Cochabamba valley at 2,500–2,600 masl (Gyarmati and Varga 1999). Cochabamba was one of the "bread baskets" of the Inca Empire. Maize was a major crop grown on estate farms and transported on the Inca highway to Paria, allowing for an Incan strategy of state integration with the use of redistributed products and labor (D'Altroy 2002: 74, 273; Gyarmati and Varga 1999; Higueras-Hare 2001; Rowe 1946; Wachtel 1982). Over 1,539 *qollqas* were identified at Paria and ethnohistoric accounts emphasize maize as the primary crop (Gyarmati 2005; Gyarmati and Castellón 2014).

This strategic location at the crossroads of the Inca Highway or *Camino Real* located near a hot spring with natural thermal waters that continue to be used as bathhouses (Gyarmati and Castellón 2014: 59–60). A modern-day hotel/bathhouse with swimming pool and bathing rooms (Fig. 5.2) is located close to the indigenous community of Balneario de Obrajes near the Inca administrative center. Sacred landscapes or places (*huacas*) associated with water were provided ritual offerings to supernatural entities who were venerated to assist in various aspects of life including journeys and health (Cobo 1990 [1653]; Staller 2008). Such perceptions were a common feature of Andean cosmology (Bray 2013; Glowacki and Malpass 2003; Moore 2014; Quilter 2014; Staller and Stross 2013). Macrobotanical evidence of

Fig. 5.1 Bolivia shaded relief map, 1993. Paria is located approximately 20 km northeast of the city of Oruro (From: https://commons.wikimedia.org/wiki/File:Bolivia_rel93.jpg. Image is in the public domain https://www.lib.utexas.edu/maps/bolivia.html {{PD-USGov-CIA}})

food and possible fermented beverage consumption, as well as hallucinogenic and medicinal plants, are present suggesting this was a location the Inca and others could rest or take respite on long journeys to Cochabamba or further south into present-day Chile. It was also a central location of labor, textile production and metallurgical arts, and ritual feasting with its preparation and consumption, as well as a sacred place to access the tripartite cosmos, ancestors, and deities. This macrobotanical analysis reveals much about the extensive interconnectedness of people from different regions

Fig. 5.2 Photo of the modern-day thermal baths of Obrejas, located near Ancient Paria (Site Ce 01) (Photo by Renée M. Bonzani in 2006)

and the various functions (residential, ceremonial, administrative, storage, housing for travelers/possible tambos) that the site might have served.

Excavations at Ancient Paria

The Paria River basin is 20 km northeast of the city of Oruro and approximately 230 km to the southeast of La Paz and 200 km to the northwest of the Cochabamba valley. A systematic survey conducted in 2004 identified 51 sites with Inca remains associated with the Late Horizon Period (A.D. 1400/1450–1535). Excavations were conducted at Ancient Paria, also known as Khota Chullpa (Ce 01), as well as Site Ce 51 (Gyarmati and Castellón 2014). Ancient Paria encompasses a continuous area measuring 90 hectares with a central plaza of 6.5 hectares (Fig. 5.3). The northern boundary includes six clusters of storehouses (*qollqas*) in the vicinity yielding a total of 1,539 with an additional 236 structures completely deteriorated or tentatively identified as storehouses bringing a possible total to 1,775 *qollqas* (Fig. 5.4) (Gyarmati and Castellón 2014: 29–58, 60).

The administrative/ceremonial center inside the central zone has the most extant architectural constructions. Many structures collapsed on themselves forming low mounds with dimensions ranging from 20 to 40 m × 8 to 10 m. These structures resemble Inca *kallankas*, or elongated rectangular buildings with interior spaces for administrative and possibly ceremonial uses. One of these, Structure BH, had extant stone walls of 60–80 cm high and had an orientation of 78° east of north.

Fig. 5.3 Present-day overview of the site of Ancient Paria (Site Ce 01), Paria River Basin, Department of Oruro, Bolivia (Courtesy of Janós Gyarmati)

Fig. 5.4 Satellite image of the archaeological site of Ancient Paria (Site Ce 01) (Courtesy of Google Earth 2006; text by Janós Gyarmati)

Excavation of Structure BH revealed a rectangular formation 35–40 m long and 10 m wide and stone foundation. Such structures generally function for public events and, in some instances, served as temporary living quarters for visiting guests or administrators/dignitaries. Two ^{14}C dates from Structure BH indicate an occupation dated to the early to mid-fifteenth century (see Gyarmati and Castellón 2014: 67, Table IV.1).

A second rectangular building, Structure BM (Surface II) is a 200 m^2 area that was excavated in the administrative/ceremonial core. This structure may be part of a possible *kancha*, in essence, living quarters for production and possible consumption. It is comprised of three or more rectangular structures placed symmetrically around the side of a compound with a central patio (Hyslop 1990: 16–17). People living within a single *kancha* may or may not have been related. Other artifacts recovered from this structure included abundant evidence of textiles, adornments, and possible *chicha* (maize beer) production, as well as bone weaving implements, spindle whorls, needles, sodalite, turquoise, various metals (copper, and possibly silver and gold), numerous ceramic fragments, and carbonized maize cobs—remains often associated with *aqllakuna* activities (Alberti Manzanares 1985; Cobo 1990 [1653]: 174). *Aqllakuna* or *mamaconas* were chosen women from high-ranking lineages of different ethnic groups associated with the Inca elite who performed various functions for the state (Bandelier 1910: 273, 288). Excavations at Structure BM revealed it was a residential house/workshop (a possible House of the *Mamaconas*) for women specializing in textile and adornment production with the occupants being possible *aqllas* (Gyarmati and Castellón 2014: 73). This structure also dates to the early to mid-fifteenth century.

Site Ce 51 located 1.5 km. south-southwest of Paria, on hills above Balneario de Obrajes and overlooking the Inca administrative center, is one of a group of associated storage houses included primarily for comparative purposes. Site Ce 51 consists of a group of two rows of at least 78 *qollqas* or storage structures (Gyarmati and Castellón 2014: 52–54). The botanical remains from one rectangular and one circular structures (1 and 2) are discussed here to provide comparative insights into Inca food management strategies. Although initially believed to have functioned for habitation during the Late Intermediate Period (A.D. 1000–1400/1450), artifacts recovered in the structure suggest it probably functioned as a mortuary tower; however, no burials were found (Gyarmati and Castellón 2014: 39–40).

Methodology of the Recovery and Analysis of Macrobotanical Remains

Paleoethnobotanical analysis of macrobotanical remains, pollen, phytoliths, and starch grains are a growing and indispensable part of archaeological investigations, particularly to understand past food practices (Bonzani 2014b, c; Bruno 2008; Dillehay et al. 2010; Hastorf 1999; Morcote Ríos 1996; Oyuela-Caycedo and Bonzani

Fig. 5.5 Portable flotation device used at Site Ce 01 with field crew (Paula and Ernest) in the process of floating samples (Photo by Renée M. Bonzani in 2006)

2005, 2014; Pearsall 2000, 2003; Piperno and Pearsall 1998; Rossen et al. 2010; Wagner 1986). As has been noted for recovering plant foods, the importance of techniques, such as flotation systems, is essential (Rossen 1999). For this analysis, excavators processed archaeobotanical remains using an easily transportable flotation device to guarantee recovery of fragile plant remains (Bonzani 2014a) (Fig. 5.5).

Laboratory analysis was carried out following procedures outlined by Pearsall (2000). Only the light fractions recovered from flotation were further analyzed, and each sample from Surface II of Paria (Structure BM) and Structures 1 and 2 from Site Ce 51 underwent complete analysis. A random sub-sample of each light fraction was also analyzed from Surface I of Paria (Structure BH) (Tables 5.1, 5.2, 5.3, 5.4, 5.5 and 5.6). Identifications were substantiated with a reference collection and secondary sources of various identification manuals were also used (Bird 1985; Castañeda 1965, 1991; Hather 2000; Honores and Rodríguez 2007; Lentz and Dickau 2005; Martin and Barkley 2000; Montgomery 1977; Rocas 1989; Ugent and Ochoa 2006).

A number of factors can affect the preservation of prehistoric plant remains. These include human cultural as well as non-human factors such as plant and animal perturbations, soil type, post-depositional geological activities, plant preservation differences, etc. The results of macrobotanical ecofactual analysis were quantified based on statistical measures to provide quantitative supporting evidence for these interpretations (Bonzani 1997, 1998; Hastorf and Popper 1988; Lennstrom and Hastorf 1992, 1995; Oyuela-Caycedo 1998; Pearsall 2000).

Table 5.1 Macrobotanical remains from Surface II (Structure BM) at Ancient Paria (Ce 01), Oruro, Bolivia (Paria Archaeological Project, 2006)

Flot number	Specimen type number	1	6	3	2	5	4	8	15	7	Totals
Provenience		Ce 01	Ce 01	Ce 01	Ce 01	Ce 01	Ce 01	Ce 01	Ce 01	Ce 01	
		Estr. BM	Estr. BM	Estr. BM	Estr. BM	Estr. BM	Estr. BM	Estr. BM	Estr. BM	Estr. BM	
		Exterior	Basural	Exterior	Cuadro 1	Cuadro 3	Caudro 6	Cuadro 15	Cuadro 21	Cuadro 27	
		Basurero	Nivel 3	Q-Area	Supf. II	Nivel 3	Supf. II	Nivel 3	Supf. II	Nivel 1	
		Nivel: 40 cm	(30–50 cm)	quemada	Nivel	(30–40 cm)	Nivel	(-40 cm)		(20–40 cm)	
		Pozo 1 y 2	Pozo 10	Nivel 2	of -45 cm	Rasgo 7	(-40 cm)		Pozo de	Exterior	
				(50–60 cm)					ceniza	muro oeste	
				Pozo 21							
Dry weight of sample (kg)		5.98	ca. 4.55	ca. 4.8	4.4	4.25	5.54	9.6	ca. 4.3	ca. 4.1	47.52
Volume floated		5	4.5	ca. 4	3	3.5	4	8	4	4	40
Weight of light fraction (g) (100% analyzed)		36.1	50.9	63.4	19.4	20.8	24.9	95.1	58.5	35	404.1
Carbonized wood >2 mm no.		147	212	94	129	29	36	123	573	96	1339

(continued)

Table 5.1 (continued)

Flot number	Specimen type number	1	6	3	2	5	4	8	15	7	Totals
Carbonized wood >2 mm weight (g)		6.7	9.3	4.1	4.7	1.1	1.2	5.2	21.3	5.4	59
Amaranthaceae *Amaranthus* sp.	44	9		2	7	5	7	16		20	66
cf. Anacardiaceae	21	15(0.2 g)					5(0.1 g)				20
cf. Asteraceae	15	1			1			5	1		8
Cactaceae cf. *Opuntia soehrensii*	14	5	3	7			4	1	1	1	22
Cactaceae *Trichocereus* sp.	13			1	4	8	1	5	1		20
Chenopodiaceae	19	43	3	8		6	9	25	4	57	155
Chenopodiaceae *Atriplex* sp.	5	1									1
Chenopodiaceae *Chenopodium* spp.[a]	17	37	25	15	11	21	10	86	16	22	243
Juncaceae cf. *Juncus* spp.	7	2		1	1	10	2	9	3	16	44
Leguminosae cf. *Astragalus* sp.	11		2	3		7	1	10	4	18	45
Leguminosae/Mimosaceae	26		1(0.1 g)								1
cf. Malvaceae	16	16									16

(continued)

Table 5.1 (continued)

Flot number	Specimen type number	1	6	3	2	5	4	8	15	7	Totals
Malvaceae cf. *Malvastrum* sp.	2	78	21	13	4	98	67	78	9	101	469
cf. Nyctaginaceae	30							5			5
Nyctaginaceae cf. *Boerhavia* sp.	29							18			18
cf. Onagraceae	3					3	16	20	1	12	52
Poaceae	4	3	3	2		4	4	23	4	10	53
Poaceae	28							20			20
Poaceae *Zea mays*	9	3[c]				3		1		4	11
Portulacaceae *Portulaca* sp.	18			2		1					3
Solanaceae	27									1	1
Solanaceae *Capsicum* sp.[b]	12	6	4		5	7	4	2		1	29
Unknown type 22	22	1									1
Unknown type 23	23						1				1

(continued)

Table 5.1 (continued)

Flot number	Specimen type number	1	6	3	2	5	4	8	15	7	Totals
Unknown type 24	24					1					1
Unknown type 25	25					1					1
Unknown type 31	31								1		1
Totals		220	62	54	33	175	131	324	45	263	1307
Bone		3					1			7	11
Insect		5					4				9
Mollusk		1									1
Unidentified carbonized fragments				1	3				3		7
Carbonized bracts/culms		1					4				5
Carbonized tuber frags (probable)			2(0.2 g)					3			5
Uncarbonized seeds							2	3	1		6

[a] *Chenopodium* cf. *quinoa* was not differentiated from other possible *Chenopodium* species
[b] All seed remains of *Capsicum* were recovered desiccated except for one seed from Flot 7 that was carbonized
[c] Includes one complete kernel
Diversity Index of carbonized seeds is 0.81 with 1 being the most diverse
Cf. = Conforms to the morphology of seeds from the taxon. However, identifications remain tentative

Table 5.2 Macrobotanical remains from Surface I (Structure BH) at Ancient Paria (Ce 01), Oruro, Bolivia (Paria Archaeological Project, 2005)

Flot number	Specimen type number	16	17	18 (2 bags)	19 (4 bags)	20 (2 bags)	20c (3 bags)
Provenience		Ce-01	Ce-01	Ce-01	Ce-01	Ce-01	Ce-01
		Estr. BH	Estr. BH	Estr. BH	Estr. BH	Estr. BH	Estr. BH
		Cuadro 2	Cuadro 2	Cuadro 3	Cuadro 9	Cuadro 12	Cuadro 12
		Supf. 1		Supf. 1	Supf. 1	Supf. 1	Supf. 1
		Nivel 0–30 cm	Nivel 1–30 cm	Nivel 80–100 cm	Nivel 60–80 cm	Nivel piso-10 cm	Nivel piso-10 cm
				Exterior del muro	Bajo del muro	Hoyo abajo del muro	Hoyo abajo del muro
Dry weight of sample (kg)		5	4.3	12.33	8.1	7.8	3.5
Volume floated		5	4	8	9.5	8	3
Weight of light fraction (g)		44.4	35.6	78.3	128.9	280.9	191.2
Weight of light fraction analyzed (g) (percent analyzed)		22.2 (50%)	17.8 (50%)	15.7 (20%)	12.9 (10%)	28.1 (10%)	191.2 (100%)
Carbonized wood >2 mm no.		23	28	83	52	210	1013[g]
Carbonized wood >2 mm weight (g)		0.7	0.9	2.2	1.3	16.9	90
Amaranthaceae *Amaranthus* sp.	44	36	42		9		

(continued)

Table 5.2 (continued)

Flot number	Specimen type number	16	17	18 (2 bags)	19 (4 bags)	20 (2 bags)	20c (3 bags)
Annonaceae cf. *Annona cherimola*	47						
cf. Asteraceae	15	1	4				
cf. Brassicaceae	6						1
Brassicaceae cf. *Lepidium* sp.	45						
Cactaceae cf. *Opuntia soehrensii*	14	24	6				
Cactaceae *Trichocereus* sp.	13	2	1				
Chenopodiaceae	19						
Chenopodiaceae *Atriplex* sp.	5	1	3			1	1
Chenopodiaceae *Chenopodium* spp.	17	11					
Chenopodiaceae *Chenopodium* cf. *quinoa*	1	73	75	14	13	4	28
Cucurbitaceae cf. *Cucurbita* sp.	10						1
Cyperaceae cf. *Scirpus* sp.	20						
Juncaceae cf. *Juncus* spp.	7	3	1		4		2
cf. Leguminosae	39		1				
Leguminosae cf. *Astragalus* sp.	11	7	6	1	1	1	1

(continued)

Table 5.2 (continued)

Flot number	Specimen type number	16	17	18 (2 bags)	19 (4 bags)	20 (2 bags)	20c (3 bags)
Leguminosae *Lupinus* cf. *mutabilis*	38		9[b]				
cf. Malvaceae	16						
Malvaceae cf. *Malvastrum* sp.	2	29	38	1	7		8
Nyctaginaceae cf. *Boerhavia* sp.	29				1[c]		
cf. Onagraceae	3	17	12				4
Poaceae	4						1
Poaceae	32	51	35			1	
Poaceae	40				1		
Poaceae cf. *Bromus* sp.	34	2	3				
Poaceae cf. *Festuca* sp.	35	4	5				
Poaceae cf. *Hordeum* sp.	33	1					
Poaceae *Zea mays*	9						1 (<0.1 g)
cf. Portulacaceae/Caryophyllaceae	42						
Portulacaceae *Portulaca* sp.	18	1			1		
Solanaceae *Capsicum* sp.	12		1[a]				
Unknown type 8	8						1

(continued)

Table 5.2 (continued)

Flot number	Specimen type number	16	17	18 (2 bags)	19 (4 bags)	20 (2 bags)	20c (3 bags)
Unknown type 37	37		2				
Unknown type 41	41				1		
Unknown type 43	43						
Unknown type 46	46						
Unknown type 48	48						
Totals		263	244	16	38	7	49
Bone		1	5				3
Insect		2	5		1	1	3
Shell							1
Unidentified carbonized fragments					2		
Carbonized bracts/culms			1	3	1		1
Carbonized possible nutshell			1 (0.1 g)				
Carbonized seed fragments							
Uncarbonized seeds			1				

(continued)

Table 5.2 (continued)

Flot number	21 (2 bags)	22 (2 bags)	23	24	25	26	27
Provenience	Ce-01	Ce 01	Ce 01	Ce 01	Ce 01	Ce 01	Ce 01
	Estr. BH	Estr. BH	Estr. BH	Estr. BH	Estr. BH	Estr. BH	Estr. BH
	Cuadro 18	Cuadro 18	Cuadro 23	Cuadro 24	Cuadro 25	Cuadro 26	Cuadro 26
	Supf. 1	Supf. 1	Supf. 1	Supf. 1	Supf. 1		Supf. 1
	Nivel 30–40 cm	Nivel 40–50 cm	Nivel 50–60 cm	Nivel 60 cm	Nivel 80 cm	Nivel 30 cm	Nivel 30 cm
	Fogon	Abajo de fogon	Muestro sobre el piso	Rasgo 4	Rasgo 4		Rasgo: capa del ceniza
				Hueco en el piso	Hueco en el piso		
Dry weight of sample (kg)	9.66	9.62	6	4	5.6	4.9	5.3
Volume floated	8.5	10.5	5	3	4	5	4.5
Weight of light fraction (g)	160.6	71.4	33.1	7.1	14.2	135.4	64.2
Weight of light fraction analyzed (g) (percent analyzed)	16.1 (10%)	71.4 (100%)	33.1 (100%)	7.1 (100%)	14.2 (100%)	13.5 (10%)	16.1 (25%)
Carbonized wood >2 mm no.	65	86	22	4	7	15	48
Carbonized wood >2 mm weight (g)	2.4	3.3	0.7	0.1	0.2	0.6	1.6
Amaranthaceae *Amaranthus* sp.	8	26	15	8	6	9	5
Annonaceae cf. *Annona cherimola*							1 (0.2 g)
cf. Asteraceae		3		1	1	1	

(continued)

Table 5.2 (continued)

Flot number	21 (2 bags)	22 (2 bags)	23	24	25	26	27
cf. Brassicaceae							
Brassicaceae cf. *Lepidium* sp.						1	
Cactaceae cf. *Opuntia soehrensii*		2	3	2			
Cactaceae *Trichocereus* sp.	4	149				1[f]	
Chenopodiaceae			53				
Chenopodiaceae *Atriplex* sp.					1		
Chenopodiaceae *Chenopodium* spp.				26	119		
Chenopodiaceae *Chenopodium* cf. *quinoa*	25	19	14	129	136		
Cucurbitaceae cf. *Cucurbita* sp.							
Cyperaceae cf. *Scirpus* sp.			1				
Juncaceae cf. *Juncus* spp.	4	13	116		1		
cf. Leguminosae							
Leguminosae cf. *Astragalus* sp.	1	3	4		3		1
Leguminosae *Lupinus* cf. *mutabilis*							
cf. Malvaceae		2					

(continued)

Table 5.2 (continued)

Flot number	21 (2 bags)	22 (2 bags)	23	24	25	26	27
Malvaceae cf. *Malvastrum* sp.	2	17	27	7	1	1	4
Nyctaginaceae cf. *Boerhavia* sp.							
cf. Onagraceae		9	6			2	
Poaceae		2					
Poaceae	2			6	9	2	6
Poaceae					1		
Poaceae cf. *Bromus* sp.					1		
Poaceae cf. *Festuca* sp.				1	8	2	
Poaceae cf. *Hordeum* sp.							
Poaceae *Zea mays*		1 (<0.1 g)		10 (<0.1 g)[d]	11 (0.1 g)[e]		
cf. Portulacaceae/Caryophyllaceae				3	4		
Portulacaceae *Portulaca* sp.			1				
Solanaceae *Capsicum* sp.	1	3		4[a]	3[a]		
Unknown type 8							
Unknown type 37							
Unknown type 41				1			1
Unknown type 43					3 (0.1 g)		
Unknown type 46							1

(continued)

Table 5.2 (continued)

Flot number	21 (2 bags)	22 (2 bags)	23	24	25	26	27
Unknown type 48							
Totals	47	249	240	198	308	19	19
Bone							
Insect	2	6	23		6	2	1
Shell							
Unidentified carbonized fragments	2	1					1
Carbonized bracts/culms	1	9					
Carbonized possible nutshell							
Carbonized seed fragments					1		
Uncarbonized seeds		4				2	

Flot number	28 (2 bags)	29 (2 bags)	30 (3 bags)	Totals	Hand-collected Structure BM
Provenience	Ce 01	Ce 01	Ce 01		Ce 01
	Estr. BH	Estr. BH	Estr. BH		Estr. BM
	Cuadro 26		Cuadro 2		Cuadro 6
	Supf. 1				Supf. 4?
	Nivel 40–50 cm	Nivel 70–90 cm	Nivel 40 cm		Nivel 2–3 cm
	Rasgo: bajo de la capa de ceniza	Pozo 4	Rasgo de piso interior del muro Norte		

(continued)

Table 5.2 (continued)

Flot number	28 (2 bags)	29 (2 bags)	30 (3 bags)	Totals	Hand-collected Structure BM
		Exterior muro Norte			
Dry weight of sample (kg)	10.4	11	8.6	116.11	NA
Volume floated	9.5	7.5	8	103	NA
Weight of light fraction (g)	58.5	54.6	152.9	1511.3	320.8
Weight of light fraction analyzed (g) (percent analyzed)	14.6 (25%)	13.7 (25%)	23 (15%)	510.7 (33%)	32.8 (10%)
Carbonized wood >2 mm no.	32	34	161	1883	0
Carbonized wood >2 mm weight (g)	1.1	1.1	6	129.1	0
Amaranthaceae *Amaranthus* sp.	25		5	194	
Annonaceae cf. *Annona cherimola*				1 (0.2 g)	
cf. Asteraceae				11	
cf. Brassicaceae				1	
Brassicaceae cf. *Lepidium* sp.				1	
Cactaceae cf. *Opuntia soehrensii*	3	1		41	
Cactaceae *Trichocereus* sp.	93		1	251	

(continued)

Table 5.2 (continued)

Flot number	28 (2 bags)	29 (2 bags)	30 (3 bags)	Totals	Hand-collected Structure BM
Chenopodiaceae				53	
Chenopodiaceae *Atriplex* sp.				7	
Chenopodiaceae *Chenopodium* spp.			14	170	
Chenopodiaceae *Chenopodium* cf. *quinoa*	15	7	13	565	
Cucurbitaceae cf. *Cucurbita* sp.				1	
Cyperaceae cf. *Scirpus* sp.				1	
Juncaceae cf. *Juncus* spp.	5	1		150	
cf. Leguminosae				1	
Leguminosae cf. *Astragalus* sp.	2	2		33	
Leguminosae *Lupinus* cf. *mutabilis*				9	
cf. Malvaceae				2	
Malvaceae cf. *Malvastrum* sp.	10		10	162	
Nyctaginaceae cf. *Boerhavia* sp.				1	
cf. Onagraceae	10		1	61	

(continued)

Table 5.2 (continued)

Flot number	28 (2 bags)	29 (2 bags)	30 (3 bags)	Totals	Hand-collected Structure BM
Poaceae				3	
Poaceae	4	1	3	120	1
Poaceae				2	
Poaceae cf. *Bromus* sp.				6	
Poaceae cf. *Festuca* sp.			1	21	
Poaceae cf. *Hordeum* sp.		1		2	
Poaceae *Zea mays*				23 (<0.4 g)	
cf. Portulacaceae/Caryophyllaceae				7	
Portulacaceae *Portulaca* sp.	1			4	
Solanaceae *Capsicum* sp.	1			13	
Unknown type 8				1	
Unknown type 37				2	
Unknown type 41				3	
Unknown type 43				3 (0.1 g)	

(continued)

Table 5.2 (continued)

Flot number	28 (2 bags)	29 (2 bags)	30 (3 bags)	Totals	Hand-collected Structure BM
Unknown type 46				1	
Unknown type 48	1			1	
Totals	170	13	48	1928 (<0.7 g)	
Bone				9	
Insect		2		54	
Shell				1	
Unidentified carbonized fragments	1			7	
Carbonized bracts/culms	1			17	
Carbonized possible nutshell				1 (0.1 g)	
Carbonized seed fragments				1	
Uncarbonized seeds	1			8	

[a]Carbonized
[b]One fragment with positive identification; eight fragments with tentative identifications
[c]Tentative identification. The specimen is smaller than other specimens from Structure BM, Flot 8
[d]Specimens consist of three cob fragments, six kernel fragments, and one cupule fragments
[e]Specimens consist of one complete kernel and 10 kernel fragments
[f]Tentative identification
[g]Number estimated by using the formula: weight of 20 fragments/20 = total weight of fragments/total number
Diversity Index of carbonized seeds is 0.86 with 1 being the most diverse
Cf. = Conforms to the morphology of seeds from the taxon. However, identifications remain tentative

Table 5.3 Macrobotanical remains from the storage structures at site Ce 51 near Ancient Paria, Oruro, Bolivia (Paria Archaeological Project, 2006)

Flot number	Specimen type number	10	9	11	12	13	14	Totals
Provenience		Ce 51	Ce 51	Ce 51	Ce 51	Ce 51	Ce 51	
		Estr. 1	Estr. 1	Estr. 2	Estr. 2	Estr. 2	Estr. 2	
		exterior	Pozo 5	Pozo 1	Pozo 1	Pozo 2	Pozo 3	
		Pozo 2	interior	Supf		Nivel: -13 cm	adobe	
		muro	base nivel	afuera del	afuero del		sobre el	
				muro	muro		cimiento	
Dry weight of sample (kg)		5.8	11.1	5.9	6.4	4	5.8	39
Volume floated		4	8	4	4	3	4	27
Weight of light fraction (g) (100% analyzed)		24.6	40.4	17.2	30	21.8	20.7	154.7
Carbonized wood >2 mm no.		0	14	36	40	50	0	140
Carbonized wood >2 mm weight (g)		0	0.3	1.1	1.1	1.5	0	4
Amaranthaceae *Amaranthus* sp.	44			3				3
Cactaceae cf. *Opuntia soehrensii*	14			1		3	1	5
Chenopodiaceae	19			30	16	26	4	76
Chenopodiaceae *Atriplex* sp.	5					1		1
Chenopodiaceae *Chenopodium* spp.[a]	17		1	17	17	17	6	58
Juncaceae cf. *Juncus* spp.	7			23	19	11	2	55
Leguminosae cf. *Astragalus* sp.	11		1	17	13	18	1	50
Malvaceae cf. *Malvastrum* sp.	2			22	26	11	3	62
cf. Onagraceae	3			10	7	28	2	47
Poaceae	4			3	4		1	8
Portulacaceae *Portulaca* sp.	18			1	1			2

(continued)

Table 5.3 (continued)

Flot number	Specimen type number	10	9	11	12	13	14	Totals
Totals		0	2	127	103	115	20	367
Bone			6					6
Insect		2	6		1	2	1	12
Mollusk			1					1
Carbonized fruit fragment						1		1
Uncarbonized seeds		2	72				2	76

[a]*Chenopodium* cf. *quinoa* was not differentiated from other possible *Chenopodium* species
Diversity Index of carbonized seeds is 0.85 with 1 being the most diverse
Cf. = Conforms to the morphology of seeds from the taxon. However, identifications remain tentative

Table 5.4 Density measures from Structures BM and BH (CE 01) and the storage structures (Ce 51)

		Number	Wght (g)	Number	Wght (g)
Structure BM (Ce 01)	Wood	1339	59	33.5	1.5
Samples analyzed (40 L)	Seeds	1307	0.3	32.7	<0.1
Structure BH (Ce 01)	Wood	1883	129.1	18.3	1.3
Samples analyzed (103 L)	Seeds	1928	<0.7	18.7	<0.1
Structure 1 (Ce 51)	Wood	14	0.3	1.2	<0.1
Samples analyzed (12 L)	Seeds	2	<0.1	0.2	<0.1
Structure 2 (Ce 51)	Wood	126	3.7	8.4	0.2
Samples analyzed (15 L)	Seeds	365	<0.1	24.3	<0.1

[a]Number or weight divided by total liters of processed fill for cultural context

Macrobotanical Evidence from Ancient Paria

The macrobotanical analysis includes Surfaces I (Structure BH) and II (Structure BM) of Paria and two structures identified at Site Ce 51. The density of wood and seeds recovered from Surface II was almost two times greater than that from Surface I and about four times higher or more for wood density than that from the structures at Site Ce 51 (Table 5.4). A total of 3,602 carbonized seeds were recovered from flotation (Tables 5.1, 5.2, 5.3, 5.4, 5.5 and 5.6). Also recovered were many pieces of carbonized wood (Surface II, n = 1,339; Surface I, n = 1,883). In contrast, only 140 fragments of carbonized wood and 367 carbonized seeds were recovered from Site Ce 51 (Tables 5.1, 5.2 and 5.3). These remains are analyzed in terms of the locations where they occur. Specific attention is given to botanical remains known to have food

Table 5.5 Ubiquity of macrobotanical remains from Structures BM and BH (Ce 01) and Structures 1 and 2 (Ce 51)

Scientific determination	Specimen type number	Number	Ubiquity %	Number	Ubiquity %	Number	Ubiquity %	Number	Ubiquity %
Provenience		Ce 01 Structure BM	Ce 01 Structure BM	Ce 01 Structure BH	Ce 01 Structure BH	Ce 51 Structure 1	Ce 51 Structure 1	Ce 51 Structure 2	Ce 51 Structure 2
			(n = 9)		(n = 15)		(n = 2)		(n = 4)
Amaranthaceae *Amaranthus* sp.	44	66	78	194	80	0	0	3	25
cf. Anacardiaceae	21	20	22	0	0	0	0	0	0
Annonaceae cf. *Annona cherimola*	47	0	0	1	7	0	0	0	0
cf. Asteraceae	15	8	44	11	40	0	0	0	0
cf. Brassicaseae	6	0	0	1	7	0	0	0	0
Brassicaceae cf. *Lepidium* sp.	45	0	0	1	7	0	0	0	0
Cactaceae cf. *Opuntia soehrensii*	14	22	78	41	47	0	0	5	75
Cactaceae *Trichocereus* sp.	13	20	67	251	47	0	0	0	0
Chenopodiaceae	19	155	89	53	7	0	0	76	100
Chenopodiaceae *Atriplex* sp.	5	1	11	7	27	0	0	1	25
Chenopodiaceae *Chenopodium* spp.	17	243	100	170	27	1	50	57	100
Chenopodiaceae *Chenopodium* cf. *quinoa*	1	0	0	565	87	0	0	0	0
Cucurbitaceae cf. *Cucurbita* sp.	10	0	0	1	7	0	0	0	0
Cyperaceae cf. *Scirpus* sp.	20	0	0	1	7	0	0	0	0
Juncaceae cf. *Juncus* spp.	7	44	89	150	67	0	0	55	100

(continued)

Table 5.5 (continued)

Scientific determination	Specimen type number	Number	Ubiquity %	Number	Ubiquity %	Number	Ubiquity %	Number	Ubiquity %
cf. Leguminosae	39	0	0	1	7	0	0	0	0
Leguminosae cf. *Astragalus* sp.	11	45	78	33	80	1	50	49	100
Leguminosae *Lupinus* cf. *mutabilis*	38	0	0	9	7	0	0	0	0
Leguminosae/Mimosaceae	26	1	11	0	0	0	0	0	0
cf. Malvaceae	16	16	11	2	7	0	0	0	0
Malvaceae cf. *Malvastrum* sp.	2	469	100	162	93	0	0	62	100
cf. Nyctaginaceae	30	5	11	0	0	0	0	0	0
Nyctaginaceae cf. *Boerhavia* sp.	29	18	11	1	7	0	0	0	0
cf. Onagraceae	3	52	56	61	53	0	0	47	100
Poaceae	4	53	89	3	13	0	0	8	75
Poaceae	28	20	11	0	0	0	0	0	0
Poaceae	32	0	0	120	73	0	0	0	0
Poaceae	40	0	0	2	13	0	0	0	0
Poaceae cf. *Bromus* sp.	34	0	0	6	20	0	0	0	0
Poaceae cf. *Festuca* sp.	35	0	0	21	40	0	0	0	0
Poaceae cf. *Hordeum* sp.	33	0	0	2	13	0	0	0	0
Poaceae *Zea mays*	9	11	44	23	27	0	0	0	0
Portulacaceae *Portulaca* sp.	18	3	22	4	27	0	0	2	50
cf. Portulacaceae/Caryophyllaceae	42	0	0	7	13	0	0	0	0

(continued)

Table 5.5 (continued)

Scientific determination	Specimen type number	Number	Ubiquity %	Number	Ubiquity %	Number	Ubiquity %	Number	Ubiquity %
Solanaceae	27	1	11	0	0	0	0	0	0
Solanaceae *Capsicum* sp.	12	29	78	13	40	0	0	0	0
Unknown type 8	8	0	0	1	7	0	0	0	0
Unknown type 22	22	1	11	0	0	0	0	0	0
Unknown type 23	23	1	11	0	0	0	0	0	0
Unknown type 24	24	1	11	0	0	0	0	0	0
Unknown type 25	25	1	11	0	0	0	0	0	0
Unknown type 31	31	1	11	0	0	0	0	0	0
Unknown type 37	37	0	0	2	7	0	0	0	0
Unknown type 41	41	0	0	3	20	0	0	0	0
Unknown type 43	43	0	0	3	7	0	0	0	0
Unknown type 46	46	0	0	1	7	0	0	0	0
Unknown type 48	48	0	0	1	7	0	0	0	0

Table 5.6 Carbonized maize (*Zea mays*) remains from Structures BM and BH (Ce 01)

Provenience	Type	Row	Length		Diameter	Weight	Cupules		Side angle measurements of kernels in degrees
		Number	cm		cm	(g)	Width (mm)	Kernel Thickness (mm)	
Cob Fragments									
Ce 01, Structure BM	cob frag	14	1.6		0.7	0.2			
Supf. ll, Cuadr. 8 (Hand-collected)									
Ce 01, Structure BM	cob frag	16	2.9		1	0.5			
Supf. ll, Cuadr. 8 (Hand-collected)									
Ce 01, Structure BM	cob frag	12	2.4		1.5	0.7			
Supf. Ll, Cuadr. 5									
Nivel: 4–40-60 cm									
Rasgo: Interior estructura (Hand-collected)									
Ce 01, Structure BH	cob frag	14	1.6		0.7	0.2			

(continued)

Table 5.6 (continued)

Provenience	Type	Row	Length		Diameter	Weight	Cupules		Side angle measurements of kernels in degrees
Cuadr. 2, Nivel: 40 cm									
Rasgo de piso: interior cerca del muro (Hand-collected)									
							Cupules on cob		
Flot 24, Ce 01, Structure BH	cob frag. 1	NA	0.98		0.7	<0.1	4	2	
Cuadro 24, Supf. 1, Nivel 60 cm							4	2	
Rasgo 4 Hueco en el piso							3.2	2	
							3	2.2	
							4	1.8	
							4	2.2	
	cob frag. 2	NA	0.46		0.5	<0.1	4	2	
							3.5	2	
	cob frag. 3	NA	0.42		0.4	<0.1	3.5	1.8	
							3.5	1.8	

(continued)

Table 5.6 (continued)

Provenience	Type	Row	Length		Diameter	Weight	Cupules		Side angle measurements of kernels in degrees
Kernel and Cupule Fragments			**Kernel**						
			Depth	**Width**	**Thickness**				
			mm	**mm**	**mm**				
Flot 1, Ce 01, Structure BM	1 complete		6.5	4	3	<0.1			
Exterior Basurero	kernel +								
Nivel: 40 cm, Pozo 1 y 2	2 frags								
Flot 7, Ce 01, Structure BM	4 kernel		3.8	5	3.5	<0.1			
Cuadr. 27, Nivel 1 (20–40 cm)	frags		4	5.5	3.1	<0.1			
Exterior muro oeste			4.2	6.5	4.5	<0.1			
Flot 5, Ce 01, Structure BM	3 highly		na	na	na	<0.1			
Cuadr. 3, Nivel 3 (30–40 cm)	fragmented		na	na	na	<0.1			
Rasgo 7	kernels		na	na	na	<0.1			
Flot 8, Ce 01, Structure BM	1 kernel		4.6	4	3	<0.1			

(continued)

Table 5.6 (continued)

Provenience	Type	Row	Length		Diameter	Weight	Cupules		Side angle measurements of kernels in degrees
Flot 20c, Ce 01, Structure BH	1 kernel		2.5	3.2	2	<0.1			
Cuadro 12, Supf. 1, Nivel piso-10 cm	frag								
Hoya abajo del muro									
Flot 22, Ce 01, Structure BH	1 kernel		2.1	3	1.7	<0.1			
Cuadro 18, supf. 1, Nivel 40–50 cm	frag								
Abajo de fogon									
							Cupule Fragment		
Flot 24, Ce 01, Structure BH	1 kernel		5	4	2.2	<0.1	2.6	1	30[a]
Cuadro 24, Supf. 1, Nivel 60 cm	frag								
Rasgo 4 Hueco en el piso	and 5 highly								

(continued)

Table 5.6 (continued)

Provenience	Type	Row	Length		Diameter	Weight	Cupules		Side angle measurements of kernels in degrees
	fragmented								
	kernels								
Flot 25, Ce 01, Structure BH	1 complete kernel	complete	4.7	3.1	2.4	0.1			30[a]
Cuadro 25, Supf. 1, Nivel 80 cm	and 10 highly		5.4	4	3	<0.1			
Rasgo 4 Hueco en el piso	fragmented								
	kernels								

[a]Indicative of a 12-rowed cob or higher

and fermented beverage uses and hallucinogenic or medicinal properties. Botanical analysis of all remains from these sites has been published (Bonzani 2014a).

Surface II (Structure BM) of Paria (Site Ce 01)

From Surface II, 1,307 carbonized seeds were recovered and include 13 families, 12 genera, and 2 species. The remains are differentiated into food crops and wild or weedy plants found above 3,000 masl, and plants adapted to lower elevations below 2,500 masl. These classifications and differentiation provide evidence of interactions and exploitation at Structure BM which appears to have been a *kancha* occupied by *aqlla* or *mamacona* involved in textile production and adornments (Gyarmati and Castellón 2014).

Food crops grown above 3,000 masl from Surface II include quinoa (*Chenopodium* spp.) (potentially two different species and other types) and amaranth (*Amaranthus* spp.). The quinoa varieties were differentiated by size and morphology to *quinoa* (*Chenopodium* cf. *quinoa*), which was commonly encountered in Surfaces II and I with a tentative species identification given to specimens from Surface I. The other variety appears to be the wild or weedy *Chenopodium pallidicaule* (i.e., *kañiwa), paik'o* (*C. ambrosioides*), or *llipecha* (*C. petiolare*). Another new quinoa variety was recently identified at the site of La Barca, Department of Oruro, in contexts dated to the Formative Period (1800 B.C.–A.D. 400) (Langlie et al. 2011; Bruno 2005).

Quinoa is an important food crop grown in elevations of 2,770–3,800 masl. Ashes of *quinoa* were commonly used as lime in coca chewing (Gade 1975: 153–156). *Quinoa* was originally domesticated in the Bolivian altiplano and its consumption extends back to the Formative Period (c.1500 B.C.) (Bruno and Whitehead 2003). *Kañiwa* is also adapted to extreme altitudes above 3,800 masl. *Paik'o* seeds are medicinally used against worms and to cure stomach disorders. It has a strong odor and grows above about 2,600 masl, while the *llipecha* weed leaves are generally consumed after boiling (Gade 1975: 156).

Kiwicha (*Amaranthus caudatus*) is a domesticated herbaceous species sparingly cultivated from 2,500 to 3,000 masl. Another weed, *hatak'o* (*A. quitensis*) grows in all kinds of fields at all elevations and its leaves are boiled and eaten as a potherb. One cultivated plant known as *kiwillo* yields edible seeds and the ash is made into lime for coca chewing, and for making a red dye (Gade 1975: 156).

Other positively and tentatively identified plants adapted to the high Andes include members of the cactus family (cf. *Opuntia soehrensii*) (Fig. 5.6) and *pasakana* (*Trichocereus peruviana* syn. *Echinopsis* spp.), *liwi liwi* (Chenopodiaceae, *Atriplex* sp.), *kora* (Malvaceae, cf. *Malvastrum* sp.), *garbancillo* (Leguminosae, cf. *Astragalus* sp.), and *chiwa* (Juncaceae, cf. *Juncus* spp.) (Gade 1999: 49). Members of the grass family *Poaceae* spp. were also recovered from Surface II, but specific genera and species could not be identified in the analysis.

Opuntia is a common cactus that grows below 3,750 masl (Gade 1975: 15) (Fig. 5.7). The archaeological specimens may represent *ayrampu* (*Opuntia*

Fig. 5.6 Photo of cactus seed, tentatively identified as *ayrampu* (Cactaceae, cf. *Opuntia soehrensii*). Example recovered from Site Ce 01, Surface II (Structure BM), Squares 1 and 2. Dimensions: 3.5 × 2.8 × 2.5 mm (L × W × TH) (Photo by Renée M. Bonzani)

soehrensii), a prostrate cactus with yellow flowers, and grows wild (Gade 1975: 192; see also Anderson 2001). Its seeds yield a reddish-purple dye used for food coloring, and to make a violet dye for textiles. Its seeds are coated with a dried red substance also used to color beverages, medicines, and foods such as gelatins, jellies, and puddings (Antúnez de Mayolo 1989: 188; Gade 1975: 192). Fruits of a species of *Opuntia* called *airampu* (*O. sulphurea, O. soehrensii*, and another unidentified type) are incorporated into *chicha* to obtain a deep reddish-purple color in the Cochabamba Valley. A red to nearly black-colored maize variety called *culli* is used to make this type of *chicha* (Cutler and Cárdenas 1981: 250). Ethnographic evidence suggests the seeds and/or fruit of this cactus were also utilized for food, dye, in *chicha*, or as a textile dye in the vicinity of Structure BM during Inca times. These data support evidence of weaving and *chicha* production by women as characteristic of Inca culture.

In the case of the Peruvian Torch Cactus *pasakana* (*Echinopsis* spp.), the macrobotanical remains were identified using photos by Morón Ríos (2006a: 65, 2006b). The fruit can be stored for up to seven days in areas with good ventilation and low humidity and is commonly consumed (Morón Ríos 2006a: 70–71, 84, 123–124). In regions where trees do not grow, the large cacti stems are used in carpentry and furniture-making (Zeballos Montes de Oca et al. 2003: 60). The genus has many uses in traditional medicine, rites, rituals, and ceremonies; the trunk or stem (*tallo*) of species of the Torch or San Pedro Cactus (*Echinopsis pachamoi*) is psychotropic, associated with alkaloids such as mescaline. Other varieties of this genus in the Departments of La Paz, Cochabamba, and Chuquisaca also contain psychotropic alkaloids. Medicinal and ceremonial uses among indigenous altiplano communities have as yet not been identified or documented (Morón Ríos 2006a: 161–164). Many

Fig. 5.7 Women with sherds from Ancient Paria (Site Ce 01). Note cactus of *ayrampu* in lower left-hand corner (Photo by Renée M. Bonzani in 2006)

more seeds of this genus were recovered from Paria, but none were recovered from structures at Site Ce 51.

Below 3,750 masl, the rushes or *chiwa* (*Juncus* spp.) are adapted to poorly drained areas (Gade 1975: 15, 145; Moerman 2000: 281–282). Although potentially utilized for fodder, fiber, or building materials, these tentative identifications as well as those for *liwi liwi*, *kora*, and *garbancillo,* indicate the environment since the Late Horizon Period has remained similar (see e.g., Bonzani 1999, n.d; Castañeda 1965: 232–233; Oyuela-Caycedo and Bonzani 2005, 2014; Pearsall 1980; Perez-Arbelaez 1978: 476–477; Schultes and Nemry 1998: 78; Whitehead 1999). Given its location near the juncture of three rivers, these plant remains are indicative of wetter areas (Brockmann 1986: 85; Cuenca Sempertegui et al. 2005: 23–27; Gentry 1993: 591).

Also identified were seeds tentatively identified as cf. *Astragalus* sp., a plant associated with medicinal uses. Based on phenotypic similarities to modern plants identified as *garbancillo* (*Astragalus micranthellus*), a tentative identification of this genus was made for these archaeological specimens (Zeballos Montes de Oca et al. 2003: 64–65). *Garbancillo*, also known as *t'ala t'ala* (*Astragalus micranthellus*), is a medicinal plant, as are other members of this genus, such as *Astragalus garbancill,* and are generally boiled in alcohol to relieve leg pain (Franquemont et al. 1990; see Ugent and Ochoa 2006: 143). However, other species of *Astragalus* are known to be toxic to animals, due to the high selenium levels (Brockmann 1986: 85; Zeballos Montes de Oca et al. 2003: 65).

Other remains positively identified from Surface II (Structure BM) are of maize including 11 kernel or kernel fragments and 3 cob fragments recovered in situ on Surface II (Fig. 5.8) and chili pepper. All seeds of *Capsicum* from Surface II were recovered in a desiccated state except for one carbonized seed (Fig. 5.9). These crops are not typically grown at the extreme elevation around Paria (3,810–3,815 masl) but are generally adapted and cultivated in lower elevations below c. 3,600 masl and below 2,800 masl with best results below 1,500 masl for chili pepper (Gade 1975: 95–107, 201; Lentz and Dickau 2005: 215). However, a maize variety called *tunqu* is cultivated around the Lake Titicaca Basin at 3,810–4,200 masl (Chávez and

Fig. 5.8 Photo of *Zea mays* cob (Poaceae). Example recovered from Site Ce 01, Surface I (Structure BH), Square 2. 14 rowed. 1.6 cm in length, 0.7 cm in diameter (Photo by Renée M. Bonzani)

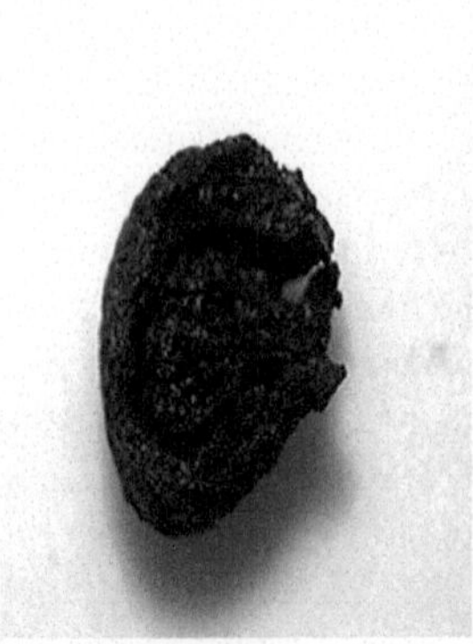

Fig. 5.9 Photo of chili pepper seed (Solanaceae, *Capsicum* sp.). Example recovered from Site Ce 01, Surface II (Structure BM), Square 27. Dimensions: 2.7 mm long × 2.4 mm wide (Photo by Renée M. Bonzani)

Thompson 2006: 416; Staller 2016; Staller et al. 2006). Maize is cultivated in elevations of 3,600 masl in the Vilcanota Valley, Peru (Gade 1975: 118), and different maize varieties are cultivated in higher elevations depending on climatic, environmental, or microenvironmental factors. This phenotypic variation among maize varieties at archaeological sites is apparent during the Middle Horizon (ca. A.D. 500–900) at Tiwanaku, Bolivia (Hastorf et al. 2006). Maize has also been well documented in the valley of Cochabamba at 2,500–2,600 masl where maize grown in this region was transported on the Inca road system to Paria for storage (Staller 2016; Covey 2009; D'Altroy 2002: 74, 273; Higueras-Hare 2001; Rowe 1946; Wachtel 1982).

Different maize varieties are used for different purposes. Not all Andean households make *chicha*, but those who brew and sell *chicha* to the different Quechua and Aymara indigenous highland communities are generally women (Gade 1975: 124, 127–128; Jennings and Bowser 2009; Moore 1989; Staller 2006, 2008, 2010; Staller and Stross 2013). Maize originated in Mesoamerica and its early spread is believed to be related to the stalk sugar which was used in the fermentation process and later production of fermented intoxicants like *chicha* (Bonzani and Oyuela-Caycedo 2006; Iltis 2000; Smalley and Blake 2003; Staller 2006, 2010).

The exact reason for the recovered desiccated pepper seeds remains to be determined. They may represent present-day inclusions but the small number of carbonized pepper seeds and their location (n = 8) suggest chiles may have been consumed during Inca times. In some instances, chili pepper remains are associated with ritual contexts, such as at La Galgada (ca. 2200–1200 B.C.) in the Río Santa valley of Peru. At La Galgada chili pepper seeds were recovered from fire pits in the mound-top chambers (Moseley 1992: 113–114). Pepper seeds were in both feasting and ritual contexts (*C. baccatum*) and domestic contexts (*C. pubescens* and *C. frutescens*) with *C. chinense* found throughout, but mainly in preceramic civic/public contexts dating c. 7,572–4,510 B.C. at Huaca Prieta and Paredones in the Chicama Valley (Chiou and Hastorf 2012; Chiou et al. 2014). Today, chili peppers are the primary condiment among indigenous and campesino Andean communities (Gade 1975: 201). The origins of *Capsicum baccatum*, *C. chinense*, *C. frutescens*,

Fig. 5.10 Photo of seeds tentatively identified as *raíz de china* (Nyctaginaceae, cf. *Boerhavia* sp.). Example recovered from Site Ce 01, Surface II (Structure BM), Square 15. Dimensions: 3.8 × 1.5 × 1.2 mm (L × W × TH) (Photo by Renée M. Bonzani)

and *C. pubescens* are in the Andes and eastern Bolivia (Chiou and Hastorf 2012; Pearsall 1992, 2008; Pickersgill 2009).

Of particular interest are the possible medicinal plants and recovery of a few other tentatively identified seed remains. Although these identifications are based on morphological similarities to photographs from Lentz and Dickau (2005: 68, 152), cf. *Boerhavia* sp. (Nyctaginaceae) (Fig. 5.10) is a herb commonly grown in cultivated fields only up to 1,400 masl indicating evidence of interaction between the altiplano and lower elevations. The genus is also used for food and medicine in Central America (Lentz and Dickau 2005: 68, 152). Furthermore, *kallawayas*, or healers in Bautista Saavedra Province, Bolivia, also use the plant (called *raíz de china* [*Boerhavia caribaea*]) with others to purify the blood, as a diuretic and to cleanse the kidneys, urethra and remove kidney stones or with other plants to make a tea to treat venereal diseases (Bastien 1987: 50, 150, 155). Recovery of similar seed remains at Paria suggests this plant was brought there for medicinal properties, particularly those concerning treatments for the urinary and reproductive systems.

Three other types, cf. Anacardiaceae, Leguminosae/Mimosaceae, and cf. Nyctaginaceae, also require further research for positive identification. They may be indicative of fruit or medicinally used trees from lower elevations. The single Leguminosae/Mimosaceae seed requires further research as it has many morphological similarities to *Inga* seeds (see Lentz and Dickau 2005: 96). In the lower valleys, shade trees such as *Inga* are often planted and the genus occurs in the forest inventories of the Bolivian Amazon region (Boom 1986). *Kusillo pakay* (*Inga adenophylla*) grows below 1,900 masl in a cultivated or volunteer situation with the pulp around the seeds eaten. *Pakay* (*Inga edulis*) is a tree cultivated in house gardens below 1,500 masl. *Pakay* is found archaeologically along the Peruvian coast (Towle 1961: 47–48) and may have been used as "*tidbit delicacies of the Incas*" (Gade 1975: 163–164).

Surface I (Structure BH) of Paria (Site Ce 01)

From Surface I, 1,928 carbonized seeds were recovered and included 18 families, 20 genera, and 5 species. These remains were broken down into crop plants and other wild or weedy plants associated with environments either above 3,000 masl, or elevations below 2,500 masl.

As with Surface II, the potential crop plants are grown above 3,000 masl recovered from Surface I include probable *quinoa* (*Chenopodium* cf. *quinoa*), other types of *Chenopodium* spp., *Amaranthus* sp., and probable *tarwi* (Leguminosae [Fabaceae], *Lupinus* cf. *mutabilis*). *Tarwi* is primarily a subsistence crop grown for its seeds and has a wide ecological niche from 2,630 to 3,830 m (Gade 1975: 96) and ranges 1,500–4,500 m (Ugent and Ochoa 2006: 156). The plant is an erect herbaceous annual probably domesticated in Peru, although definitive archaeological evidence of the species remains lacking there, while possible specimens of this genus were recovered from Ancón (Ugent and Ochoa 2006: 156). The first identified reference to *tarwi* was in 1539 by Fr. Vicente Valverde (1865 [1539]) in a letter to the Spanish Crown. Seeds must be boiled and then soaked in running water for 3–4 days to get rid of their bitter flavor. They are often served in salads with capsicum pepper, onions, and herbs, or eaten as a confection. It is often found alongside maize or as ground cover for its nitrogen-fixing properties (Gade 1975: 168).

Other positively and tentatively identified plants on Surface I, which grow in the high Andes were also recovered with many being the same as those recovered from Surface II. These include: possible *airampu* (cf. *Opuntia soehrensii), pasakana* (*Trichocereus* sp. or *Echinopsis* sp.), *liwi liwi* (*Atriplex* sp.), *kora* (cf. *Malvastrum* sp.), *garbancillo* (cf. *Astragalus* sp.), and *chiwa* (cf. *Juncus* spp.) (see Table 5.2). Other recovered carbonized remains include *mata conejo* (Brassicaceae, cf. *Lepidium* sp.), a plant toxic to animals in various animals in the area (Cuenca Sempertegui et al. 2005: 68) and cf. *Scirpus* sp. As previously mentioned, many more *pasakana* seeds were recovered only from the structures at Paria with the majority being from below a hearth and an ash lens layer identified in Surface I (Gyarmati and Castellón 2014: 66).

An attempt was made to identify possible types of carbonized grass seeds recovered from Paria. Identifications are based on reference collections made on the Condarco family farm near Paria on the Jacha Uma River in 2008. Six different types of grasses (types 4, 32, 33, 34, 35, and 40) were recovered from the archaeological site. Two types of grasses (types 32 and 35) were also recovered from modern llama dung analyzed for seed content and in a more extensive dung analysis from various Andean animal species (Bonzani 2009; Hastorf and Wright 1998). Three, including *cebadilla* [cf. *Bromus* sp., type 34]; *paja brava*, (aka *iruya* and *iru wichu*) [cf. *Festuca* sp., type 35]), and wild barley (cf. *Hordeum* sp., type 33]) were tentatively identified. For further information on these grasses and their ethnobotanical uses and recovery at other sites see Bonzani 2014a, b, c; Brockmann 1986: 85; Cuenca Sempertegui et al. 2005: 23–27, 54; Moore et al. 2007; Morón Ríos 2006b: 18–19; Planella and Blanca Tagle 2004; Renvoize 1998; Sánchez et al. 2004.

Three food crops were identified from Surface I, maize (*Zea mays* L.) including 19 kernel fragments, a cupule and 4 cob fragments, chili pepper (*Capsicum* sp.), and 1 probable squash specimen (cf. *Cucurbita* sp.) (see Table 5.6). These crops are generally adapted to elevations below 3,600 masl for maize, below 2,800–1,500 masl for chili pepper, and below 2,600 masl and rarely over 3,000 masl for squash (*Cucurbita* spp.) (Gade 1975: 95–107, 201, 216–217; Lentz and Dickau 2005: 215). Squash was domesticated early on in South America, as well as in Mesoamerica and eastern North America (Dillehay et al. 2007; Pearsall 1992, 2008; Piperno and Pearsall 1998; Simon 2000, 2009; Smith 1992, 1997). Since a specific genus and species determination could not be made for the one seed, it is possible it may represent either a lower growing taxon of *Cucurbita* (i.e., *Cucurbita maxima*) or the cold-tolerant taxon of *Cucurbita ficifolia* suggesting this plant was cultivated at Paria.

One tentatively identified small-sized *Boerhavia* sp. (Nyctaginaceae) seed was recovered from Surface I (see Tables 5.1 and 5.2). One specimen of the Annonaceae family was also tentatively identified as *chirimoya* (cf. *Annona cherimola*). This fruit has large seeds surrounded by an edible sweet, white flesh. It is from a tree adapted to below 1,800 masl (Gade 1975: 157–158). Archaeological use of *chirimoya* extends from the Preceramic Period (2100 B.C. at Huarmey) up through the Colonial Period (XVI–XVII centuries) (Ugent and Ochoa 2006: 50–51). This information suggests that *chirimoya* fruits were being transported to Paria from lower elevations such as Cochabamba (below 2,500 masl).

Structures of Site Ce 51

The botanical remains recovered from the two structures (1 and 2) at Site Ce 51 include 367 carbonized seeds and represent 9 families, 8 genera, and 1 possible species. None of the common remains ethnohistorical and archaeologically known to have been stored in *qollqa*, such as maize in the circular structures and potatoes or other tubers in the rectangular structures were recovered (Cobo 1990 [1653]; D'Altroy 1992: 66). Structure 1, the rectangular Inca Late Horizon Period *qollqa*, yielded only 2 carbonized seeds and 14 carbonized wood fragments, verifying its use only for storage where human activities of burning/cooking would not be occurring. Further, rectangular *qollqa* such as Structure 1 at Ce 51 are usually associated with storage of tubers such as potatoes, while circular ones were used for maize (D'Altroy 1992: 66; Morris and Thompson 1985).

The Late Intermediate Period Structure 2, yielded many of the same remains as those in Surface II (Structure BM), located in the center of Ancient Paria. These remains include *Amaranthus* sp., cf. *Opuntia soehrensii, Atriplex* sp., *Chenopodium* spp., cf. *Juncus* spp., cf. *Astragalus* sp., cf. *Malvastrum* sp., and *Portulaca* sp., as well as others. Particularly noteworthy was the recovery of *Chenopodium, Amaranthus,* and cf. *Opuntia* seeds that were used for food or for food/beverage/textile dyes or colorants. The potential that people belonging to local pre-Inca ethnic groups occupied or at least were storing seeds from useful plants in Structure 2 for part of its existence during the Late Intermediate Period may be borne out by the recovered botanical remains. However, none of the botanical remains from Structure 2 indicates growth habitats from elevations different from those of the site at 3,810–3,815 masl.

These results support the hypothesis that occupants of Paria proper had access to maize transported from lower elevations or other locations and distributed through elite Inca contacts. These results concur with the model developed by Hastorf (1993; D'Altroy and Hastorf 2001) using the Xauxa populace of the Upper Mantaro Valley in the central Peruvian highlands. This model predicts a shift toward the intensification in the use of fewer agricultural products, such as maize (*Zea mays*) and quinoa (*Chenopodium* sp.), with increasing Inca political control.

Discussion

The remains from the structures at Ancient Paria and Site Ce51 provide insights into how empires manage food resources as part of their political economy. The macrobotanical remains examined from these two sites indicate the existence of redistribution or exchange networks with other altitudinal zones during Inca times and the Late Horizon Period. These remains provide direct evidence for the movement of foodstuffs between different environmental settings. These botanical remains include two and possibly three crop plants, maize, chili pepper, and squash, although none of these crops were recovered from Site Ce 51. Maize, chili pepper, and squash are not typically grown at these extreme elevations, but instead are known to be generally adapted to lower elevations (Gade 1975: 95–107, 201; Lentz and Dickau 2005: 215). One species of squash (*Cucurbita ficifolia*) grows at high elevations and is cold-tolerant (Pearsall 2008), and maize has been grown in the Lake Titicaca Basin above 3,810 masl (Chávez and Thompson 2006). The maize remains may represent those grown by the Inca in the valley of Cochabamba at 2,500–2,600 masl, which were later transported on the Inca Highway to Paria for storage and use as food or to make *chicha* (D'Altroy 2002: 74, 273; Gyarmati and Varga 1999; Higueras-Hare 2001; Rowe 1946; Wachtel 1982). Indeed, as indicated in Gyarmati and Castellón (2014: 10; de la Mercedes Del Río 2005: 140), the Sora *caciques*' (or chiefs) testimonials from 1593 only mention potato and *quinoa* as crops grown in Paria. These remains verify the use of maize for food or fermented beverages at Ancient Paria during Inca times.

A few other tentatively identified plant remains (cf. *Annona cherimola*, cf. *Boerhavia* sp., cf. Anacardiaceae, Leguminosae/Mimosaceae, and cf. Nyctaginaceae) may also indicate altitudinal exchange networks and the use of medicinal plants, fruit trees, and fibers. Particularly in the case of *raíz de china* (cf. *Boerhavia* sp.), it is likely that this plant was brought in from lower elevations for its medicinal properties, specifically for the treatments of urinary and reproductive systems (Bastien 1987: 50, 150, 155). In the case of *chirimoya* (cf. *Annona cherimola*) and the seed from the Leguminosae/Mimosaceae families that may be *pakay* (*Inga* sp.), these trees are normally adapted to below 1,800–1,500 masl, respectively (Gade 1975: 157–158, 163–164). It appears fruits from trees grown at lower elevations were transported to Paria during its occupation, even from locations below that of Cochabamba (below 2,500 masl). If this were the case, the incorporation of local indigenous/ethnic groups into the Inca Empire may have been, at least partially, due to the ability of the Inca state to supply products that could not otherwise be as easily attained (Menzel 1959). On the other hand, members of the same *ayllu*, or kinship are known to often maintain separate communities in different ecological zones (Allen 1988; Bastien 1978; Murra 1972, 1980; Platt 1986; cf. Van Buren 1996). These members maintain reciprocal exchange networks in which lower growing crops such as maize and peppers might be exchanged to kinship members for products found at higher elevations (i.e., llama and alpaca wool, *quinoa,* and even potatoes). It is also possible that these plants or their seeds were carried by travelers on the routes along the *Camino Real.*

Plants with hallucinogenic properties include the San Pedro cactus or *pasakana* (*Echinopsis* spp.) and along with this cactus other possible medicinally used plants recovered from the site include *raiz de china* (cf. *Boerhavia* sp.), *garbancillo* (cf. *Astragalus* sp.), and *iru wichu* (cf. *Festuca* sp.). Many are associated with food preparation, as possible dyes for food or *chicha,* and have either hallucinogenic or medicinal properties. This illustrates the production and serving of food and possibly fermented beverages by the Inca to visitors to Ancient Paria. Given its close location to a hot spring (and possible *huaca*), contact with the supernatural with the assistance of hallucinogens may have been occurring. The site may have been a *huaca* or sacred place itself. Beyond that, inhabitants most likely were treated with medicinal plants for a variety of possible health ailments while at Paria. Many of these plants are known to treat medical health problems associated with urinary, reproductive, and digestive problems, as well as, for pain and pulmonary infections. Perhaps visits to the nearby hot spring and thermal waters were also encouraged.

In summary, Ancient Paria, also referred to as Khota Chullpa was more than simply an administrative center to store maize and potatoes. The archaeological and macrobotanical remains from the site indicate that it served numerous functions including as a central location of labor, textile production, and the metallurgical arts, a location for ritual feasting with preparation and consumption, and a probable destination and resting stop for travelers on the royal Inca highway. The movement and storage of foodstuffs were the key activities that occurred at the site. These data provide empirical evidence of practices related to plants and plant foods often discussed in ethnohistorical and other accounts.

Conclusions

As shown here, I document the various plant remains and their probable social functions through an analysis of macrobotanical remains from two sites. The administrative center of Ancient Paria stored medicinal and recreational plants in addition to foodstuffs, and therefore, the center had a much greater social function than simply serving as a center for mundane supplies. Excavations at Ancient Paria (Site Ce 01) were conducted on two structures: Surface I (Structure BH) which is a probable *kallanka,* a rectangular building used for administrative and possible ceremonial functions, and Surface II (Structure BM), a probable *kancha* utilized in the production of textiles, metallurgical arts, and possible *chicha.* A second site (Site Ce 51) which may have served mortuary functions in the Late Intermediate Period were also conducted for comparative purposes.

The recovered macrobotanical remains from Structures BH and BM indicate an extensive interconnectedness by peoples from different ecological zones. Macrobotanical remains of maize, pepper, and possible squash and cherimoya, all indicate crops most likely grown at lower elevations and brought to Ancient Paria for consumption and possible redistribution. Site Ce 51 did not contain any of these plant remains. Structure BH does have maize remains and *Chenopodium, tarwi,* and pepper, and numerous *pasakana* seeds were recovered perhaps indicating feasting and ritual/ceremonial practices in this structure. Structure BM also yielded maize, *Chenopodium,* pepper, possible squash, and other food resources, indicating food and possible beverage production. Macrobotanical remains of possible medicinal and dye plants, tied to other archaeological evidence, indicate that textile and metallurgy manufacturing were also being conducted in Structure BM.

Although we cannot say with certainty, given its location by a hot spring and thermal waters, various botanical remains of medicinal and psychotropic plants, it seems to have also been a place of respite, a place to rest on the long journey south from Cuzco or heading north on the Inca highway from the furthest southern expanses of the Empire. And, perhaps it was a place to redistribute goods, feast, and drink *chicha* on occasion, commune with the supernatural, and recuperate before the long journey over the Cordillera de Azanaques of the Andes Mountains and down the, at times, treacherous trails to the fertile, warmer valleys of Cochabamba, or even on into the lowlands of the Amazon Basin on the way to more adventures that lie ahead.

Acknowledgements The author would like to thank János Gyarmati and Álvaro and Carola Condarco Castellón for the opportunity to work with them at the site of Ancient Paria. Botanical analyses were conducted under the permit for the Paria Archaeological Project granted for a survey in 2004, and fieldwork and excavations in 2005/06 by the Bolivian Dirección Nacional de Arqueología (DINAR AUT. No. 27/04, 030/05, COM. EXT. 228/06). The fieldwork was funded by the Heinz Latin American Archaeology Program of the University of Pittsburgh. A more extensive version of this chapter with photographs was previously published in Gyarmati and Castellón (2014). I also thank Ernest Presher and Paula and Rinaldo Condarco during the 2006–08 field seasons. Also, thanks to Dr. Marc Bermann of the Department of Anthropology, University of Pittsburgh, for his assistance in returning to Bolivia in the 2007/08 field seasons. Dr. Richard Jefferies of the Department of Anthropology, University of Kentucky, also kindly assisted with the photography of the botanical remains. Most of all, the author offers her sincere thanks to the extended Condarco family for their help and great hospitality while she stayed on their farm to conduct this research. It was a great delight and honor to be able to work at this Inca site on the Inca Royal Road in the Bolivian altiplano.

References

Alberti Manzanares P (1985) La Influencia Economica y Politica de las Acllacuna en el Inkanato. Revista de Indias 45(176):557–585

Allen CJ (1988) The hold life has: Coca and cultural identity in an Andean community. Smithsonian Institution Press, Washington

Anderson EF (2001) The cactus family. Timber Press, Portland, Oregon

Antúnez de Mayolo KK (1989) Peruvian natural dye plants. Econ Bot 43(2):181–191

Bandelier AF (1910) The islands of Titicaca and Coati. Hispanic Society of America, New York

Bastien JW (1978) Mountain of the condor: metaphor and ritual in an Andean Ayllu. West Publishing Company, New York

Bastien JW (1987) Healers of the Andes: Kallawaya herbalists and their medicinal plants. University of Utah Press, Salt Lake City

Bauer BS (2002) The development of the Inka state. University of Texas Press, Austin

Bauer BS, Alan Covey R (2002) Processes of state formation in the Inka heartland (Cuzco, Peru). Am Anthropol 104:846–864

Bird JB (1985) The preceramic excavations at Huaca Prieta, Chicama valley, Peru. Anthropological Papers of the American Museum of Natural History, vol 62, part 1. American Museum of Natural History, New York

Bonzani RM (1997) Plant diversity in the archaeological record: a means toward defining hunter-gatherer mobility strategies. J Archaeol Sci 24:1129–1139

Bonzani RM (1998) Learning from the present: the constraints of seasonality on foragers and collectors. In: Oyuela-Caycedo A, Scott Raymon J (eds) Advances in the archaeology of the Northern Andes. Monograph 39. The Institute of Archaeology, University of California, Los Angeles, pp 20–35

Bonzani RM (1999) Medicinal use of plants in the peasant community of San Jacinto, Northern Colombia. Caldasia 21(2):203–218. Bogotá

Bonzani RM (2009) Botanical analysis for the formative period site of Pukara 2, Bolivia. Ms in possession of the author

Bonzani RM (2014a) Macrobotanical remains recovered during the archaeological investigations of the Paria basin. In: Gyarmati J, Castellón CC (eds) Paria La Viexa: prehispanic settlement patterns in the Paria basin, Bolivia, and its Inka provincial center. Museum of Ethnography, Budapest, Hungary, pp 119–132

Bonzani RM (2014b) Appendix G: macrobotanical remains from Gaván-complex sites. In: Spencer CS, Redmond EM (eds) A pre-Hispanic chiefdom in Barinas, Venezuela. Excavations at Gaván-complex sites, vol 2. American Museum of natural history anthropological papers, number 100. American Museum of Natural History, New York, New York, pp 862–896

Bonzani RM (2014c) Carbonized botanical remains from sites in Purén and Lumaco. In: Dillehay TD (ed) The teleoscopic polity. Contributions to global perspectives in archaeology, vol 38, Appendix 2. Springer International Publishing, Switzerland, pp 321–338

Bonzani RM, Oyuela-Caycedo A (2006) The gift of the variation and dispersion of maize: social and technological context in Amerindian societies. In: Staller J, Tykot R, Benz B (eds) Histories of maize: multidisciplinary approaches to the prehistory, biogeography, domestication, and evolution of maize. Routledge, Taylor & Francis Group, New York, pp 343–356.

Boom BM (1986) A forest inventory in Amazonian Bolivia. Biotropica 18(4):287–294

Bray TL (2002) To dine splendidly. Imperial pottery, commensal politics, and the Inka state. In: Bray TL (ed) The archaeology and politics of food and feasting in early states and empires. Kluwer Academic Publishers, New York, pp 93–142

Bray TL (2013) Water, ritual, and power in the Inca empire. Lat Am Antiq 24(2):164–190

Brockmann CE (ed) (1986) Perfil Ambiental de Bolivia. Instituto Internacional para el Desarrollo y Medio Ambiente. Agencia de los Estados Unidos para el Desarrollo Internacional, La Paz, Bolivia, Washington, DC

Bruno MC (2005) ¿Domesticado o Silvestre? Resultados de la Investigación de Semillas de *Chenopodium*, Chiripa, Bolivia (1500–100 A.C.). Textos *Antropológicos* 15(2):39–50. Carreras de Antropología y Arqueología, Universidad Mayor de San Andrés, La Paz, Bolivia

Bruno MC (2008) Etnobotánica y las Investigaciones sobre la Agricultura Prehistoric en la Cuenca del Lago Titicaca. In: Casanovas CR (ed) Arqueología de las Tierras Altas, Valles Interandinos, y Tierras Bajas de Bolivia. Memorias del I Congreso de Arqueología de Bolivia. Instituto de Investigaciones Antroplógicas y Arqueológicas, Universidad Mayor de San Andrés, La Paz, Bolivia, pp 27–34

Bruno MC, Whitehead WT (2003) *Chenopodium* cultivation and formative period agriculture at Chiripa, Bolivia. Lat Am Antiq 14(3):339–355

Castañeda RR (1965) Flora del Centro de Bolivar. Universidad Nacional de Colombia, Instituto de Ciencias Naturales, Bogotá, Colombia

Castañeda RR (1991) Frutas Silvestres de Colombia, 2nd edn. Instituto Colombiano de Cultura Hispánica, Bogotá, Colombia

Chávez SJ, Thompson RG (2006) Early maize on the Copacabana peninsula: implications for the archaeology of the lake Titicaca Basin. In: Staller J, Tykot R, Benz B (eds) Histories of maize: multidisciplinary approaches to the prehistory, biogeography, domestication, and evolution of maize. Routledge, Taylor & Francis Group, New York, pp 415–428

Chiou K, Hastorf CA (2012) *Capsicum* spp. at the preceramic sites of Huaca Prieta and Paredones, Chicama valley, Peru. Report written for Tom Dillehay, Vanderbilt University 2012. UCB Archaeobotany Laboratory Report #74. Berkeley, CA

Chiou KL, Hastorf CA, Bonavia D, Dillehay TD (2014) Documenting cultural selection pressure changes on Chile Pepper (*Capsicum baccatum* L.) seed size through time in Coastal Peru (7,600 B.P.-Present). Econ Bot 68(2):190–202

Cieza de León P (1973 [1553]) La Crónica del Perú. Promoción Editorial Inka S.A., Lima

Cieza de León P (1985 [1553]) Crónica del Perú. Segunda parte. Pontificia Universidad Católica, Lima

Cobo FB (1990 [1653]) Inka religion and customs (trans: Hamilton R). University of Texas Press, Austin

Condarco CC (2009) Poblamiento y Desarrollo Cultural en el Altivalle de Paria del Período Formativo al Período Tardío/Colonial (ca. 1500 a.C. 1535 d.C.), Oruro-Bolivia. Doctorado en Arqueologia y Patrimonio, Geografía, Historia y Filosofía, Area de Arqueologia, Universidad Pablo de Olavide, Sevilla, Spain

Condarco CC, Edgar HM, Mile VR (2002) Tras Las Huellas del Tambo Real de Paria. Fundación PIEB, La Paz, Bolivia

Covey RA (2006) How the Inkas built their heartland: state formation and the innovation of imperial strategies in the sacred valley, Peru. The University of Michigan Press, Ann Arbor

Covey RA (2009) Inka agricultural intensification in the imperial heartland and provinces. In: Marcus J, Williams PR (eds) Andean civilization: a tribute to Michael E. Moseley. Monograph 63. Cotsen Institute of Archaeology, University of California, Los Angeles, pp 365–377

Cuenca Sempertegui A, Bahoz EG, Canelas EL, Cáceres IM (2005) Más Allá de las Pajas y Espinas: Biodiversidad en el Municipio de Oruro (Comunidades Cochiraya – Iroco – Chuzekery). Latinas Editores, Oruro, Bolivia

Cutler HC, Cárdenas M (1981) Chicha, una Cerveza sudamericana Indígena. In: La Tecnología en el Mundo Andino. Runakunap kawsayninkupaq rurasqankunaqa. Universidad Nacional Autónoma, México, pp 247–259

D'Altroy TN (1992) Provincial power in the Inka empire. Smithsonian Institution Press, Washington

D'Altroy TN (2002) The Inkas. Blackwell Publishers, Oxford

D'Altroy TN, Hastorf CA (eds) (2001) Empire and domestic economy. Kluwer Academic, New York

de la Mercedes Del Río M (2005) Etnicidad, Territorialidad y Colonialismo en Los Andes: Tradición y Cambio entre Los Soras de Los Siglos XVI y XVII. IFEA, IEB, La Paz, Lima

Dillehay TD, Rossen J, Ugent D, Karathanasis A, Vásquez V, Netherly PJ (2010) Early holocene Coca chewing in Northern Peru. Antiquity 84:939–953

Dillehay TD, Rossen J, Andres TC, Wilson DE (2007) Preceramic adoption of squash, peanut, and cotton in Northern Peru. Science 316:1890–1893

Franquemont C, Franquemont E, Davis W, Plowman T, King SR, Sperling CR, Niezgoda C (1990) The ethnobotany of Chinchero, an Andean community in Southern Peru. Fieldiana Bot (Nuevas Ser) 24:1–126

Gade DW (1975) Plants, man and the land in Vilcanota valley of Peru. Dr. W. Junk B.V., The Hague

Gade DW (1999) Nature and culture in the Andes. The University of Wisconsin Press, Madison, Wisconsin

Gentry AH (1993) A field guide to the families and genera of woody plants of Northwest South America (Colombia, Ecuador, Peru). Conservation International, Washington, DC

Glowacki M, Malpass M (2003) Water, *Huaca*, and ancestor worship: traces of a sacred Wari landscape. Lat Am Antiq 14(4):431–448

Gyarmati, János 2005 Paria Archaeological Project (PAP). Report of Fieldwork Season, 2005Gyarmati J (2005) Paria archaeological project (PAP). Report of fieldwork season 2005. Ms on file at the Museum of Ethnography, Budapest, Hungary

Gyarmati J, Castellón CC (2006) Reporte Preliminar del Proyecto Arqueológico Paria, Oruro (PAP) 2004–2006. Ms on file at the Museum of Ethnography, Budapest, Hungary

Gyarmati J, Castellón CC (2014) Paria La Viexa: prehispanic settlement patterns in the Paria Basin, Bolivia, and its Inka provincial center. Museum of Ethnography, Budapest, Hungary

Gyarmati J, Varga A (1999) The Chacaras of war. An Inka state estate in the Cochabamba valley, Bolivia. Néprajzi Múzeum, Budapest, Hungary

Hastorf CA (1993) Agriculture and the onset of political inequality before the Inka. Cambridge University Press, Cambridge

Hastorf CA (1999) Recent research in paleoethnobotany. J Archaeol Res 7(1):55–103

Hastorf CA, Popper VS (eds) (1988) Current paleoethnobotany: analytical methods and cultural interpretations. University of Chicago Press, Chicago

Hastorf CA, Whitehead WT, Bruno MC, Wright M (2006) The movements of maize into middle horizon Tiwanaku, Bolivia. In: Staller J, Tykot R, Benz B (eds) Histories of maize: multidisciplinary approaches to the prehistory, biogeography, domestication, and evolution of maize. Routledge, Taylor & Francis Group, New York, pp 429–448

Hastorf CA, Wright MF (1998) Interpreting wild seeds from archaeological sites: a dung charring experiment from the Andes. J Ethnobiol 18(2):211–227

Hather J (2000) Archaeological Parenchyma. Archetype Publications, London

Higueras-Hare A (2001) El Periodo Intermedio (Horizonte Medio) en los Valles de Cochabamba: una Perspetiva del Analisis de Asentamientos Humanos y Uso de Tierras. Boletín de arqueología PUCP 5:625–646

Honores AMF, Rodríguez Rodríguez EF (2007) Etnobotánica del Perú Pre-Hispano. Ediciones Herbarium Truxillense (HUT), Universidad Nacional de Trujillo, Trujillo, Perú

Hyslop J (1984) The Inka road system. Academic Press, Orlando

Hyslop J (1990) Inka settlement planning. University of Texas Press, Austin

Iltis HH (2000) Homeotic sexual translocations and the origin of maize (*Zea mays*, *Poaceae*): a new look at an old problem. Econ Bot 54:7–42

Jennings J, Bowser BJ (eds) (2009) Drink, power, and society in the Andes. University Press of Florida, Gainesville

Langlie BAS, Hastorf CA, Bruno MC, Bermann M, Bonzani RM, Condarco WC (2011) Diversity in Andean *Chenopodium* domestication: describing a new morphological type from La Barca, Bolivia 1300–1250 B.C. J Ethnobiol 31(1):72–88

Lennstrom HA, Hastorf CA (1992) Testing old wives' tales in paleoethnobotany: a comparison of bulk and scatter sampling schemes from Pancan, Peru. J Archaeol Sci 19:205–229

Lennstrom HA, Hastorf CA (1995) Interpretation in context: sampling and analysis in paleoethnobotany. Am Antiq 60(4):701–722

Lentz D, Dickau R (2005) Seeds of Central America and Southern Mexico: the economic species. Memoirs of the New York Botanical Garden, vol 91. The New York Botanical Garden, New York
Martin AC, Barkley WD (2000) Seed identification manual, 2nd edn. The Blackburn Press, Caldwell, New Jersey
Menzel DA (1959) The Inka occupation of the South coast of Peru. Southwest J Anthropol 15(2):125–142
Moerman DE (2000) Native American ethnobotany, 3rd edn. Timber Press, Portland, Oregon
Montgomery FH (1977) Seeds and fruits of plants of Eastern Canada and Northeastern United States. University of Toronto Press, Toronto
Moore JD (1989) Pre-Hispanic beer in coastal Peru: technology and social context of prehistoric production. Am Anthropol 91(3):682–695
Moore JD (2014) A prehistory of South America: ancient cultural diversity on the least known continent. University Press of Colorado, Boulder, Colorado
Moore K, Bruno M, Capriles J, Hastorf C (2007) Integrated contextual approaches to understanding past activities using plant and animal remains from Kala Uyuni. In: Bandy M, Hastorf C (eds) Kala Uyuni: an early political center in the Southern lake Titicaca Basi. Archaeological Research Facility, Berkeley, pp 113–133
Morcote Ríos G (1996) Evidencia Arqueobotanica de Cultigenos Presentes en Grupos Muiscas de la Sabana de Bogotá en los Siglos VIII y XI. In: Enciso B, Therrien M (eds) Bioantropología de la Sabana de Bogotá, Siglos VIII al XVI D.C. Instituto Colombiano de Antropología-ICAN, Colcultura, Bogotá, pp 59–83
Morón Ríos EH (2006a) Pasakana: Fruto del Ecosistema Altiplánico. Latinas Editores, Oruro, Bolivia
Morón Ríos EH (2006b) Halofitas: Plantas Tolerantes a la Salinidad. Deposito Legal No. 1-040/2007, Oruro, Bolivia
Morris C, Thompson DE (1985) Huánuco Pampa: an Inka city and its hinderland. Thames and Hudson, London
Moseley ME (1992) The Inkas and their ancestors. The archaeology of Peru. Thames and Hudson, New York
Murra JV (1972) El "Control Vertical" de un Máximo de Pisos Ecológicos en la Economía de las Sociedades Andinas. In: de Zuòiga IO (ed) Visita de la Provincia de León de Huánuco en 1562, Tomo II. Lima, Perú, pp 429–468
Murra JV (1980) The economic organization of the Inka state. JAI Press, Greenwich, Connecticut
Oyuela-Caycedo A (1998) Seasonality in the tropical lowlands of Northwest South America: the case of San Jacinto 1, Colombia. In: Rocek TR, Bar-Yosef O (eds) Seasonality and sedentism: archaeological perspectives from old and new world sites. Peabody Museum Bulletin 6, Peabody Museum of Archaeology and Ethnology, Harvard University, Cambridge, MA, pp 165–180
Oyuela-Caycedo A, Bonzani RM (2005) San Jacinto 1: a historical ecological approach to an archaic site in Colombia. University of Alabama Press, Tuscaloosa
Oyuela-Caycedo A, Bonzani RM (2014) San Jacinto 1: Ecologia Historica, Origenes de la Ceramica, e Inicios de la Vida Sedentaria en el Caribe Colombiano. Spanish Edition. Universidad del Norte, Barranquilla, Colombia
Pärssinen M (2002) Confederaciones Interprovinciales y Grandes Señores Interétnicos en el Tawantinsuyu. Boletín de Arqueología 6:23–41
Pärssinen M (2005) Caquiaviri y la Provincia Pacasa: Desde el Alto-Formativo Hasta la Conquista Española (1–1533). Producciones CIMA, La Paz, Bolivia
Pearsall DM (1980) Pachamachay ethnobotanical report: plant utilization at a hunting base camp. In: Rick JW (ed) Prehistoric hunters of the high Andes. Academic Press, New York, pp 191–231
Pearsall DM (1992) The origins of plant cultivation in South America. In: Wesley Cowan C, Watson PJ (eds) The origins of agriculture. Smithsonian Institution Press, Washington, DC, pp 173–205
Pearsall DM (2000) Paleoethnobotany: a handbook of procedures, 2nd edn. Academic Press Inc., New York

Pearsall DM (2003) Plant food resources of the Ecuadorian formative: an overview and comparison to the central Andes. In: Scott Raymond J, Burger RL (eds) Archaeology of formative Ecuador. Dumbarton Oaks Research Library and Collection, Washington, DC, pp 213–257
Pearsall DM (2008) Plant domestication and the shift to agriculture in the Andes. In: Silverman H, Isbell WH (eds) Handbook of South American archaeology. Springer, New York, pp 105–120
Perez-Arbelaez E (1978) Plantas Utiles de Colombia. Litografía Arco, Bogotá
Pickersgill B (2009) Domestication of plants revisited: Darwin to the present day. Bot J Linnaean Soc 161:203–212
Piperno DR, Pearsall DM (1998) The origins of agriculture in the lowland tropics. Academic Press, San Diego
Planella MT, Blanca Tagle A (2004) Inicios de Presencia de Cultígenos en la Zona Central de Chile, Períodos Arcaico y Agroalfarero Temprano. Chungura, Revista de Antropología Chilena Volumen Especial: 387–399
Platt T (1986) Mirrors and maize: the concept of *Yanantin* among the Macha of Bolivia. In: Murra JV, Wachtel N, Revel J (eds) Anthropological history of Andean polities. Cambridge University Press, Cambridge, pp 228–259
Quilter J (2014) The ancient central Andes. Routledge World Archaeology, New York, New York
Renvoize SA (1998) Graminaes de Bolivia. Royal Botanical Gardens, KEW, London
Rice PM (2012) Torata Alta: an Inka administrative center and Spanish colonial Reducción in Moquegua, Peru. Lat Am Antiq 23(1):3–28
Rocas AN (1989) Semillas de Plantas Leñosas: Morfología Comparada. Limusa Noriega Editores, Mexico
Rossen J (1999) The Flote-Tech flotation machine: Messiah or mixed blessing? Am Antiq 64:370–372
Rossen J, Planella MT, Stehberg R (2010) Archaeobotany of Cerro del Inga, Chile, at the Southern Inka Frontier. In: Malpass MA, Alconini S (eds) Distant provinces in the Inka Empire: toward a deeper understanding of Inka imperialism. University of Iowa Press, Iowa City, Iowa, pp 14–43
Rowe JH (1946) Inka culture at the time of the Spanish conquest. In: Steward J (ed) Handbook of South American Indians, volume 2. Bureau of American Ethnology Bulletin 143, Washington DC, pp 183–330
Sánchez M, Quiroz D, Massone M (2004) Domesticación de Plantas y Animales en la Araucanía: Datos, Metodologías y Problemas. Chungura, Revista de Antropología Chilena Volumen Especial: 365–372
Schultes RE Nemry von Thenen de Jaramillo-Arango MJ (trans). Sánchez M, Quiroz D, Massone M (1998) The journals of Hipólito Ruiz: Spanish botanist in Peru and Chile 1777–1788. Timber Press, Portland, Oregon
Simon ML (2000) Regional variations in plant use strategies in the midwest during the late Woodland. In: Emerson TE, McElrath DL, Fortier AC (eds) Late Woodland societies: tradition and transformation across the midcontinent. Lincoln, University of Nebraska Press, pp 37–75
Simon ML (2009) A regional and chronological synthesis of archaic period plant use in the midcontinent. In: Emerson TE, McElrath DL, Fortier AC (eds) Archaic societies: diversity and complexity across the midcontinent. SUNY Press, Albany, New York, pp 81–114
Smalley J, Blake M (2003) Sweet beginnings: stalk sugar and the domestication of maize. Curr Anthropol 44:675–703
Smith BD (1992) Rivers of change: essays on early agriculture in Eastern North America. Smithsonian Institution Press, Washington, DC
Smith BD (1997) The initial domestication of *Cucurbita Pepo* in the Americas 10,000 years ago. Science 276(5314):932–934
Staller JE (2006) The social, symbolic, and economic significance of *Zea mays* L. in the late horizon period. In: Staller J, Tykot R, Benz B (eds) Histories of maize: multidisciplinary approaches to the prehistory, linguistics, biogeography, domestication, and evolution of maize. Routledge, Taylor & Francis Group, New York, pp 449–467

Staller JE (2008) Dimensions of place: the significance of centers to the development of Andean civilization: an exploration of the *Ushnu* concept. In: Staller JE (ed) Pre-Columbian landscapes of creation and origin. Springer, New York, New York, pp 269–312

Staller JE (2010) Maize cobs and culture history of Zea mays L. Springer, Berlin, Heidelberg

Staller JE (2016) High altitude maize (*Zea mays* L.) cultivation and endemism in the lake Titicaca Basin. J Bot Res 1(1):8–21

Staller JE, Stross B (2013) Lightning in the Andes and Mesoamerica: pre-Columbian, colonial and contemporary perspectives. Oxford University Press, NY; Oxford

Staller J, Tykot R, Benz B (eds) (2006) Histories of maize: multidisciplinary approaches to the prehistory, linguistics, biogeography, domestication, and evolution of maize. Routledge, Taylor & Francis Group, New York

Towle MA (1961) The ethnobotany of pre-Columbian Peru. Viking Fund Publications in Anthropology 30. Wenner-Grenn Foundation for Anthropological Research, New York

Trimborn H (1967) Das Verwunschene Dorf in Carangas. Archäologische Studien in den Kordilleren Boliviens, Baessler Archiv, Beiträge zur Völkerkunde neue Folge, Beiheft 5:52–62

Ugent D, Ochoa CM (2006) La Etnobotánica del Perú: Desde la Prehistoria al Presente. Consejo Nacional de Ciencia, Tecnología e Innovación Tecnológica-CONCYTEC, Lima, Perú

Vaca de Castro Cavellero C (1908 [1543]) Ordenanzas de Tambos. *Revista Histórica*, T. 3. Lima, Perú, pp 427–492

Valverve V Fr (1865 [1539]) Carta Dirigida al Emperador Carlos V. Colección de Documentos Inéditos Relatives al Descubrimiento, Conquista y Colonicación de las Posesiones Españolas en América y Oceanía. Tomo III. Madrid

Van Buren M (1996) Rethinking the vertical archipelago: ethnicity, exchange, and history in the South-Central Andes. Am Anthropol 98(2):338–351

Wachtel N (1982) The *Mitimas* of the Cochabamba valley: the colonization policy of Huayna Capac. In: Collier GA, Rosaldo RI, Wirth JD (eds) The Inka and Aztec states, 1400–1800: anthropology and history. Academic Press, New York, pp 199–235

Wagner G (1986) The corn and cultivated beans of the Fort Ancient Indians. Mo Archaeol 47:107–135

Whitehead WT (1999) Paleoethnobotanical evidence. In: Hastorf CA (ed) Early settlement at Chiripa, Bolivia. Contributions of the Archaeological Research Facility 57. Archaeological Research Facility, Berkeley, pp. 95–103

Zeballos Montes de Oca M, Emilia García E, Beck SG (2003) Contribución al Conocimiento de la Flora del Departamento de Oruro. Herbario Nacional de Bolivia. Artes Gráficas Latina. UAP, La Paz, Bolivia

Chapter 7
Identification of *Chicha de Maiz* in the Pre-Columbian Andes Through Starch Analysis: New Experimental Evidence

Crystal A. Dozier and Justin Jennings

Abstract Maize beer, or *chicha de maiz*, has held an important political, social, and ritual role within Andean society for thousands of years. However, definitive identification of maize beer production in the archeological record has proven difficult. This chapter surveys different methodologies that have been used to identify *chicha* and introduces the first documentation for malted maize starch, a crucial step in many *chicha* productions. While the boiling regime of this experimental *chicha* gelatinized most of the starch, archeological residues containing possibly fermented maize starch has been recovered in several contexts within South America. This chapter serves to advance the utility of ancient starch analysis for identifying *chicha* production in the pre-Columbian Andes.

Introduction

Maize beer or *chicha de maiz* is one of the most extensively researched indigenous brewing traditions of the Americas (Bruman 2000; Cutler and Cardenas 1947; Hayashida 2008; Jennings 2004; Moore 1989; Nicholson 1960; Vallejo et al. 2013). Although the beverage has played important political, social, and ritual roles within Andean society for thousands of years, definitive identification of maize beer production in the archaeological record has proven difficult. This chapter surveys different methodologies that have been used to identify ancient *chicha* production and discusses the first experimental documentation of malted maize starch, a crucial step in making many varieties of *chicha* (see Leibowicz and Jacob, Chap. 5 and

C. A. Dozier (✉)
Department of Anthropology, Wichita State University, 1845 Fairmount Street, 67260 Wichita, KS, USA
e-mail: crystal.dozier@wichita.edu

J. Jennings
Department of Art and Culture, Royal Ontario Museum, Toronto, Canada

Department of Anthropology, University of Toronto, Toronto, Canada

J. E. Staller (ed.), *Andean Foodways*, The Latin American Studies Book Series,
https://doi.org/10.1007/978-3-030-51629-1_7

Laffey and Llanque, Chap. 8 for more insights regarding *chicha* production and its roles in rites and rituals). Boiling gelatinized the starch in our experimental chicha, but archaeological residues from vessels used earlier in the brewing process could potentially be analyzed for malted maize starch, offering for the first time a direct means to identify *chicha* production in the pre-Columbian Andes.

There is a wide diversity of brewing technologies for *chicha*, and the term is sometimes used to describe both fermented beverages made from starches other than maize, such as manioc or quinoa (Cutler and Cardenas 1947; Weismantel 1988, 96) as well as sweetened beverages which may not be alcoholic (Cutler and Cardenas 1947; Smalley and Blake 2003: 695). We employ the shortened term *chicha*—a Nahuatl term for fermented intoxicants that were brought into the Andes during the Spanish Conquest (Staller 2006: 449; Bonavia 2013: 258–259)—within this article to characterize maize beer. There are other indigenous terms and traditions from across the Americas, such as *tiswin* (Buskirk 1986), *aswa* (Weismantel 1988), or *tesgüino* (Bruman 2000), that are not addressed here.

The maize (*Zea mays* L.) fermentation process is of particular interest to archaeologists throughout the Americas, given that maize cultivation was and remains widespread among many indigenous cultures. There is growing consensus that initial domestication of maize from wild teosinte occurred in the Southern Mexican highlands around the Balsas River drainage at c. 6000 BC (Bryant 2007; Doebley and Stec 1991, 1993; Matsuoka et al. 2002; Piperno et al. 2009; Staller 2010). Smalley and Blake (2003) argue that exploitation of sugary pith for fermentation was the primary driver for the domestication and spread of early maize, a beer-before-bread argument paralleled in other parts of the world (Iltis 2000; Braidwood et al. 1953; Hornsey 2003; Hayden, Canuel, and Shanse 2013). More research into indigenous brewing technologies should help archaeologists better understand how maize was domesticated (Smalley and Blake 2003; Staller 2010).

Alcoholic beverages quickly became mainstays of feasting regimes and mechanisms of power for early state societies (Dietrich et al. 2012; Guerra-Doce 2015; Hayden 2014). The importance of maize beer in the ancient Andes is difficult to overstate (Bray 2003; Moore 1989; Jennings 2004, 2014; Jennings and Bowser 2009). This is particularly true of the Inca Empire, where maize beer and cloth production were the keystones of an Andean political economy based on complementarity and reciprocity (Murra 1960, 1975 [1972], 1980; Cummins 2002; Ramírez 2005; Staller 2006). Those that offered their labor to the Inca state expected to be served copious amounts of maize beer at rituals and rites, and the need to brew "millions of gallons" of beer for these events significantly shaped imperial infrastructure (Morris 1979: 32). Beer production appears to have been central to governance in earlier Andean states as well—Paul Goldstein (2003) suggests that Tiwanaku created a *chicha* economy in the first millennium AD. Brewing equipment and keros (serving vessels) are widely associated with communal gathering places, and were a mainstay of coastal and highland domestic assemblages (Gero 1990, 1992; Moore 1989; Segura Llanos 2001). If these vessels were used for maize beer, then the beverage has been foundational to many Andean societies for at least the last three thousand years (Weismantel 2009).

Confidently tracking maize beer production in the region prior to the Inca Empire, however, has proven rather difficult for archaeologists. Brewing equipment is not specific to *chicha*, and the production sequence overlaps considerably with recipes for stews and other foods made with maize (Hayashida 2008; Jennings and Duke 2018; Jennings and Valdez 2016; Parker and McCool 2015). On very rare occasions, superb preservation provides convincing evidence for the architectural spaces, tools, and waste associated with each step in maize beer production (Moore 1989; Prieto 2011). Yet in most cases, we can only capture part of this production sequence. These data, when combined with evidence for elevated maize consumption via isotope analysis for some groups (Burger and Van der Merwe 1990; Tykot et al. 2006; Hastorf 1991; Hastorf et al. 2006) or the presence of maize kernels, phytoliths, or starches associated with culinary and ritual activity areas in archaeological sites (Hastorf et al. 2006; Logan et al. 2012), can be used to develop a strong, but still circumstantial, case for *chicha* production. Direct measures are needed to demonstrate that archaeological maize remains were consumed as *chicha*. This chapter discusses the results of a starch study that may enable archaeologists to detect malted maize in the record.

Brewing *Chicha*

Ethnographically recorded *chicha* manufacture indicates wide variation between regions and recipes. However, as Fig. 6.1 demonstrates, the production stream of *chicha* can be simplified into three main steps: preparation of the flour (malting/grinding), boiling (mashing), and fermentation. Women have been responsible for the production of the vast majority of *chicha* in the Andes since at least the time of the Incas (Hayashida 2008; Jennings and Chatfield 2009; Weismantel 2009). Male brewers, however, are known from earlier periods on the North Coast of Peru (Hayashida 2009: 234), most notably in Miguel Cabello de Balboa's (1586) recording of the story of a king named Nymlap whose *chicha* was brewed by a male retainer.

Brewing maize follows the same basic biotechnologies as the other fermentable cereal grains, such as barley, wheat, and rice. The first two steps in brewing techniques for grains require the enzymatic conversion of complex carbohydrates (starch) into sugars through malting and mashing. After unwanted hulls and shells are removed from the desired grain, there are several techniques used to stimulate enzymatic production of amylase within the grain. Amylase is the enzyme responsible for hydrolysis of starch granules; hydrolysis frees simple sugars from the semicrystalline structure of starch granules.

Controlled germination, or malting, occurs when grains are soaked in water until the radicle (roots) sprouts. This stimulates the β-amylase enzyme that occurs naturally in raw, ungerminated grains and produces the α-amylase enzyme (Wilkins and Atiyeh 2012). In the Andes, maize sprouts are typically dried on woven mats in the sun after the rootlets are twice the size of the grain. The grains, called malt at this stage in the brewing literature, can then be dried and stored until needed for further processing.

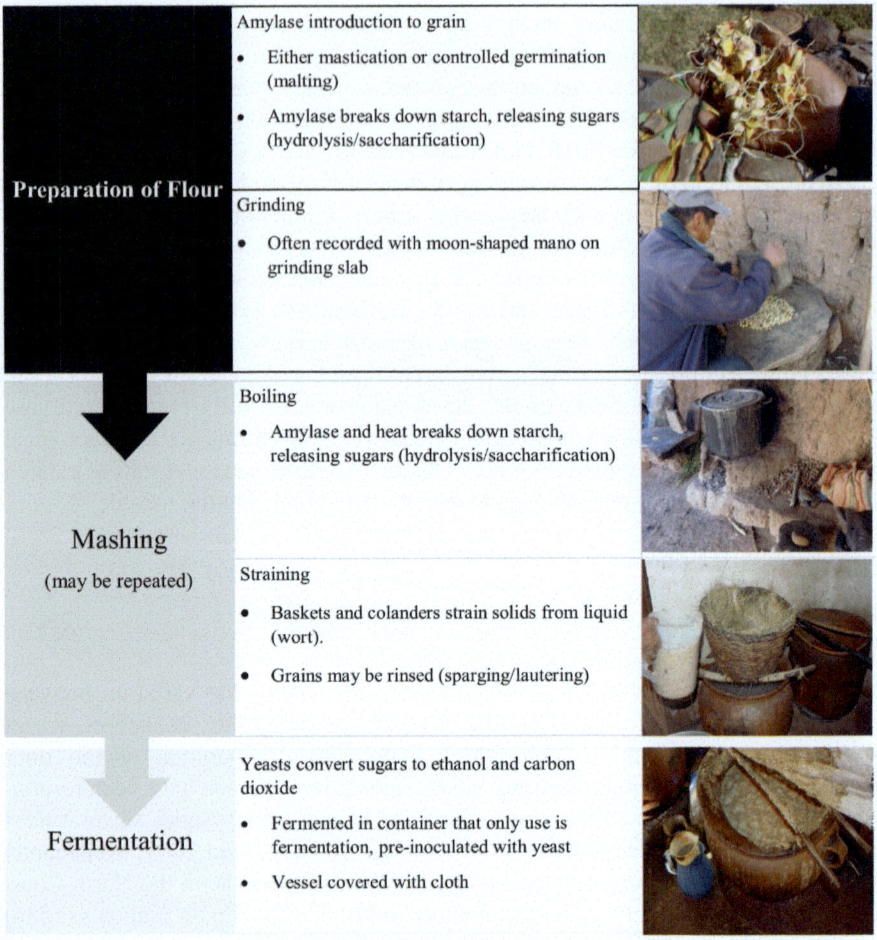

Fig. 6.1 Generalized chicha production stream, as illustrated with contemporary photos from eastern Cuzco, Peru (Courtesy of Bradley Parker and Weston McCool, Alcohol and Society Ethnoarchaeological Project)

When desired, *chicha* producers grind the malted maize kernels into flour, typically via a large, half-moon mano (*maran uña*) rocked from side to side over a stone slab (*maran*). The resulting flour is called *jora* in the Andes (Cutler and Cardenas 1947).

A second means to create *jora* is by chewing unsprouted maize kernels. Humans produce amylase enzymes in their saliva, which can be used to introduce the enzyme into maize to convert starch into sugar. Cutler and Cardenas (1947) report that *jora* was often masticated into small flour ball packages, *muko*, which could then be stored for extended periods of time. *Muko* had higher value than *jora* alone (Cutler and Cardenas 1947; Jennings and Chatfield 2009; Weismantel 2009), although it may not produce beverages with as high alcohol content as malting (Hayden, Canuel, and

Shanse 2013, 120; Jennings 2004; Driver and Massey 1957, 268). Despite masticated *jora*'s notoriety (Steinkraus 1979, 42–43), malting has been more often recorded in the field as the main source of amylase introduction (Cutler and Cardenas 1947; Nicholson 1960; Hayashida 2008). This may reflect, however, a bias in the literature toward coastal and big batch manufactures where malting is more common.

After the *jora* is produced, it is mixed with water. In order to speed up the enzymatic creation of simple sugars (saccharification) from the starches, the water is often heated and held (depending on the circumstances) between 50 and 70 °C for 30 min to 4 h. This process is called *mashing* (Wilkins and Atiyeh 2012). Although the term mashing is not universally used in the ethnographic literature (Cutler and Cardenas 1947; Hayashida 2008; Nicholson 1960), it is common for *chicha* to be boiled in open-necked vessels for several hours or even overnight in the production stream (see Fig. 6.1). Water is added to maintain a desired volume. In extremely warm temperatures, mashing is not observed in *chicha* production (Cutler and Cardenas 1947). The liquefaction of the *wort* (a pre-inoculated brewing substrate of starch-filled water) is caused by the gelatinization of the starches, indicating that the sugar molecules have been released into the substrate.

Fermentation occurs after the wort has cooled. The wort is transferred to narrow-necked vessels that provide a low-oxygen or oxygen-free environment (see Fig. 6.1). Fermentation is the anaerobic metabolism of cells; in this context, yeasts consume the sugar-rich wort and produce carbon dioxide and ethanol. *S. cerevisiae* has been identified in Peruvian *chicherias*, albeit with many genetic differences from European varieties (Vallejo et al. 2013). Other researchers have identified a new yeast, *Candida theae*, associated with *chicha* production (Chang et al. 2012). Fermentation requires 24 h to 8 days to reach completion, which yields a beer that is 3–6% alcohol by volume (ABV). Fermentation vessels in the Andes are exclusively used for *chicha* production. Small amounts of *chicha* are left from the prior batch to encourage yeast inoculation (Cutler and Cardenas 1947; Hayashida 2008; Nicholson 1960). Since the time of the Incas, *chicha* has usually been served in small, individual serving cups. Batches much be consumed fairly soon after they are produced because *chicha* has a higher pH than other fermented beverages, such as wine, and therefore is more susceptible to unwanted bacterial contamination (Wilkins and Atiyeh 2012; Jennings 2004). Ethnographic reports indicate that *chicha* is usually consumed within a week of production and can be consumed before fermentation is complete (Cutler and Cardenas 1947; Nicholson 1960; Hayashida 2008; Bruman 2000).

While ethnographic and historical analogues cannot be assumed to perfectly model pre-Hispanic *chicha* production regimes (Hayashida 2009), the biochemical requirements for converting maize grains into ethanol necessitate an operational sequence that must be followed: the starches in maize need to be broken down into sugars, and these sugars need to be broken down into ethanol and carbon dioxide (Jennings et al. 2005). Recipes undoubtedly varied widely across space and time in the Andes. This basic sequence, however, remained the same and, as will be demonstrated in our discussion of starches later in the chapter, the universal need for saccharification may be critical for developing a clear archaeological marker for maize beer production.

Archaeological Evidence of *Chicha* and Problems of Equifinality

Documenting the existence of maize beer in the archaeological record is surprisingly difficult. Maize was introduced into the Andes from Mesoamerica by at least 3000 B.C, when it appears to have been consumed as a grain and fermented intoxicant (Staller and Thompson 2002; Tykot and Staller 2002; Zarrillo et al. 2008). Maize was a minor contributor to the diet in most regions for the next four thousand years (Burger and Van der Merwe 1990; Logan et al. 2012; Tykot et al. 2006), before finally becoming a staple crop across the Andes by the end of the first millennium AD (Goldstein 2003; Moseley et al. 2005). Maize would continue to be consumed as both a food and beverage through the Inca and colonial periods (Jennings and Duke 2018). The documentation of maize starches, phytoliths, and macro-remains at sites, coupled with isotopic values in some human skeletons suggesting maize consumption, provides us with clear evidence for maize consumption in the ancient Andes, but does not speak directly to *chicha* production.

In a similar vein, evidence for brewing cannot be taken as evidence for exclusively maize-derived beer production. People in the ancient Andes made alcohol from a wide variety of plants that included molle (*Schinus molle*), potatoes (*Solanum tuberosum*), the figleaf gourd (*Cucurbita ficifolia*), and quinoa (*Chenopodium quinoa*) (Goldstein et al. 2009). The production sequences of these rival beverages were different, but brewers often used similar equipment and went through similar stages of production because of the biochemical demands of making ethanol (Jennings and Valdez 2016). "*Chicha* yeasts," like *Candida theae*, may not be restricted to maize beer (Chang et al. 2012), and the existence of compound beers made from different plants—as is seen today in the Andes—makes it more difficult to untangle production regimes (Goldstein et al. 2009). The diversity of fermented beverages suggests evidence for brewing should not be necessarily equated with evidence for the fermentation and production of *chicha* or maize beer.

As discussed by Moore (1989), Hayashida (2008), Parker and McCool (2015), among others, there are indices of brewing that can be used to build a case for maize *chicha* production in archaeological settings. Along the North Coast of Peru, where preservation is outstanding, it is possible on rare occasions to recover the actual cobs, kernels, and dregs left behind in each stage of production (Moore 1989: 686–688). When combined with evidence for brewing equipment, the argument for maize beer brewing is exceptionally strong (e.g., Prieto 2011). However, preservation is considerably poorer in the vast majority of cases, making claims for maize beer production in these locations more circumspect. This is especially the case in domestic settings where brewing was interspersed with other kitchen duties (Jennings and Chatfield 2009). Even in dedicated facilities, it can be difficult to identify what beverage was being prepared because of tenuous links between botanical remains and brewing assemblages (Moseley et al. 2005; Goldstein et al. 2009; Jennings and Valdez 2016).

Unlike Old World brewing traditions (Henry et al. 2009; Rojo-Guerra et al. 2006), little work has been published to identify the microfossil transformations that occur to maize during the brewing process (Pagán-Jiménez 2013; Pagán-Jiménez et al. 2015; Vinton et al. 2009). The experiment presented here assesses the morphological changes and availability of starch through experimental *chicha* production. If discovered in archaeological settings, the altered starch grains could provide direct evidence for maize beer brewing in a wide variety of settings where only a few indices of *chicha* were previously available.

Identification of *Chicha* Through Starch Analysis

Archaeological use of starch granule analysis has been growing over the past fifteen years (Henry 2015; Louderback et al. 2015). Starch granules are semicrystalline structures that can resist decay for millennia. Starch is produced in plants in two varieties: transport and storage starch. Storage starch is of primary interest to archaeologists as storage starch granules are much larger and can be morphologically distinct to species (Gott et al. 2006; Torrence 2006a). Identifiable starch granules have been recovered from soil samples (e.g., Beck and Balme 2006; Horrocks 2005; Torrence 2006b), artifacts (e.g., Del Pilar Babot and Apella 2003; Barton 2007; Duncan et al. 2009), dental calculus (e.g., Hardy et al. 2009; Henry and Piperno 2008; King et al. 2017), and coprolites (e.g., Vinton et al. 2009) from archaeological contexts. Contamination remains a primary concern in ancient starch analysis (Crowther et al. 2014; Dozier 2016; Laurence et al. 2011).

In addition to taxonomic identification, cooking processes alter starch granules in consistent ways (Crowther 2012; Del Pilar Babot 2003; Henry et al. 2009). Malting (controlled sprouting) and fermentation damage to starch granules have been recorded among a variety of Old World cereal grains (Crowther 2012; Henry et al. 2009; Wang et al. 2017). Spanish archaeologists have used malted cereal starch granules to argue for the presence of malting floors in beer production (Rojo-Guerra et al. 2006). Little experimental work has been published, however, on the effect of various cooking alterations to maize starch granules (Del Pilar Babot et al. 2014).

Del Pilar Babot (2003) was one of the earliest researchers utilizing starch granule damage as an indicator of cooking, although her excellent experiments did not include fermentation as a method of cooking. Vinton et al. (2009) found that experimentally fermented maize produced furrowing damage similar to what was observed in Inca and pre-Inca coprolites from the Atacama Desert, though they did not expand on the exact methodologies and equipment they used to produce the experimental *chicha*. Neither did they compare the experimental *chicha* starch after ingestion to understand the enzymatic processes that dictate maize starch degradation within human digestion tracks. As such, it is difficult to determine if the damage observed archaeologically can be attributed to *chicha* production or to other enzymatic damage from consumption or taphonomic degradation (Collins and Copeland 2011). Pagán-Jiménez (2013) has continued such work with altered maize starch in an unpublished technical report, which tested various methods of *chicha de maiz* production. He found several kinds of fissuring damage unique to *chicha* production, although he did not explore malted maize, which is one of the most common methods for producing *chicha* (Cutler and Cardenas 1947). Pagán-Jiménez et al. (2015) have tentatively identified modified maize starch in elaborate ceramic containers from Ecuador. None of the aforementioned work, however, satisfactorily identifies morphological changes to starch due to *chicha* production with *jora* produced from malted maize.

New Experimental Evidence

Dozier replicated a fermentation regime that approximated the basic steps (malting, mashing, and fermentation) available throughout pre-Columbian history to make it widely applicable for archaeologists. The maize used was a purple Peruvian popcorn variety obtained from the Barry Hill Farm in Ohio. Five pounds (2.3 kg) of dried kernels were rinsed, then soaked with tap water for two days in plastic tubs to initiate the malting process. The rinse water was analyzed under light microscopy and was found to contain a few starch granules (both maize and non-maize) and no pollen. The soaking grains were rinsed and refreshed with new water every 12 h. The soaked kernels were then rinsed and drained, and kept with a moist paper towel covering them. The sprouting kernels were kept by a window and rinsed 1–2 times a day to prohibit fungal growth. Sprouting was halted after five days, when a majority of the rootlets reached twice the length of the maize kernels, a common rule of thumb for both modern and ethnographically recorded *chicha* producers (Cutler and Cardenas 1947: 55; Hayashida 2008: 163). The sprouting rate exceeded 85 percent, and both sprouted and non-sprouted kernels were kept in the experiment. The sprouted kernels were then completely drained, and spread on a single layer to dry on paper towels. Kernels were dried for ten days before being ground with a stone *mano* and slab *metate*. Figure 6.2 illustrates the malting sequence; Fig. 6.3 illustrates the grinding set-up, which was used in an additional test of airborne starch (Dozier 2016).

The goal of this replication experiment was to more clearly understand the taphonomic changes that maize starch granules undergo during the boiling and fermentation process. This was methodologically inspired by Henry et al. (2009). The *jora* was split into two *chicha* samples with different boiling regimes to capture variation in maize starch morphological changes. Each chicha sample, containing 914 g of *jora* and 24 cups (5.6 l) of water, was heated with an electric burner for consistency in heat distribution. *Chicha* 1 boiled for an hour; *Chicha* 2 was heated just below boiling, but above the sanitation threshold of 160 °F (71 °C). Figure 6.4 displays wort temperatures of both *chichas* and specifies when starch samples were taken through the boiling (mashing) procedure. After mashing, the worts had a specific gravity of 1.045 and 1.036, respectively.

After heating, each of the worts were poured into food-grade, five-gallon plastic fermenters and inoculated with 5.5 g of commercial *Saccharomyces cerevisiae* (Dawnstar, Nottingham brewing yeast). Since no South American commercial yeasts are available, it was suggested to use Nottingham yeast strain, like other *chicha* homebrewers (Gordon Spect, personal communication 2015). The *chicha* samples were

Fig. 6.2 Malting procedure

fermented for six days before starch sampling. Final specific gravities were 1.024 and 1.014 for boiled (*Chicha* 1) and heated (*Chicha* 2) samples, respectively; the *chichas* therefore achieved 2.75–2.89% ABV. An additional sample was taken from the final fermented *chichas*.

Each *chicha* sample was 11 ml, which was then centrifuged at 1000 rpm for two minutes. The concentrated samples were then applied directly to a slide without further processing. Each sample was examined under a brightfield, polarized, and cross-polarized light under x200 power on an N200 Nikon Optiphot. Figure 6.5 illustrates examples of changes in maize starch granule morphology through the course of *chicha* heating and fermentation.

Fig. 6.3 Grinding set-up

Fig. 6.4 Mashing temperature. Black dots indicate where samples were taken

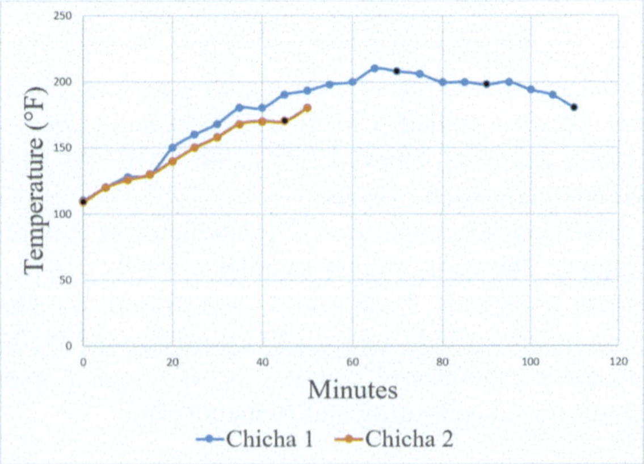

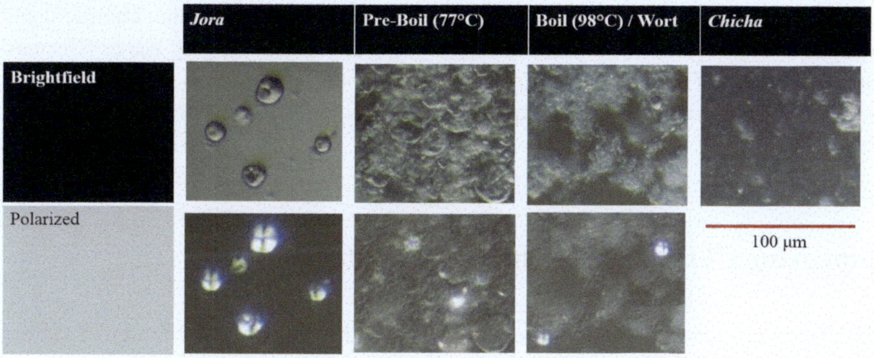

Fig. 6.5 Morphological changes in starch through fermentation

As illustrated in Fig. 6.5, maize starch granules underwent extensive morphological changes through the *chicha* production stream. Maize granules from the *jora* exhibit distinct birefringence and slightly fuzzy extinction crosses, probably due to the slight crushing damage at the hilum. Starch granules became gelatinized very early in the mashing process, swelling and losing distinctive structures (extinction cross and birefringence) at 170 °F (77 °C), after just 45 min of heating. Unmalted Old World cereal starch granules from ten domesticated species, took anywhere from 5–30 min of boiling for complete gelatinization to occur. On the other hand, whole gelatinization happened within 1–5 min of boiling for ground samples (Henry et al. 2009, 917–19). In industrial situations, Jenkins and Donald (1998) report maize starch gelatinizes between 144 and 183 °F (62–84 °C). The exact temperature for boiling is not noted in their methods; partial boiling was identified in this experiment by persistent bubbles from the wort, which occurred at 198 °F (92 °C) for *chicha* sample 1. *Chicha* sample 2 was removed from heat at 180 °F (82 °C)—after which gelatinization processes obscured any distinctive features of the starch granules. No starch granules were recovered from any fermented *chicha* samples Dozier analyzed. It is possible that a few stray starch granules could survive the sprouting, mashing, and fermentation process in good condition. When maize gelatinization becomes less complete, the more dense the starch concentration becomes (Li et al. 2015), and it should be noted that Pagán-Jiménez (2013) was able to recover maize starch granules from *chicha* that had been boiled for even longer than this experiment.

Special attention was paid to reduce airborne starch contamination (Laurence et al. 2011; Crowther et al. 2014), although the facilities employed in this study were found to contain latent starch by Laurence et al. (2011). No non-maize starch granules were observed during analysis. This fermentation experiment suggests that starch is unlikely to be recovered in significant quantities from *chicha* boiling/mashing, fermentation, or serving vessels in archaeological contexts. While Jennings and Bowser (2009: 9) call for residue analysis on *kero* drinking cups associated with *chicha* consumption, finding concentrations of undamaged maize starch in one of these vessels would thus be indicative of the consumption of an unfermented maize beverage rather than *chicha*. Archaeologists will need to consider other chemical and microbotanical regimes of residue analysis for brewing and serving wares (Barnard and Eerkens 2007). Starch grain analysis, nonetheless, could be used to identify the residue of *jora* in areas and equipment associated with the first stages of beer production.

Improving *Chicha* Field Sampling Strategies

The *chicha* production stream indicates that malted maize may be found in a variety of contexts: on grinding equipment, storage equipment, and perhaps even in dental calculus. The experiments presented here allow for more direct evidence of *chicha* production, particularly for sites, such as at Marayniyoq in Peru (Valdez et al. 2010), which have been assumed or implicated in *chicha* production but where no direct evidence exists. Recent work indicates that previously excavated and even curated collections still retain economic starch (Louderback et al. 2015). Starch granules are transported during cooking activities by wind (Dozier 2016) and water vapor (Thoms et al. 2015), which may allow for some preservation of starch granules even though the cooking regime would normally gelatinize starch granules. The results of the above study therefore suggest that non-ceramic artifacts should be considered for starch analysis. The damage from malting the maize, as shown in this experiment, may be indicative and could remain on processing surfaces such as *manos*, *metates*, drying mats, or within storage vessels, complementary to Moore's (1989) assessment for wheat.

As mentioned above, *muko*, the masticated balls of malted maize flour, seem to have had economic importance within Andean societies. These high-starch packages may leave traces within dental calculus, from which starch has been recovered in a variety of archaeological contexts. King et al. (2017) claim to have recovered fermented maize starches from dental calculus from the site of Paquime (Casas Grandes) in Northern Mexico; presumably the same can be recovered from Andean contexts. Recent work has indicated that the long boiling regimes, such as those associated with *chicha* production, can lead to slight enrichment in heavy stable oxygen isotopes in bone (Brettell et al. 2012; Gagnon et al. 2015), complicating analyses that look to parse immigration/migration patterns from diet, though opening a door to exploring differential fermented/boiled food consumption (Hastorf 1991).

The experiments explored here showcase archaeological methodologies in order to better understand brewing technologies, though many unanswered research questions remain. A more detailed sampling of the boiling process may more accurately describe the gelatinization sequence for malted, non-malted, and nixtamalized maize (the process, apparently unknown in the ancient Andes but utilized in Mesoamerica, of soaking maize with an alkaline). Green corn varieties of *chicha* may also have different gelatinization scheme. Non-heated *chicha de maiz* has been noted for more tropical areas of the Andes (Cutler and Cardenas 1947); the experiment here did not fully explore how maize starch morphologically changed through fermentation rather than heating.

Future work should evaluate how the crushing at the hilum of maize starch grains during malting occurs and the degree to which that crushing is diagnostic to malting rather than other enzymatic processes. Starch granule analysis can also be used to understand other fermented beers of the Andes made with other starchy plants, such as potato, gourds, or quinoa (Goldstein et al. 2009). Unfortunately, traditional *chicherias* are disappearing across South America as commercial beverages increasingly replace indigenous production; researchers of *chicha* desperately need to record diversity in *chicha* production and yeast, and to directly test for microfossils and chemical residues that may be indicative of *chicha* within the archaeological record.

Discussion and Conclusions

Chicha de maize is and was an incredibly important ritual, domestic, and economic product of the Andes. The archaeological identification of *chicha* production has been complicated by a lack of approaches or techniques to identify *chicha* microscopically or chemically. This research provides the utility of starch granule analysis for identifying *chicha* in archaeological contexts. An experimentally brewed *chicha* indicates that the malting sequence leaves morphologically distinct damage on the starch granules, though the mashing sequence may gelatinize (destroy) the majority of starch granules. These results challenge archaeologists to sample artifacts earlier in the production stream: grinding materials, drying mats, and even dental calculus. Other research (Pagán-Jiménez 2013; Thoms et al. 2015) indicates that some starch granules can survive extreme cooking conditions, so testing ceramics associated with boiling, fermentation, or consumption may still provide evidence of *chicha de maiz*.

As with most of archaeology, multiple lines of evidence are most convincing when trying to identify a particular tradition. Fermentation technologies, as an organic process, are especially enhanced by a multi-evidentiary approach (Biwer and VanDerwarker 2015). Researchers, like Lantos et al. (2015), exemplify this approach by combining starch analysis with ceramic and chemical analysis in order to argue for *chicha* production in Argentina. However, these authors were able to identify only that a maize-based liquid (not necessarily a maize-based beer) was consumed from certain vessels.

With the multitudinous roles that *chicha* plays within ritual, political, and domestic life, archaeological methodologies to identify its production are incredibly important. Ancient starch analysis provides a potential avenue for such identification, with respect to the limitations of the science. It is our hope that the research offered in this chapter further advances microscopic archaeology, taking us a step closer to understanding ancient Andean *chicha* production.

Acknowledgements We thank John E. Staller for inviting us to contribute to this volume. Thanks go to Vaughn Bryant and especially Alston Thoms for their support and critical suggestions in developing the replication experiment. The Palynological Research Laboratory, the Archaeology Ecology Laboratory, and the Anthropology Research Collections at the Department of Anthropology at Texas A&M University provided analysis equipment and space. Many thanks to Michael Waters for use of the N200 Nikon Optiphot. We also thank Bradley Parker and Weston McCool for permission to include their excellent photos of current *chicha* production in Peru.

References

de Balboa MC (1586) Miscelanea Antartica. New York Public Library, New York

Barnard H, JW Eerkens (2007) Introduction. In: Theory and practice of archaeological residue analysis. British archaeological reports international series, vol 1650. Archaeopress, Oxford, pp 1–7

Barton H (2007) Starch residues on museum artefacts: implications for determining tool use. J Archaeol Sci 34(10):1752–1762. https://doi.org/10.1016/j.jas.2007.01.007

Beck W, Balme J (2006) Starch as an indicator of activity areas. In: Barton H, Torrence R (eds) Ancient starch research. Left Coast Press, Walnut Creek, pp 171–173

Biwer ME, VanDerwarker AM (2015) Paleoethnobotany and ancient alcohol production: a mini-review. Ethnobiol Lett 6(1):28–31

Bonavia D (2013) Maize: origin, domestication, and its role in the development of culture. Cambridge University Press, Cambridge

Braidwood RJ, Sauer JD, Helbaek H, Mangelsdorf PC, Cutler HC, Coon CS, Linton R, Steward J, Leo Oppenheim A (1953) Symposium: did man once live by beer alone? Am Anthropol 515–526

Bray TL (ed) (2003) The archaeology and politics of food and feasting in early states and empires. Kluwer Academic/Plenum Publisher, New York

Brettell R, Montgomery J, Evans J (2012) Brewing and stewing: the effect of culturally mediated behaviour on the oxygen isotope composition of ingested fluids and the implications for human provenance studies. J Anal At Spectrom 27(5):778–785. https://doi.org/10.1039/C2JA10335D

Bruman HJ (2000) Alcohol in ancient Mexico. University of Utah Press, Salt Lake City

Bryant VM (2007) Microscopic evidence for the domestication and spread of maize. Proc Natl Acad Sci 104(50):19659–19660. https://doi.org/10.1073/pnas.0710327105

Burger RL, Van der Merwe NJ (1990) Maize and the origins of highland Chavin civilization: an isotopic perspective. Am Anthropol 92(1):85–95

Buskirk W (1986) The Western Apache: living with the land before 1950. University of Oklahoma Press, Norman, OK

Chang C-F, Lin Y-C, Chen S-F, Barriga EJC, Barahona PP, James SA, Bond CJ, Roberts IN, Lee C-F (2012) *Candida theae* sp. nov., a new anamorphic beverage-associated member of the Lodderomyces clade. Int J Food Microbiol 153(1–2):10–14. https://doi.org/10.1016/j.ijfoodmicro.2011.09.012.

Collins MJ, Copeland L (2011) Ancient starch: cooked or just old? Proc Natl Acad Sci 108(22):E145–E145. https://doi.org/10.1073/pnas.1103241108

Crowther A (2012) The differential survival of native starch during cooking and implications for archaeological analyses: a review. Archaeol Anthropol Sci 4(3):221–235. https://doi.org/10.1007/s12520-012-0097-0

Crowther A, Haslam M, Oakden N, Walde D, Mercader J (2014) Documenting contamination in ancient starch laboratories. J Archaeol Sci 49:90–104. https://doi.org/10.1016/j.jas.2014.04.023

Cummins TBF (2002) Toasts with the Inca: Andean abstraction and Colonial image on Quero vessels. University of Michigan Press, Ann Arbor

Cutler HC, Cardenas M (1947) Chicha, a native South American beer. Bot Mus Leafl Harv Univ 13(3):33–60

Del Pilar Babot M (2003) Starch grain damage as an indicator of food processing. In: Phytolith and starch research in the Australian-Pacific-Asian regions: the state of the art. Pandanus Press, pp 69–81

Del Pilar Babot M, Apella MC (2003) Maize and bone: residues of grinding in Northwestern Argentina. Archaeometry 45(1):121–132. https://doi.org/10.1111/1475-4754.00099

Del Pilar Babot M, Lund J, Olmos AV (2014) Taphonomy in the kitchen: culinary practices and processing residues of native tuberous plants of the south-central Andes. Intersecc en Antropol 15:35–53

Dietrich O, Heun M, Notroff J, Schmidt K, Zarnkow M (2012) The role of cult and feasting in the emergence of neolithic communities. New evidence from Göbekli Tepe, south-eastern Turkey. Antiquity 86(333):674–695

Doebley JF, Stec A (1991) Genetic analysis of the morphological differences between maize and teosinte. Genetics 129(1):285–295

Doebley JF, Stec A (1993) Inheritance of the morphological differences between maize and teosinte: comparison of results for two F2 populations. Genetics 134(2):559–570

Dozier CA (2016) Airborne starch dispersal from stone grinding: experimental results and implications. J Archaeol Sci: Rep 8:112–115. https://doi.org/10.1016/j.jasrep.2016.05.057

Driver HE, Massey WC (1957) Comparative studies of North American Indians. Trans Am Philos Soc 47(2):165–456

Duncan NA, Pearsall DM, Benfer RA (2009) Gourd and squash artifacts yield starch grains of feasting foods from preceramic Peru. Proc Natl Acad Sci 106(32):13202–13206. https://doi.org/10.1073/pnas.0903322106

Gagnon CM, Fred C, Andrus T, Ida J, Richardson N (2015) Local water source variation and experimental Chicha de Maíz brewing: implications for interpreting human hydroxyapatite Δ18o values in the Andes. J Archaeol Sci: Rep 4:174–181. https://doi.org/10.1016/j.jasrep.2015.09.008

Gero JM (1990) Pottery, power, and . . . parties! Archaeology 43(2):52–56

Gero JM (1992) Feasts and females: gender ideology and political meals in the Andes. Nor Archaeol Rev 25:15–30

Goldstein D, Goldstein RC, Williams PR (2009) You are what you drink: a sociocultural reconstruction of pre-Hispanic fermented beverage use at Cerro Baul, Moquegua, Peru. In: Jennings J, Bowser B (eds) Drink, power, and society in the Andes. University Press of Florida, Gainesville, pp 133–166

Goldstein PS (2003) From stew-eaters to maize-drinkers. In: Tamara LB (ed) The archaeology and politics of food and ffeasting in early states and empires, Boston, MA, Springer, US, pp 143–72. https://doi.org/10.1007/978-0-306-48246-5_6

Gott B, Barton H, Samuel D, Torrence R (2006) Biology of starch. In: Barton H, Torrence R (eds) Ancient starch research. Left Coast Press, Walnut Creek, pp 35–46

Guerra-Doce E (2015) The origins of inebriation: archaeological evidence of the consumption of fermented beverages and drugs in prehistoric Eurasia. J Archaeol Method Theory 22(3):751–782. https://doi.org/10.1007/s10816-014-9205-z

Hardy K, Blakeney T, Copeland L, Kirkham J, Wrangham R, Collins M (2009) Starch granules, dental calculus and new perspectives on ancient diet. J Archaeol Sci 36(2):248–255

Hastorf CA (1991) Gender, space, and food in prehistory. In: Gero JM, Conkey MW (eds) Engendering archaeology: women and prehistory. Blackwell, Cambridge, pp 132–159

Hastorf CA, Whitehead WT, Bruno MC, Wright M (2006) The movements of maize into middle horizon Tiwanaku, Boliva. In: Staller J, Tykot R, Benz B (eds) Histories of maize: multidisciplinary approaches to the prehistory, linguistics, biogeography, domestication, and evolution of maize. Academic Press, Burlington, pp 429–448

Hayashida FM (2008) Ancient beer and modern brewers: ethnoarchaeological observations of Chicha production in two regions of the North Coast of Peru. J Anthropol Archaeol 27(2):161–174

Hayashida FM (2009) Pots, brewers, and hosts: women's power and the limits of central Andean feasting. In: Jennings J, Bowser BJ (eds) Drink, power, and society in the Andes. University Press of Florida, Gainesville, pp 232–256

Hayden B (2014) Competitive feasting before cultivation? Curr Anthropol 55(2):230–231
Hayden B, Canuel N, Shanse J (2013) What was brewing in the Natufian? An archaeological assessment of brewing technology in the epipaleolithic. J Archaeol Method Theory 20(1):102–150
Henry AG (2015) Formation and taphonomic processes affecting starch granules. In: Marston JM, D'alpoim Guedes J, Warinner C (eds) Method and theory in paleoethnobotany. University Press of Colorado, Boulder, pp 35–50
Henry AG, Hudson HF, Piperno DR (2009) Changes in starch grain morphologies from cooking. J Archaeol Sci 36(3):915–922. https://doi.org/10.1016/j.jas.2008.11.008
Henry AG, Piperno DR (2008) Using plant microfossils from dental calculus to recover human diet: a case study from Tell Al-Raqā'I, Syria. J Archaeol Sci 35(7):1943–1950
Hornsey IS (2003) A history of beer and brewing. Royal Society of Chemistry, Cambridge
Horrocks M (2005) A combined procedure for recovering phytoliths and starch residues from soils, sedimentary deposits and similar materials. J Archaeol Sci 32(8):1169–1175. https://doi.org/10.1016/j.jas.2005.02.014
Iltis HH (2000) Homeotic sexual translocations and the origin of maize (Zea Mays, Poaceae): a new look at an old problem. Econ Bot 54(1):7–42. https://doi.org/10.1007/BF02866598
Jenkins PJ, Donald AM (1998) Gelatinisation of starch: a combined SAXS/WAXS/DSC and SANS study. Carbohydr Res 308(1–2):133–147. https://doi.org/10.1016/S0008-6215(98)00079-2
Jennings J (2004) La chichera y el patrón: Chicha and the energetics of feasting in the prehistoric Andes. Archaeol Pap Am Anthropol Assoc 14(1):241–259
Jennings J (2014) A glass for the Gods and a gift to my neighbor: the importance of alcohol in the pre-Columbian Andes. In: Pierce G, Toxqui A (eds) From Aguardente to Zambumbia. A social and cultural history of alcohol in Latin America. University of Arizona Press, Tucson, pp 25–45
Jennings J, Antrobus KL, Atencio SJ, Glavich E, Johnson R, Loffler G, Luu C (2005) "Drinking beer in a blissful mood": alcohol production, operational chains and feasting in the ancient world. Curr Anthropol 46(2):275–303
Jennings J, Bowser BJ (eds) (2009) Drink, power, and society in the Andes. University Press of Florida, Gainesville
Jennings J, Chatfield M (2009) Pots, brewers, and hosts: women's power and the limits of central Andean feasting. In: Jennings J, Bowser BJ (eds) Drink, power, and society in the Andes. University Press of Florida, Gainesville, pp 200–231
Jennings J, Duke G (2018) Food, cuisine, status, and feasting. In: Alconini S, Alan Covey R (eds) Oxford handbook of the Inca. Oxford University Press, New York (in press)
Jennings J, Valdez LM (2016) Ingredients matter: maize versus molle brewing in ancient Andean feasting. In: Fitzpatrick S (ed) Psychoactive substances in the ancient world. University of Florida Press, Tallahasse, pp 286–318
King DJ, Searcy MT, Yost CL, Waller K (2017) Corn, beer, and marine resources at Casas Grandes, Mexico: an analysis of prehistoric diets using microfossils recovered from dental calculus. J Archaeol Sci: Rep 16(Supplement C):365–379. https://doi.org/10.1016/j.jasrep.2017.10.013
Lantos I, Spangenberg JE, Giovannetti MA, Ratto N, Maier MS (2015) Maize consumption in pre-hispanic South-Central Andes: chemical and microscopic evidence from organic residues in archaeological pottery from Western Tinogasta (Catamarca, Argentina). J Archaeol Sci 55:83–99 (March). https://doi.org/10.1016/j.jas.2014.12.022
Laurence AR, Thoms AV, Bryant VM, McDonough C (2011) Airborne starch granules as a potential contamination source at archaeological sites. J Ethnobiol 31(2):213–232. https://doi.org/10.2993/0278-0771-31.2.213
Li Z, Liu W, Zhengbiao Gu, Li C, Hong Y, Cheng Li (2015) The effect of starch concentration on the gelatinization and liquefaction of corn starch. Food Hydrocol 48:189–196. https://doi.org/10.1016/j.foodhyd.2015.02.030
Logan AL, Hastorf CA, Pearsall DM (2012) "Let's drink together": early ceremonial use of maize in the Titicaca Basin. Lat Am Antiq 23(3):235–258

Louderback LA., Field J, Janetski JC (2015) Curation practices and extraction methods in relation to starch grain yields from ground stone artifacts. J Archaeol Sci Rep 4:535–40 (December). https://doi.org/10.1016/j.jasrep.2015.10.011
Matsuoka Y, Vigouroux Y, Goodman MM, Jesus SG, Buckler E, Doebley JF (2002) A single domestication for maize shown by multilocus microsatellite genotyping. Proc Natl Acad Sci USA 99(9):6080–6084. https://doi.org/10.1073/pnas.052125199
Moore JD (1989) Pre-Hispanic beer in coastal Peru: technology and social context of prehistoric production. Am Anthropol 91(3): 682–695 (New Series). https://doi.org/10.2307/680873
Morris C (1979) Maize beer in the economics, politics, and religion of the Inca empire. In: Gastineau CF, Darby WJ, Turner TB (eds) Fermented food beverages in nutrition. Academic Press, New York, pp 21–34
Moseley ME, Nash DJ, Williams PR, deFrance SD, Miranda A, Ruales M (2005) Burning down the brewery: establishing and evacuating an ancient imperial colony at Cerro Baúl, Peru. Proc Natl Acad Sci USA 102(48):17264–17271
Murra JV (1960) Rite and crop in the Inca state. In: Diamond S (ed) Culture in history: essays in honor of Paul Radin. Columbia University Press, New York, pp 393–407
Murra JV (1975 [1972]) El control vertical de un máximo de pisos ecológicos en los sociedades andinas. In: Murra JV (ed) Formaciones Económicas y Politicas del Mundo Andino. Instituto de Estudios Peruanos, Lima, pp 59–115
Murra JV (1980) Economic organization of the Inca state. JAI Press, Greenwich, Connecticut
Nicholson GE (1960) Chicha maize types and Chicha manufacture in Peru. Econ Bot 14(4):290–299. https://doi.org/10.1007/BF02908039
Pagán-Jiménez JR (2013) Evaluando algunos mecanismos de conservación/degradación en almidones modernos por medio de ensayos y experimentos controlados que replican ciertas formas antiguas de procesamiento y cocción de órganos almidonosos. Elaboración de dos tipos de chicha de maíz: chicha fermentada con saliva y otra con levadura Unpublished Reporte preliminar. Paleoethnobotanica de las culturas ancestrales del ecuador: estudio de almidones en contextos antiguos. Instituto Nacional de Patrimonio Cultural, Quito-Ecuado.
Pagán-Jiménez JR, Saavedra López PR, Guachamín-Tello AM (2015) Análisis de residuos microbotánicos (almidones) en varios objetos cerámicos relacionados con la confección y el consumo de bebidas prehispánicas, colección Avilés Marcillo, Instituto Nacional de Patrimonio Cultural, Quito, Ecuador. Unpublished Technical Report. Instituto Nacional de Patrimonio Cultural, Quito-Ecuado
Parker BJ, McCool W (2015) Indices of household maize beer production in the Andes. J Anthropol Res 71(3):359–400. https://doi.org/10.3998/jar.0521004.0071.303
Piperno DR, Ranere AJ, Holst I, Iriarte J, Dickau R (2009) Starch grain and phytolith evidence for early ninth millennium BP maize from the Central Balsas River Valley, Mexico. Proc Natl Acad Sci 106(13):5019–5024
Prieto GO (2011) Chicha production during the Chimu period at San José de Moro, Jequetepeque Valley, North Coast of Peru. In: Zori CM, Johnson I (eds) From state to empire in the prehistoric Jequetepeque Valley, Peru. BAR international series, vol 2310. Oxford Press, Oxford, pp 105–128
Ramírez SE (2005) To feed and be fed: the cosmological bases of authority and identity in the Andes. Stanford University Press, Stanford
Rojo-Guerra MÁ, Garrido-Pena R, García-Martínez-de-Lagrán Í, Juan-Treserras J, Matamala JC (2006) Beer and bell beakers: drinking rituals in Copper age inner Iberia. In: Proceedings of the prehistoric society, vol 72. Cambridge University Press, pp 243–265
Segura Llanos R (2001) Rito y economía en Cajamarquilla: Investigaciones arqueológicas en el conjunto arquitectónico Julio C. Tello. Pontificia Universidad Católica del Perú, Lima
Smalley J, Blake M (2003) Sweet beginnings. Curr Anthropol 44(5):675–703
Staller JE (2006) The social, symbolic, and economic significance of Zea mays L. in the late horizon period. In: Staller J, Tykot R, Benz B (eds) Histories of maize: multidisciplinary approaches to the prehistory, linguistics, biogeography, domestication, and evolution of maize. Routledge, Taylor & Francis Group, New York, pp 449–467

Staller JE (2010) Maize cobs and culture history of Zea mays L. Springer, Berlin, Heidelberg
Staller JE, Thompson RG (2002) A multidisciplinary approach to understanding the initial introduction of maize into coastal ecuador. J Archaeol Sci 29(1):33–50
Steinkraus KH (1979) Nutritionally significant indigenous foods involving an alcohoic fermentation. In: Gastineau CF, Darby WJ, Turner TB (eds) Fermented food beverages in nutrition. Academic Press, New York, pp 35–59
Thoms AV, Laurence AR, Short L, Kamiya M (2015) Baking Geophytes and tracking microfossils: taphonomic implications for earth-oven and paleodietary research. J Archaeol Method Theory 22(4):1038–1070. https://doi.org/10.1007/s10816-014-9216-9
Torrence R (2006a) Description, classification, and identification. In: Barton H, Torrence R (eds) Ancient starch research. Left Coast Press, Walnut Creek, pp 115–143
Torrence R (2006b) Starch in sediments. In: Barton H, Torrence R (eds) Ancient starch research. Left Coast Press, Walnut Creek, pp 170–176
Tykot RH, Burger RL, Van Der Merwe NJ (2006) The importance of maize in the initial period and early horizon Peru. In: Staller J, Tykot R, Benz B (eds) Histories of maize: multidisciplinary approaches to the prehistory, linguistics, biogeography, domestication, and evolution of maize. Academic Press, Burlington, pp 187–198
Tykot RH, Staller JE (2002) The importance of early maize agriculture in coastal Ecuador: new data from La Emerenciana. Curr Anthropol 43(4):666–667
Valdez LM, Bettcher KJ, Ernesto Valadez J (2010) Production of maize beer at a Wari site in the Ayacucho Valley, Peru. Arqueol Iberoam 5:23–35
Vallejo JA, Miranda P, Flores-Félix JD, Sánchez-Juanes F, Ageitos JM, González-Buitrago JM, Velázquez E, Villa TG (2013) Atypical yeasts identified as Saccharomyces cerevisiae by MALDI-TOF MS and gene sequencing are the main responsible of fermentation of Chicha, a traditional beverage from Peru. Syst Appl Microbiol 36(8):560–564. https://doi.org/10.1016/j.syapm.2013.09.002
Vinton SD, Perry L, Reinhard KJ, Santoro CM, Teixeira-Santos I (2009) Impact of empire expansion on household diet: the Inka in Northern Chile's Atacama Desert. PLoS ONE 4(11):e8069. https://doi.org/10.1371/journal.pone.0008069
Wang J, Liu Li, Georgescu A, Le VV, Ota MH, Tang S, Vanderbilt M (2017) Identifying ancient beer brewing through starch analysis: a methodology. J Archaeol Sci: Rep 15:150–160. https://doi.org/10.1016/j.jasrep.2017.07.016
Weismantel MJ (1988) Food, gender, and poverty in the Ecuadorian Andes. University of Pennsylvania Press, Philadephia
Weismantel MJ (2009) Have a drink: Chicha, performance, and politics. In: Jennings J, Bowser BJ (eds) Drink, power, and society in the Andes. University Press of Florida, Gainesville, pp 257–278
Wilkins M, Atiyeh H (2012) Fermentation. In: Food and industrial bioproducts and bioprocessing. Elsevier, British Columbia, pp 185–203
Zarrillo S, Pearsall DM, Raymond JS, Tisdale MA, Quon DJ (2008) Directly dated starch residues document early formative maize (Zea mays L.) in tropical ecuador. Proc Nat Acad Sci 105(13):5006–11. https://doi.org/10.1073/pnas.0800894105

Chapter 8
Sustainable Resources in Pre-hispanic Coastal Ecuador: Their Associated Iconography and Symbolism

César Iván Veintimilla-Bustamante and Mariella García-Caputi

Abstract The transition from hunting and gathering to sedentism began on the Ecuadorian Coast c. 10,000 years ago. Changes in adaptation, year-round settlements, and an increased dependence upon cultigens transformed prehistoric Andean cultures and created widespread modifications within the coastal ecology. Coastal Ecuador is characterized by extreme ecological diversity, particularly with regard to terrestrial plants, animals, birds, and maritime resources. Cultural changes in adaptation resulted in dramatic increase in population densities. Interdisciplinary research presented here incorporates zoological, botanical, and faunal remains identified archeologically, with descriptions and analysis of their representations on pre-Hispanic material culture and symbolically depicted in ancient iconography. The interdisciplinary research presented here enriches our knowledge and understanding of the rise of complex social organization associated with a more agricultural way of life. Our evidence indicates the cultivation and domestication of plants and animals, as well as the increased exploitation of maritime resources, provided the basis for early social complexity along coastal Ecuador.

Introduction

Particular plants and animals have always been critical to the daily life of societies since the dawn of humanity. They were generally directly associated with adaptation to the natural environment along with hunting, fishing, and gathering. Hunters and gatherers lived in an intimate relationship with nature, interacting with the annual

"Food is a language that speaks materially of social and symbolic dimensions" (Marcelo Alvarez 2002)

C. I. Veintimilla-Bustamante (✉) · M. García-Caputi
Museo Nacional del Ecuador, Quito, Ecuador
e-mail: ci20milla@hotmail.com

M. García-Caputi
e-mail: mgarcia.caputi@gmail.com

J. E. Staller (ed.), *Andean Foodways*, The Latin American Studies Book Series,
https://doi.org/10.1007/978-3-030-51629-1_8

cycles of plants and animals, and extracting food resources for their daily life. In the management of the natural environment, various Andean cultures in Ecuador developed a symbiotic relationship with certain plants and animals in the surrounding landscape that eventually led to the cultivation and domestication of various species during the Archaic and Early Formative Periods, c. 9000 and 1500 B.C. Cultivation is characterized by conscious human selection for certain favorable traits or characteristics, as well as unconscious selection, that is, genetic and phenotypic changes which occur as a consequence of plant management (Staller 2015b). According to Harlan (1992), cultivation refers to the management of plants in the natural environment (e.g., pruning, watering, soil removal, etc.) to select for certain characteristics that benefit human adaptation. Domestication involves evolutionary and genetic changes of wild plants and their biogeography through human and natural selection. Domesticated plants generally depend on humans for their reproduction in the natural environment (Harlan 1992; Staller 2015a, b).

Southwestern coastal Ecuador is situated along the zone between the tropical moist forests to the north and the dry xerophytic marine forests further south. These boundaries precisely match the limits of two of the most important Neotropical ecosystems: Chocó-Darién at the north and Tumbes to the south. This is a region of the Andes renowned for its complex ecologies, complex biodiversity, and endemism. This part of the Andean coast is west of the lowest region of the cordillera, which geologists generally refer to as the Huancabamba Depression (Staller 2010). This is the only part of the Andes where plant and animal species can move from the upper Amazon down into the western lowlands and coast (Staller 2010, 2014). Consequently, SW coastal Ecuador and the southern highlands in general are characterized by a high incidence of endemism (Staller 2010, 2015b, 2016). Between these two macro-geographical zones there are also environments and micro-environments with highly complex diversity.

Archaeological and anthropological evidence from SW coastal Ecuador has documented some of the earliest evidence of agriculture, ceramic technology, and complex sociocultural development in the Pre-Columbian New World (Lathrap et al. 1975; Veintimilla 2004; Pearsall 1988a, 1988b; Pearsall et al. 2016). Valdivia culture (3500–1500 B.C.) represents the earliest ceramic culture in the Andes, and prehistorians generally associate ceramic technology with sedentism and the cultivation and domestication of food crops (Piperno y Pearsall 1998; Veintimilla 2004; Pearsall et al. 2016).

There is widespread evidence of human–wildlife interactions in various ecological habitats. This is verified by organic remains and zooarchaeological analysis from prehistoric sites (Byrd 1976; Stahl 1985; Chase 1988; Wing 1988). Faunal remains were identified that correspond to the ecological niches they inhabited at various coastal sites dating back to the Pre-ceramic period (Gutiérrez 2009). Early complex cultures that cultivated and managed their surrounding landscapes expressed the importance of many natural species and cultigens in their iconography and imagery on ceramics and particularly effigy vessels. Our objective is to describe, analyze, and identify food crops, fauna, and other staples represented and recreated on prehistoric ceramics and stone objects to identify their sociocultural meaning. These vessels and material objects are from collections in the archaeological repository of the national Museum of Anthropology and Contemporary Art (MAAC) in Guayaquil.

This research explores the various roles of plants and animals in ancient economies, subsistence and religious ideologies as well as provides insights into plants, animals, and foodstuffs in ceramic and iconographic depictions. The focus is on representations of flora and fauna on ceramics and other objects to gain a clearer understanding of their importance to both subsistence and belief systems. A total of 27 plants species were identified and classified as either food crops or psychotropic species, many of which are central to rituals and rites. Additionally included are several species of animals important to subsistence with metaphoric and symbolic associations with spiritual realms.

Material Culture and Research Objectives

Material culture should not be viewed as a mere reflection of conceptual systems and social practices, it is actively involved in the formation and structuring of such practices (Barceló 1989; García-Caputi 2016). This research focuses on the identification of coastal plants and other biota represented by various cultures. The sample consists of 159 archaeological objects pertaining to various cultures from the MAAC repository. These results are derived from previous research carried out in 2015 by César Veintimilla. Iconographic analysis regarding the symbolism and representation of coastal plants is linked to archaeobotanical collections uncovered in archaeological research from various regions. Many samples also represent or symbolize faunal resources, reinforcing previous assertions that early sedentary coastal cultures of Ecuador practiced plant cultivation as well as exploited maritime resources for their subsistence. Early evidence of social complexity on the Ecuadorian coast is apparent from artificial earthen mound constructions and elite burials (Staller 2015c). Our identification of such sustainable resources is assessed and analyzed in terms of their social value, their importance or meaning as economic resources, as well as their symbolic and possible ideological significance to providing access to the tripartite cosmos among pre-Hispanic cultures.

The archaeological materials selected include (1) ceramic vessels classified in terms of their "phytomorphology" or plant morphology, (2) effigy vessels and objects with motifs or representations of plants, and (3) ceramics with incised decorations or that incorporate zoomorphic species in their motifs. The consumption and dependence upon such plants and animals is supported by paleoethnobotanical or faunal evidence (e.g., Stothert 1988; Stothert et al. 2003; Stahl 1985; Byrd 1976; Gutiérrez 2009).

The collection analyzed covers 5000 years of prehistory from the Early Formative to the Integration Period. The objects in the MAAC collections are geographically diverse with examples from the southern provinces of Santa Elena and Guayas to the northern provinces of Manabí and Esmeraldas (Fig. 8.1).

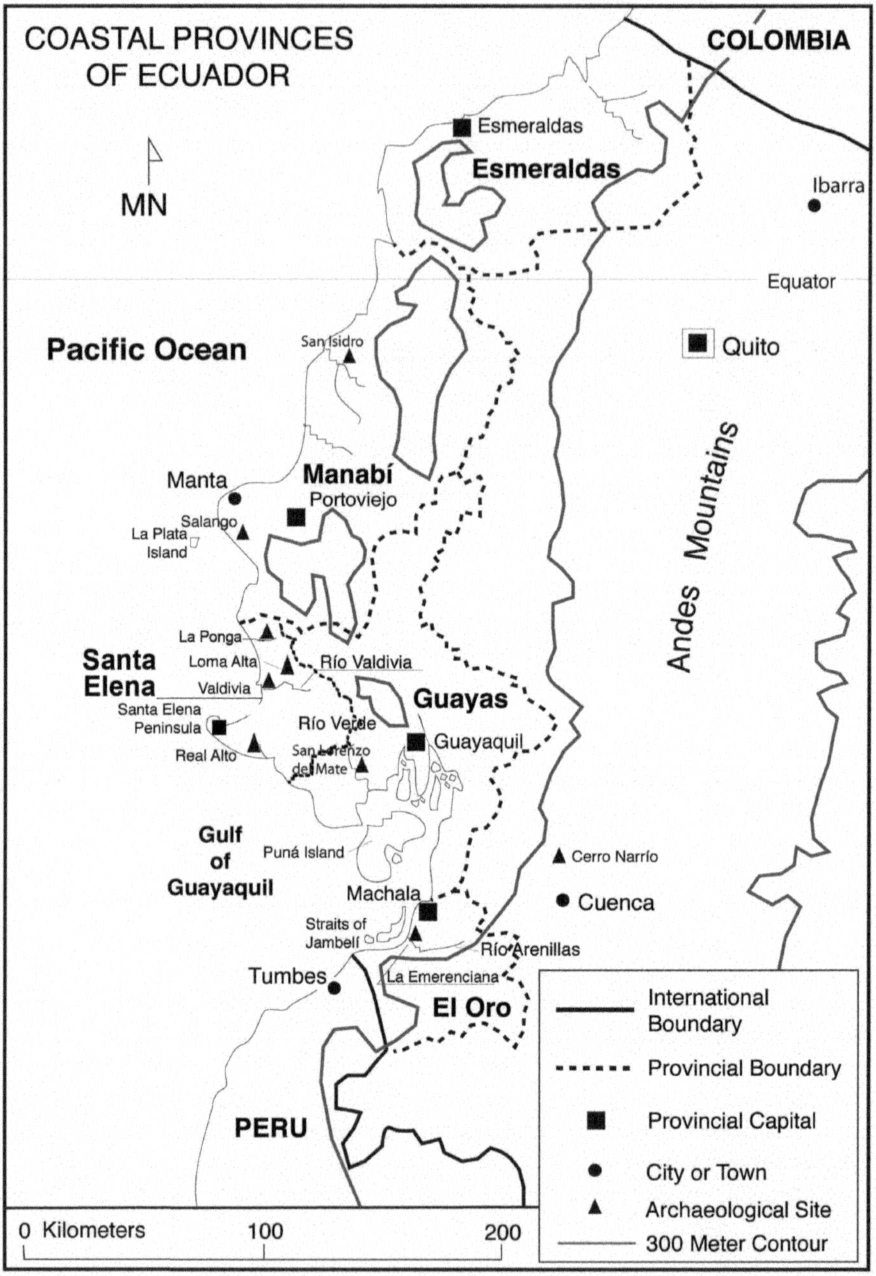

Fig. 8.1 Map of the coast of Ecuador showing the main archaeological sites mentioned in the text

Evidence of Plant Consumption

Paleoethnobotanical evidence involved in the analysis and identification of botanical remains from archaeological contexts elucidate evidence of interrelationships established between humans and plants over the time (Pearsall 2015). Botanical remains recovered from archaeological contexts fall into two categories: (1) "macrofossils," such as seeds, fruits, grains, cobs, wood, organic remains, and some plant materials visible to the naked eye and (2) "microfossils" invisible to the naked eye, excavated from sediments and often deposited in ceramic vessels from domestic contexts. Microscopic analysis is used for the identification of phytoliths, starch, and/or pollen (Piperno and Pearsall 1998; Pearsall 2015; Pearsall et al. 2016). Each of these types of paleobotanical evidence requires specific techniques and procedures for retrieval and analysis, as well as limitations; however, the identification of archaeological specimens is through comparative analysis. Empirical plant evidence complements data used to identify prehistoric use, including iconographic representations of fruits and tubers as well as plant decorations on ceramics and figurines. In tropical regions, the humidity and heat can have a deleterious effect on the botanical evidence. Therefore, most botanical remains are limited to carbonized macrofossils often deposited under anaerobic conditions in marshes and lagoons. This affirms the importance of multidisciplinary evidence in their identification.

Pre-hispanic Plant Resources in Coastal Ecuador: Contexts and Associations

The earliest evidence of human occupation along the Ecuadorian coast corresponds to the Pre-ceramic Las Vegas culture (ca. 9300 and 5300 B.C.). This culture was centered on the Santa Elena Peninsula, and excavations at two sites produced the earliest dates (Stothert et al. 2003; Pearsall et al. 2016). Ethnobotanical research identified squash phytoliths (*Cucurbita* ssp.) in sediments directly dated between 10,000 and 8,000 B.C. (Piperno and Stothert 2003). Maize has been documented with calibrated radiocarbon dating to 5985 B.C. Arrowroot or "lerén" (*Calathea allouia*) and the bottle squash (*Lagenaria siceraria* L.) have also been documented in early Las Vegas deposits. They represent an early adaptation to the dry tropical forest, with a mixed economy characterized by plant cultivation, and the exploitation of different terrestrial, riverine, and aquatic resources (Stothert 1988; Piperno and Pearsall 1998) (Table 8.1).

The Formative Period of coastal Ecuador is divided into Early, Middle, and Late. The Early Formative Period Valdivia culture (3500–1450 B.C.) was the earliest sedentary culture along the coast. Evidence of early permanent settlements in the Santa Elena Peninsula was documented at Real Alto and Loma Alta, and at the Valdivia type site (Pearsall 1979, 1988a, b, 1996, 1997) (see Fig. 8.1). Evidence of corn or maize (*Zea mays* L.) phytoliths as well as macrofossils associated with

Table 8.1 Chronological chart of the pre-Hispanic periods in Ecuador and northern Peru

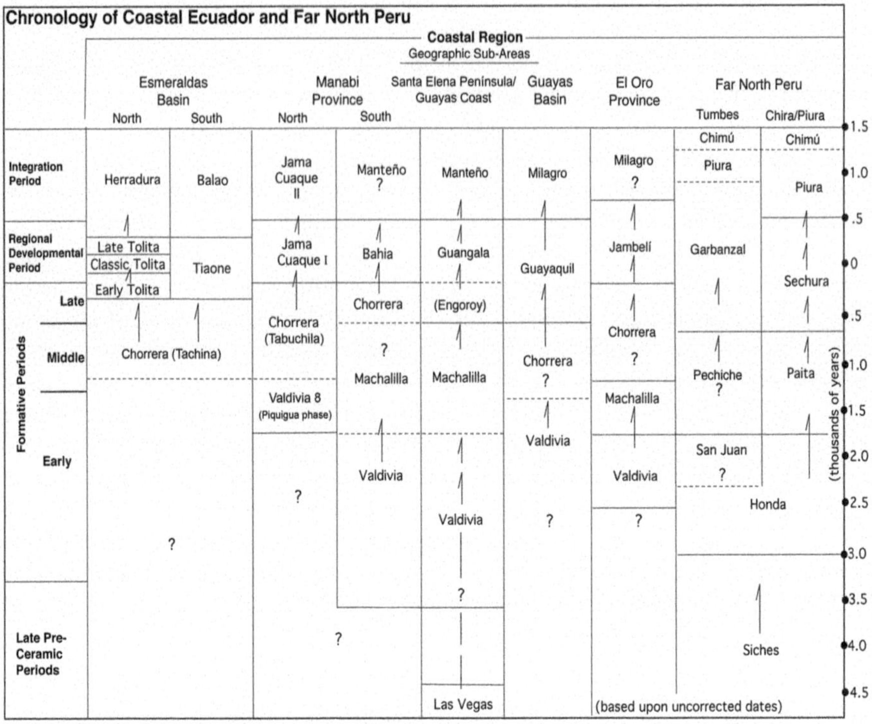

cotton (*Gossypium barbadense*), achira (*Canna edulis*), canavalia beans (*Canavalia plagiosperma*), malanca (*Maranta arundinacea*), and squash (*Cucurbita* ssp.) were identified at Real Alto. Chili peppers (*Capsicum* ssp.) manioc or cassava (*Manihot esculenta*) was also identified through starch analysis on lithic tools and in domestic structures at Real Alto and Loma Alta (Pearsall et al. 2016).

The Late Formative Period Chorrera culture (1300–300 B.C.) expanded its contacts with other cultures through both long-distance interaction and the introduction of ceramic technology into various coastal regions as well as the southern highlands. Regional styles associated with this tradition include Engoroy (Santa Elena Peninsula), Tabuchila (northern Manabi and Guayas), and Los Ríos and Cerro Narrio in the southern highlands of Azuay Province. The Chorrera-related ceramic traditions extended in El Oro Province and down to the Peruvian frontier (Staller 2010) (see Fig. 8.1). During this period there were significant transformations in the form of consolidating agricultural adaptations associated with sedentism and plant cultivation that resulted in changes in settlement patterns and increased the construction of pre-Hispanic dams, "*albarradas*" and "*camellones.*"[i]

During the Late Formative Period, maize, cassava, and various bean species became important food and drink resources. Late Formative evidence of domesticated plants and sedentism was also identified at various sites in Manabí Province, such as San Isidro, Dos Caminos, and El Mocoral (Pearsall 2004). Ethnobotanical evidence and macrobotanical remains, include palms (*Phytelephas aequatorialis*), the source of vegetable ivory or cade (*tagua*). An important industrial plant endemic to the coast is the ivory palm tree whose leaves are used in roof traditional houses. Also present on the coast is the aquatic herb *Scirpus* ssp., which is commonly used to make floor mats. The common bean (*Phaseolus vulgaris* L.), cherimoya (*Annona* ssp.), and fruits of the *Sapotaceae* family also appear on ceramics and at various sites indicating consumption.

After the Late Formative Period c. 500 B.C., pre-Columbian coastal cultures show evidence of complexity and cultural differentiation. Maize remains a primary staple, as do achira, sago, cassava, beans, and the other crops initially identified on ceramics in earlier periods. These reappear later during the Regional Developmental Period with the Guangala, Bahía, Jama-Coaque I, Jambelí, and La Tolita cultures (see Table 8.1).

During the Integration Period (c. A.D. 500–1500) cultures along the coast underwent greater economic, political, and social complexity (see Table 8.1). Maize remains important, but its primary form of consumption is as maize beer or *chicha.* As in many areas of the Andes, maize beer played important roles in Andean rituals (Staller 2006, 2016). The evidence suggests increased biodiversity of different landraces of maize associated with ethnic identity. Greater biodiversity is also apparent in cotton, cassava, squash, and various kinds of bean varieties. Tubers continue to play an important role in the ancient diet (Staller 2010).

Agricultural intensification results in an incremental increase in population densities along the coast, western lowlands, and coastal mountains. There was also marked differentiation between urban and rural regions. Religious and political elite controlled the production and redistribution of food surpluses. Metallurgy became more elaborate, pottery styles more abstract, and there is evidence of expanded long-distance interaction (Marcos 1986; Veintimilla 2004, 2013).

The Museum of Anthropology and Contemporary Art Collections (MAAC)

Archaeological collections in the Museum of Anthropology and Contemporary Art Collection (MAAC) consist primarily of ceramic effigy vessels or artifacts representing plants or that have plant-like attributes. The MAAC sample incorporates all cultural periods, including Valdivia ceramics with incised decorations, later La Tolita and Jama-Coaque ceramics, as well as objects that include or depict stems, leaves, tubers, and possible fruit. Most samples pertain to the Chorrera and Jama-Coaque cultures from Manabí Province. The Jama River Basin and San Isidro have highly diverse habitats with many archaeological occupations, some occupied for more than 3000 years (Zeidler and Pearsall 1994). Most pre-Columbian artifacts were acquired from collectors, rather than archaeological research. Consequently, their contextual associations are either unknown or problematic. Most objects are associated with burials, tombs, rituals, or religious ideologies. Few are from domestic contexts.

Morphologies were compared with illustrations and iconographic documentation of plants in the literature and by cultural association (see, e.g., Yacovleff and Herrera 1935; Lathrap et al. 1975; Weinstein 1996; Gutiérrez 2011). Artifacts of the MAAC collections analyzed in this research were assessed by their dimensions, surface textures, and association with the natural world. Evidence of plant peduncles, scars, divisions, and segments, among others were assessed using the *Catalog of Vascular Plants of Ecuador* (Jorgensen and León Sánchez 1999). This catalog was critical for the identification of the biogeography and cultural association of the plant taxa as well as if a species was native, introduced, or cultivated. We completed our interpretations and identifications using both a photographic archive of the inventory and the objects themselves. We used museum graphics, archaeological, botanical, and ethnobotanical criteria in our analysis. This data includes ceramic objects and fruit and tuber pictures, morphological descriptions, cultural affiliations, ethnobotanical data (macro- and micro-evidences), and references (Veintimilla 2016). We identified species primarily on the basis of contemporary understanding of their consumption and/or cultural importance. However, contextual verification from archaeological research provided direct evidence of chronology and forms of consumption and/or ritual importance.

The MAAC collection consisted of a total of 159 artifacts spanning the prehistoric sequence. The analyzed collection consists of 126 effigy vessels. 20 bowls (*cuenco* types), 6 Valdivia pots, and 7 vessels with quicklime called *lliptas* (small vessels containing vegetable ash or shell residue used to enhance coca leaf properties). These vessels contained iconographic and/or symbolic representations of 27 plant species (Table 8.2). A total of 92 vessels were classified as food crops, 60 were utilitarian plants, and 7 lacked any clear distinguishable plant morphologies or phenotypic attributes (Table 8.3). Many of the same species were represented throughout the pre-Columbian sequence (Table 8.4).

Food Crops

Our identification of food crops was based on plant properties and characteristics, and these were subsequently subdivided and classified into cereals, legumes, tubers or roots, vegetables and fruits (Table 8.3).

Cereals: Maize (Zea mays L.)

The origin, biogeography, and evolution of maize, the most important New World grass species in the *Poaceae* family, has been controversial and widely published in the scientific literature (Pearsall 2004; Pearsall et al. 2004; Staller 2010, 2015a,

Table 8.2 Iconographic interpretation of plant species

	Total
Achira or platanillo (*Canna* ssp.)	3
Chili (*Capsicum* spp.)	1
Cotton (*Gossypium barbadense*)	2
Ayahuasca (*Banisteriopsis caapi*)	6
Bijao o lerén (*Calathea allouia*)	2
Cacao (*Theobroma cacao*)	1
Yam (*Ipomoea batata*)	6
Ceiba (*Ceiba* ssp.)	1
Chamico or monte hediondo (*Datura* ssp.)	5
Custard Apple (*Annona cherimola*)	5
Coca (*Erythroxylum coca*)	26
Bean (*Phaseolus* ssp. o *Canavalia* ssp.)	1
Beans (*Phaseolus* ssp.)	3
Soursop (*Annona* ssp.)	1
Legume (*Anadananthera*)	2
Corn (*Zea mays*)	12
Corn (*Zea mays*), cassava (*Manihot esculenta*) and bean (*Phaseolus* ssp.)	1
Red Mammee apple (*Pouteria* ssp.)	1
Calabash Tree (*Crescentia cujete*)	18
Pineapples (*Ananas comosus*)	1
Sapodilla (*Pouteria* ssp.)	3
Cassava (*Manihot esculenta*)	1
Pumpkin (*Cucurbita moschata*)	28
Pumpkin (*Cucurbita pepo*)	14
Pumpkin (*Cucurbita* ssp.)	1
Pumpkin (*Lagenaria siceraria*)	7
Unidentified phytomorph	7
Overall total	159

2016). Maize was originally domesticated in Mesoamerica. It is directly derived from a variety of teosinte (*Zea* ssp. *parviglumis* L.) through a long process of human and natural selection which began approximately 8000 years ago (Doebley 1990; Pipemo and Pearsall 1998; Pearsall et al. 2016; Staller 2010).

This iconographic research identified maize through incised decorations and modeled forms on Valdivia ceramics. Evidence of motifs consisting of incised line forms as well as oblique and uniform lines joined from a central axis suggest stylized spikes or tassels (Lathrap et al. 1975); modeled forms included grains and ears of corn on the edges of vessels (Veintimilla 2016: Figs. 9.1, 9.2e). The most realistic

Table 8.3 Classification of food species

	Common Names	Total
Cereals	Maize	13
Total Cereals		13
Fruits	Cacao	1
	Custard Apple	5
	Soursop	1
	Red Mammee apple	1
	Pineapples	1
	Sapodilla	3
Total Fruits		12
Peppers	Chili	1
		50
Total Vegetables		51
Legumes	Beans	4
Total Legumes		4
Tubers	Achira or platanillo	3
	Bijao o lerén	2
	Yam	6
	Cassava	1
Total Tubers		12
Total		92
Non-Food Plants	Common Name	Total
Economic	Cotton	2
	Ceibo	1
	Calabash Tree	18
Total Economic		21
Psychotropic	Ayahuasca	6
	Chamico or monte hediondo	5
	Coca	26
	Legumes	2
Total Psychotropic		39
Total	60	

representations of maize are on Jama-Coaque ceramics, evident by a zoomorphic agouti figurine eating a maize cob (Gutiérrez 2011). Other examples include anthropomorphic figurines holding partially shredded cobs, often associated with annual harvest rituals (Veintimilla 2016: Figs. 3–4). A Jama-Coaque effigy bottle also had stylized designs incised on the shoulder of corn cobs representation (Veintimilla 2016: Fig. 5) (Fig. 8.2).

Table 8.4 Cultural affiliations of food species

	Total
Bahía	8
Chorrera	79
Guangala	1
Undetermined	1
Jama-Coaque	51
Machalilla	1
Manteño	4
Tolita (Tumaco)	8
Valdivia	6
Overall total	159

Fig. 8.2 Jama-Coaque zoomorphic sculpture showing an agouti eating a maize cob

Maize initially appears in the late Pre-ceramic period Las Vegas sites. During the Early Formative Period maize, phytoliths were documented at Real Alto, Loma Alta, and the Valdivia type site. Such ethnobotanical evidence was later corroborated by phytoliths and starch granules on lithic tools (*manos* and *metates*) dated to Phase 3 Valdivia (c. 2800–2400 B.C.) at Structure 20 at Real Alto (Pearsall et al. 2016). These results suggest maize was commonly associated with domestic contexts (Pearsall 2016; Pearsall et al. 2004).

During the Middle and Late Formative Periods, maize macrobotanical remains were documented at the Machalilla (1430–830 B.C.) site of La Ponga, and with Chorrera culture along the Jama River Basin, in Manabí Province (Pearsall 2004). During the subsequent Regional Development and Integration periods, maize phytoliths, grains and macrobotanical remains suggest it had an important role.

Root Crops: The Fruit of the Earth

Effigy vessels decorated with tubers or roots attached to a stem are considered works of art. Some have two fruits projected on their sides surrounded by an applet representing a trifoliate plant, such as a vine or pod. These morphological elements probably represent cassavas (*Manihot esculenta* Crantz), maize cobs (*Zea mays* L.), and beans (*Phaseolus vulgaris* L.). These food crops were all associated with lowland and coastal cultures and also have strong cultural associations with religious symbolism (Veintimilla 2016: Fig. 16).

Legume Species

There are some 20 species of beans *Canavalia* ssp. in the New World (Gentry 1993). Jack beans (C. *ensiformis* L.) and Piper beans (C. *plagiosperma*) are from the Andes and adapted to xerophytic and saline environments. Common beans (*Phaseolus vulgaris*) are one of the best known New World species. In Ecuador they are documented beginning with the Early Formative. This annual herb is adapted to a multitude of environments from the coast to the highlands, the Upper Amazon, and tropical eastern lowlands. It extends from Mesoamerican to southern Peru (Kaplan and Kaplan 1996).

The lima bean (*Phaseolus lunatus* L.) is long and flattened and originally comes from middle elevations of the western cordillera. Certain species are also adapted to the western lowlands of Ecuador and Peru. Pre-Hispanic evidence suggests the species extends to Moquegua in southern coastal Peru (Staller 2015b). Lima beans were recovered from domestic contexts at Regional Development and Integration Period sites (Veintimilla 2004).

Multidisciplinary evidence (genetic, biochemical, and geographic) indicates that the domesticated beans come from two genetic sources: Mesoamerica and the Andean coast and cordillera (Piperno and Pearsall 1998). The species biogeography extends from Mexico, Central America, and the Caribbean into Colombia and south along the Andean lowlands. The common bean is present early in coastal Ecuador and Peru. These common beans *P. vulgaris* and *P. lunatus* originated from these regions of the Andes. Varieties of wild bean species in these regions further support such assertions (Piperno and Pearsall 1998; Pearsall et al. 2016).

Earliest evidence of legume (*Canavalia plagiosperma* Piper) consumption is documented at Real Alto and associated with Valdivia Phases 1 and 2 (c. 4400–3000 B.C.) (Pearsall 1979, 1988; Damp et al. 1981). Phytoliths from the genus *Phaseolus vulgaris* were recovered at San Isidro, (Manabí Province in association with Chorrera and Jama-Coaque cultures) (Pearsall 2004). Research by Veintimilla (2004) in the "*albarradas*" of the Santa Elena Peninsula have identified *P. vulgaris* and *P. lunatus* in Regional Developmental and Integration Period contexts (see Table 8.1).

These species were initially identified on Chorrera and Bahia effigy bottles and bowls flexed in the center and demarcated by a dark edge from where the bottle spike is projected (Veintimilla 2016: Fig. 6). Plant decorations on small bottles appear as trifoliate vines and pods symbolizing their hallucinogenic or psychotropic properties and effects in Jama-Coaque ceramics (Veintimilla 2016: Fig. 7). However, manioc or cassava and some legume species also contain chemical properties such as cyanide and other poisonous substances usually removed by repeated soaking and/or cooking the seeds (see Fig. 8.3).

Tubers and Roots

Four species of tubers were identified from the MAAC collection associated with the genus *Calathea* such as *leren* or *bijao* (*Calathea allouia* Lind), achira or platanillo (*Canna edulis* L.), sweet potato (*Ipomoea batatas* L.), and cassava (*Manihot esculenta* Crantz) (see Fig. 8.3). On the basis of their biological characteristics and representation among different cultures, we decided to carry out a comparative study of formal characteristics and attributes embodied in each object, comparing these with iconographic representations presented (see Yacovleff and Herrera 1935).

Manioc or cassava is primary a starchy root crop used as food in the tropical lowlands. Manioc originated in north and northeastern South America in modern-day Colombia, Venezuela, and northwestern Brazil (Piperno and Pearsall 1998). Another region of origin is Mesoamerica (Piperno and Pearsall 1998). Cassava is adapted to sandy soils where the roots can easily grow and expand.

Cassava first appears in Ecuador with early Valdivia culture where it has been identified through starch grains extracted from lithic tools, particularly grinding tools, i.e., *manos* and *metates*, and soils associated with domestic dwellings at Real Alto and Loma Alta (Pearsall et al. 2016). Indirect evidences of cassava comes from manioc graters and ceramic vessels associated with *chicha* processing as well as representations on the bases of ceramic pots and sitting anthropomorphic figures. Cassava, corn, beans, squash, and sweet potato were some of the most important food crops of Andean cultures on both sides of the cordillera. Manioc or cassava is symbolized by elongated, thick roots at the top, thin and conical roots at the base (Yacovleff and Herrera 1935:270; Fig. 9). Effigy vessels representing manioc tubers

Fig. 8.3 Jama-Coaque sculpture representing manioc tubers, corn cobs on the sides, and a legume possibly a bean specie

(*Manihot esculenta*, Crantz), maize (*Zea mays* L.), and beans (*Phaseolus* spp. L.) are identified from various ceramic traditions (Veintimilla 2016: Fig. 10) (Fig. 8.3).

The genus *Calathea* (Marantaceae family) has sweet edible roots. This sweet corn root is a robust herb in the arrowroot family, native to the lowland forests on both sides of the cordillera. Several *Calathea* species have edible roots; however, C. *allouia* is consumed most frequently. *Leren* was one of the earliest cultivated Andean plants. Carbon residues of *Calathea allouia* phytoliths were identified in c. 6000 B.C., with Pre-ceramic Las Vegas culture and at Valdivia sites such as Real Alto and Loma Alta, as well as carbon residues Late Formative Chorrera and Jama-Coaque ceramics.

The genus *Canna (Canna indica L.)* is comprised of 10 New World tropical species that grow at about 2000 m. altitude, three have edible rhizomes. *Achira* or *platanillo* (*Canna indica* L.) is also consumed in Middle America, but originated in the northern Andes, in the humid tropical forests of Colombia, Ecuador, and Peru. Ethnobotanical evidence for *achira* includes phytoliths from Valdivia ceramics in domestic contexts at Real Alto (Pearsall 1979, 1988). We identified *achira* or *platanillo* on the handles of Late Formative Period Chorrera, La Tolita and Jama-Coaque ceramics (Table 8.1).

Sweet potatoes *or yams (Ipomoea spp. L.)* are the most represented species after cassava and were widely consumed among tropical lowland cultures. They prefer warm and humid climates, well drained, sunny soils and annual precipitation between 750 and 1250 mm. Chroniclers during the sixteenth and seventeenth centuries stated that yams were associated with numerous dishes and cuisines in Manabí and Guayas Provinces (Estrella 1990). The largest were cultivated around Catarama, Puerto Viejo, and Daule, near Guayaquil (Estrella 1990).

Fig. 8.4 Sweet potatoes tubers represented at the base or legs of a seat of Jama-Coaque sculpture

Sweet potatoes were identified archaeologically on the basis of starch granules on lithic tools generally used to process food and pollen recovered from estuaries in the Santa Elena Peninsula. It has been inferred that these lithic tools were used to process tubers and this is supported by representations on ancient ceramics (Veintimilla 2016: Fig. 8). Sweet potatoes often represent the base or legs of bowls with the seats of high-status individuals on Chorrera, La Tolita, and Jama-Coaque ceramics (Fig. 8.4).

Vegetables: Pumpkins, Squash, and Chili Pepper

There is considerable phenotypic variability among pumpkins, squashes, as well as fruit species. Fruit species identified in our samples came primarily from two different families: *Cucurbitaceae*, particularly the genus (*Cucurbita* and *Lagenaria*) and *Bignoniaceae,* specifically the calabash tree (*Crescentia cujete* L.). Identification of these species is complicated by their similar morphology. Our analysis combined established formal stylistic and cultural criteria along with specific phenotypic attributes, particularly surface characteristics and textures. We also analyzed plant morphological details including the existence peduncles, scars, divisions, and segments, among others. For example, the samples interpreted as *calabaza* or winter squash (*Cucurbita pepo* L.) and butternut squash (C. *moschata* L.) were differentiated by height, one is usually taller than wide (Veintimilla 2016: Fig. 11). The species *Lagenaria siceraria* (pumpkin or bottle squash) has a double body and peduncle, while calabash trees have a smooth surface and the scar is defined as a hole usually in the form of an opening or rim on ceramic effigy vessels.

The genus *Cucurbita* is comprised of five New World species, originally cultivated and domesticated in the tropical lowlands and adapted to dry xerophytic environments, with moderate rainfall. They do not tolerate high levels of precipitation. The biogeography of the species C. *moschata* extends from the American Southwest to the coastal lowlands along the Andes and tropical forests of South America (Piperno and Pearsall 1998).

The genus *Cucurbita* has a long history in coastal Ecuador: Ethnobotanical evidence indicates its cultivation extends back to Pre-ceramic Las Vegas culture (Stothert et al. 2003). *Cucurbita* phytoliths were identified in sediments dated between 10,000 and 8,000 B.P. (Stothert 2003; Pearsall et al. 2016). *Lagenaria* phytoliths were documented in Pre-ceramic sediments dated to 5985 B.P. These early dates are possible evidence a pumpkin species (possibly *Cucurbita moschata*) evolved from the wild species *Cucurbita ecuadorensis*. Numerous Late Formative Chorrera, Bahía, and La Tolita and Jama-Coaque effigy vessels are squash or pumpkin shaped (Veintimilla 2016: Figs. 12–14).

*Chili Peppers (***Capsicum** *ssp.)*

Some of the earliest effigy vessels of chili peppers come from the coast and are associated with the Late Formative Chorrera culture. Some of these vessels were decorated with a creamy white background that accentuated and simulated a stylized silhouette of chili peppers in the red parts of the vessel exterior. Pre-Hispanic chili consumption on the coast was documented through ethnobotanical evidence of starch granules on lithic tools at Structure 1 at Real Alto (Pearsall et al. 2016). Chilies grow in all environments, from the xerophytic maritime regions of Santa Elena Peninsula to dense tropical forests further to the north. Three chili species are native to the Andes: *Capsicum pubscens* or "rocoto," which was domesticated in the central Andes (Peru-Bolivia), *Capsicum annuum*, originated in the lowlands of northern South America and C. *baccatum*, spread east into the tropical eastern Andean lowlands (Pickersgill 1984; Andrews 1996; Pipemo and Pearsall 1998).

Fruit Species in Coastal Ecuador

Fruits including cherimoya (*Annona cherimola*), guanabana (*Annona muricata*), pineapples (*Ananas comosus* L.), and eggfruits from mamey sapote (*Pouteria sapota*) were widely consumed among pre-Columbian cultures along the coast and continue to be popular among contemporary cultures. These plants are adapted to dry and humid tropical forests. Botanical evidence of cherimoya, guanabana, and sapotes were reported at San Isidro in the Jama River Basin (Pearsall 2004).

Fig. 8.5 Chorrera spout whistling bottle in the shape of a cacao bean

Cacao (*Theobroma cacao* L.) is native to the Amazon Highlands and Orinoco River Basins (Cope 1979; West 1996; Lanaud et al. 2012). Ethnohistoric, ethnographic, and botanical research indicate it was cultivated in the high, medium, and low watersheds of the Guayas River and humid regions along the gulf coast (Estrella 1990). Ethnobotanical remains of cacao starch granules were extracted from the handles of stirrup-spout bottles at Santa Ana, La Florida, and Zamora Chinchipe. Cacao was often given as a ritual offering (Valdez 2019) (see Fig. 8.1). The MAAC collections include a single spout whistling bottle in the shape of a cacao bean with an elongated morphology, and divisions or sections that join at the end point on one side, where a small hole emulates the peduncle from where the fruit grows (Veintimilla 2016: Fig. 5) (Fig. 8.5).

Domesticated Plants: Cotton and Calabash

Economic plants depicted on pre-Columbian coastal ceramics from Ecuador include a species of cotton (*Gossypium barbadense* L.) domesticated in the Andes and native to the Neotropics. It is generally adapted to dry, open, and sunny climates, and extends to desert coast and western lowlands of Peru. This species was initially domesticated along the coast at sites such as Real Alto (4400 B.C.) and Ancon (4200 B.C.) in coastal Peru (Piperno and Pearsall 1998; Valverde 2004). Numerous Chorrera effigy vessels in our collection represent various fruit and/or cotton seeds. Two triangular vessels were demarcated by sections with two ends pointing out to the apex and part of the peduncle (Veintimilla 2016: Fig. 17) (Fig. 8.6).

Fig. 8.6 Chorrera effigy vessel representing a cotton seed

The cultivation of cotton is well documented on the basis of carbonized seeds recovered in domestic structures associated with Early Valdivia culture at Real Alto (Damp and Pearsall 1994). Prehistoric cotton has been inferred from fabric imprints on clay remains and from the presence of shell hooks and spindle whorls. The evidence suggests it was cultivated for cloth in Valdivia culture as early as 3000 B.C.

The ceiba (*Ceiba* ssp.) and calabash (*Crescentia cujete* L.) trees are also represented on Chorrera ceramics. Ethnobotanical evidence from phytoliths and carbonized wood indicates that both plants are associated with sites in the *albarradas* of Santa Elena Peninsula, and south along the coast of El Oro Province (Staller 2001), and in Early Huancavilca occupations (Veintimilla 2004). Ceiba (*Ceiba* ssp.) phytoliths were identified in sediments from sites in the Santa Elena Peninsula Manabí and at El Oro (Staller 2001; Veintimilla 2004; Pearsall et al. 2016). The ceiba tree has strong associations with ancient mythology and religious ideology and is commonly associated with the sacred center or *axis mundi* among ancient New World cultures (Demarest 2004).

Psychotropic Plant Species

MAAC samples include 23 archaeological examples generally, globular and single spout bottles with plant motifs created with the "pulled sugar" technique ("*pastillaje*"). These bottles are, smoothed (not polished), with everted rims and annular bases. Many have long spouts with representations of leaves, possible stems, fruits, or buttons on the body of the vessels, suggesting associations with rites or offerings in mortuary contexts. Gutiérrez (2011:180) defines such decorations and vessels as "special elements with a ritual and sacred symbolism."

We identified three psychotropic species central to ancient rituals that were used to gain access to the spiritual realms by shamans; these were special plants for Andean peoples (Veintimilla 2016: Figs. 19–19) (See Table 8.3).

*Coca (*Erythroxylum coca *Lam.)*

Coca leaves were symbolized by one or more circular designs with a sort of flower bud or fruit from which a plant with petiolate leaves and clear venation arises in a stylized way. Its leaves, the main element emphasized in designs, covey the impression that coca was important to Andean culture.

Coca is native to the Amazon regions of Colombia, Ecuador, Peru, and Bolivia. In Ecuador, coca has been identified on the coast and Amazonia, particularly in the lower regions of Cañar, Azuay, Chimborazo Provinces and Napo, between an altitude of 500 and 1000 masl. (Jorgensen and León-Sánchez 1999). The shrub reaches between 1.5 and 3 m in height, has lanceolate or oval flowers, small white flowers, and the fruit is a red oval drupe of 1 cm. diameter. It is the alkaloid cocaine that has stimulating, anesthetic, therapeutic, and properties to mitigate cold and hunger that makes it important among highland Andean cultures. Coca is widely used by indigenous populations throughout the cordillera, and the coca leaf properties were enhanced using quicklime (vegetable ash or shell) which was placed into small containers, called *lliptas*. Leaves are chewed by simply adding lime to the bolus with a toothpick previously moistened with saliva and immersed in the pot containing lime.

Evidence of early coca use is evident on two Valdivia anthropomorphic figurines. The earliest (Phase IIb—2950–2600 B.C.) has a jaguar headdress and hole in the upper part of its head along with quicklime remains (OGSECH-012/FS 4378), which is how shaman used it (García Caputi 2008, 2016). Another Phase 4—2600–2400 B.C. figurine has coca bolus in the cheek, further supporting early coca use (Plowman 1984). Dental tartar in certain human skeletons as well as objects of ritual paraphernalia used in ecstatic shamanism, such as "*lliptas,*" spatulas, and sticks, support its early consumption. Anthropomorphic figurines indicating coca consumption are also in Bahía, Guangala, La Tolita, and some of Jama-Coaque ceramics (Veintimilla 2016: Fig. 20).

Ayahuasca or yagé (Banisteriopsis caapi)

A single spout bottle has representations of a liana of the *Malpighiaceae* family, called Ayahuasca or yagé (*Banisteriopsis caapi*) (Schultes 1972). Ayahuasca is a psychotropic plant with more than 40 varieties. Indigenous Andeans consider it

Fig. 8.7 Jama-Coaque sculpture in the shape of a house with stems and shoots on the ceiling, representing a liana, possibly ayahuasca

sacred, particularly Colombian and Ecuadorian cultures in the Amazon. A potion or tea made from the liana of ayahuasca is consumed by shaman and its powerful hallucinogenic effects are believed to provide a basis to foresee future events and access the underworld and celestial realms (Reichel-Dolmatoff 1978; Schultes 1972). Effigy bottles depicting or representing ayahuasca were decorated with two opposing circles decorated with plant elements including a lump similar to a peduncle and a sort of undulating stem that twists through the walls of the vessel. Additionally, small circles at the ends and/or in the body simulate plant nodules or buds (Veintimilla 2016: Fig. 21). Some ceramics were shaped as houses with stems and buds depicted on the roof. Our example has an elite or shaman seated at the doorway (Veintimilla 2016: Fig. 22). (Fig. 8.7).

Ayahuasca consumption is not well documented along prehistoric coastal Ecuador. However, consumption as a beverage strongly suggest either mate or ceramic bowls were used. Ayahuasca is native to the humid forests of coastal Ecuador, particularly Esmeraldas, northern Manabí, and Los Ríos Provinces, extending south El Oro Province along the southern cordillera around Cuenca in Azuay Province at altitudes ranging from 0 to 500 masl (Jorgensen and León-Sánchez 1999; Staller 2000) (see Fig. 8.1). The MAAC collections only have Jama-Coaque ceramic artifacts with liana of ayahuasca. This culture inhabited the humid coastal forests of Manabí, south of Esmeraldas and the west of Santo Domingo de Los Tsachilas.

Some ceramics were decorated with plants with leaves that emanate from a flower bud, seed, or fruit pods (Veintimilla 2016: Figs. 7, 23). This stylistic feature generally inferred the plant was probably a legume, possibly *Anadenanthera* spp. known as *yopo* or *vilca*. The seeds are used to make a powder with powerful hallucinogenic affects (Schultes 1972). Gutiérrez (2011) suggests that *yopo* was used by pre-Hispanic Ecuadorian cultures since the late Early Formative Period. Valdivia ceramic bowls and plates contained the *yopo* powder and mortars with plant representations to grind the seeds. This species is native to dry tropical forests of Loja Province and grows between 1000 and 2000 masl (Jorgensen and León-Sánchez 1999).

*Monte Hediondo or Chamico (***Datura** *sp.)*

Another psychotropic plant identified in the ceramic iconography is nightshade *Datura* ssp., known as "*monte hediondo*" or "*chamico,*" which is often represented on globular pots, flattened at the ends, with ring baselines (with or without a lid) and surfaces decorated with sharp projections (Veintimilla 2016:Fig. 24). *Datura* is a perennial herb, native to Central and South America. It is a potent hallucinogen containing tropano alkaloids, scopolamine, and other natural components that can produce psychomotor excitations (Valverde 2004). *Datura* seeds were uncovered archaeologically at Chorrera and Manteño-Huancavilca sites in the Santa Elena Peninsula (Veintimilla 2004). The MAAC Bahía and Jama-Coaque ceramic collections also include examples characteristics of this plant (Fig. 8.8).

Fig. 8.8 Chorrera globular pot, flattened at the ends, with ring baselines and surfaces decorated with sharp projections representing a datura fruit

Ceramic Representations and Iconography of Diet and Agricultural Fertility

A large body of literature on pre-Hispanic coastal Ecuador indicates that ancient cultures were intrinsically tied to their natural environment or mother earth (*Pachamama*). Agricultural rituals associated with sowing and harvesting food crops continue to be celebrated among indigenous communities. The solstices (June and December) and equinoxes (March and September) are marked with rituals related to the harvest (*Inti Raymi*) and associated with solar (*Inti Tayta*) and lunar (*Mama Quilla*) cycles, as well as with planting and sowing crops (*Kulla Raymi*), and shamans continue to carry out rituals to evoke rain. Fishermen along the coast still take voyages to honor the Virgin or San Pedro, their patron saint. These clearly reflect syncretism of aboriginal and colonial religious beliefs and customs. Fisherman pray and make ritual offerings for both productive catches and their safety. The thorny oyster (*Spondylus* ssp.) has served as a climatic marker since the Early Formative. The worship and veneration to the Lord of the Waters of Colonche, a saint attributed with good dry seasons, are carried out every May 29 and 30. Cultures around the Colonche hills venerate this spiritual entity and bring offerings of horticultural crops, fruit and grass, and meat protein from beef cattle, goats and horses.

Fertility rituals are evident since the Early Formative. Ceramics and figurines provide evidence of ecstatic shamanism, as ceramic seats of power have been documented in ceremonial and domestic contexts. The figurine collections include a pregnant female with a jaguar *(Panthera onca)* headdress suggesting shamanic transformation (Fig. 8.9). Over 50 late Valdivia Venus figurines were compared with early figurines of Buenavista-Chacras tradition uncovered in early ceramic contexts at Real Alto (FS 3736). The Valdivia figurines had animal skin headdresses and headdresses of the Harpy eagle (*Harpia harpyje*) suggesting transformation in altered states of consciousness.

There is clear evidence of long-term continuity in rituals associated with calculating solar and lunar cycles, and symbolic associations with the underworld (jaguar) and celestial realm (harpy eagle). Ideologically, Valdivia culture was focused upon the fertility of the earth, characteristic of agricultural societies. Late Valdivia figurines are of the Venus style and more symbolically focused on shamanic transformation (Staller 2000, 2001). Other important symbols associated with fertility are the thorny oyster[ii] (*Spondylus princeps*) and Strombus conch (*Lobatus galeatus*) dyad (Fig. 8.9). They initially appear at Real Alto on the floor of a meeting house (Garcia Caputi 2016).

During the Regional Developmental Period, anthropomorphic figures dressed in ritual ponchos were sometimes embroidered with small snails to reproduce a sound resembling rain when moving (see Table 8.1). Evidence of ritual specialists associated with agricultural fertility are also symbolized by figures with sticks or wooden posts associated with planting. "Shamans are mediators that access rain divinities, to sprout the seeds and complete the agricultural cycle, as well as control climatic forces capable of destroying these processes" (Ontaneda and Valdez 2016).

Fig. 8.9 Middle Valdivia ceramic figurine representing a pregnant female with a jaguar headdress suggesting shamanic transformation

Our research suggests hunting rituals were carried out by shamans who accessed the spiritual realm to propitiate such activity implying direction, knowledge, and agility. All of these are forces that give great prestige to hunters. An iconographic example on a Jama-Coaque bottle has a hunter seated on a deer. The hunter's headdress is ornamented with five snakes, he has a nose ring and tubular earrings, and a four-round collar segmented around the neck. The hunter holds snakes in his hands and his legs are folded and ornamented. The deer is lying below him with a semi-open snout showing its teeth (Fig. 8.10).

There are a wide variety of edible and utilitarian animal species central to sedentism along the coast in the MAAC collection. Approximately, 100 artifacts depict animals, birds, and maritime species, providing a basis for assessing their social and ideological dimensions. Various MAAC artifacts provide evidence of the cultural importance and associations with traditional customs as well as ancient religious ideology. Zoomorphic artifacts indicate several common cultural features including meat consumption of various terrestrial, maritime, and bird species, as well as domesticated species. Some are utilitarian objects providing insights into perceptions and customs related to calculating the agricultural cycles. Thus, these ceramics provide insights into cultural concepts, perceptions, and ideological beliefs among ancient coastal cultures.

We methodologically selected three types of objects, including those related to subsistence, objects that served as either agricultural tools, or those associated with

Fig. 8.10 Jama-Coaque sculpture representing a hunter seated on a deer. The hunter's head dress is ornamented with five snakes, a nose ring, tubular earrings, and four-round collar segmented around the neck

hunting and food preparation. We analyzed *Anadara grandis* or *Anadara tuberculosa* clam shells to determine if they served as masks, such as the Thorny Oysters (*Spondylus princeps*[4]) or Strombus conch (*Lobatus galeatus*[5]) which were trumpets, and carved shells tools, spoons, containers or ploughs or hoes formed by the columella of these gastropods. Shells used as ornaments and utilitarian objects include a small selection from the Valdivia culture as well as agricultural tools and axes associated with Bahía, Guangala, Milagro Quevedo, and Manteño cultures. Objects used to prepare and process food such as grinding stones or *manos* (Manteño and Milagro Quevedo) and *metates* (Manteño) were also included.

Valdivia occupations at Loma Alta intentionally drilled gastropods to extract food, but ethnographic evidence indicates that they were also whistles (Stahl 1985). Oysters and clams served as important food resources at Loma Alta and, particularly, La Emerenciana in El Oro Province (Stahl 1985; Staller 2013, 2015c). Ancient cultures also used shells to manufacture various elements. They apparently were not processed and discarded on the beaches; therefore, we have numerous examples in the MAAC collection. Ethnographic evidence suggests rodents and large land gastropods (*Strophocheilus* sp.) were a plague to ancient agriculture (Stahl 1985). However, snakes depicted in Valdivia iconography were perceived as a benefit because they protected crops by feeding on rodents. Valdivia collections include objects depicting snakes or corn cobs on bowls and ceramic vessels. Two artifacts

Table 8.5 Cultural affiliation of zoomorphic and maritime species

Cultural affiliation	Iconographic representation	Tools
Valdivia	2 Monoliths 2 Thorny Oyster (*Spondylus princeps* L.) masks	4 (one stone ax, one spoon, and two hoe picks formed by the spines of large gastropods) 1 Thorny Oyster *Spondylus* ssp. shell bivalve
Chorrera	20	
Guangala	6	1 ax
Bahía	3	1 ax
Tolita	5	3 ichthyomorphic graters
Jama-Coaque	6	
Manteño	3	1 stone ax 1 metate 1 grinding hand
Milagro-Quevedo		1 stone ax 1 grinding stone (*mano*)

were ritual mortars, one representing a parrot, in this case a macaw or "*guacamayo*," (*Ara* spp.), and another, a golden-mantled howler monkey (*Alouata palliate*). During the Early Formative Period, large middens of ark clams (*Anadara tuberculosa*), as well as deer bones (*Odocoileus virginianus*), were found throughout various coastal regions (Gutiérrez 2009).

The MAAC collection included Chorrera (n = 20), Guangala (n = 6) and Jama-Coaque (n = 6); Tolita (n = 3), as well as Bahia (n = 3) and Manteño (n = 3) ceramics and artifacts as well as two Valdivia figurines. *Spondylus* shell masks (n = 2) were important ritual objects as was an unidentified bivalve (Tables 8.5). We included 15 zoomorphic objects from the *10,000 years of the Ancient Ecuador* exhibition at MAAC. These artifacts provided insights into the ecological diversity along the coast as well as the presence of permanent settlements and the beginnings of social complexity.

Faunal Food Species and the Associated Symbolism

Although faunal representations associated with archaeological excavations allow us to infer consumption, variability in cultural representations of different traits proved a challenge to this research and our preconceived concepts associated with such species. We describe four symbolic aspects of ancient cultures including shamanic transformations, such as figurines with animal headdresses, including the jaguar (Early Valdivia) commonly associated with the underworld. Fifty figurines (both complete and fragmented) have cranial deformations and protruding nose that resembles a harpy eagle beak (Late Valdivia). The harpy eagle, along with the jaguar and the

Fig. 8.11 Bahia sculpture suggesting anthropo-zoomorphic transformations

snake, represents the human awakening with their spiritual realms. Later examples suggested anthropo-zoomorphic transformations particularly objects associated with Bahia culture[iii]. Felines were the most prominent species represented, particularly the jaguar with a hanging tongue and of priests and shamans (Fig. 8.11).

Anthropomorphic faunal representations include both zoomorphic and anthropomorphic representations, such as the jaguar ceramic sculpture and specific ornaments along with signs of power and clairvoyance from La Tolita (MAAC: GA-35-1924-81). Also, in the collection is a caiman-man (Idem. MAAC: GA-12-1771-81) telluric deity of the underworld from where the caiman emerges as a divinity originating from all plants (MAAC catalog "Shamans and divinities of pre-Columbian Ecuador" 2016). Mythological or fantastic beings often combine traits of different animals in cultural objects. These are clear examples that they precede the "Bay monsters" of the Late Formative Period Chorrera whistle bottles (Fig. 8.12). Such artifacts usually have two resonance chambers that combine the body of a shrimp or prawn (water), probably *Trapichenseus* (species *leander* or *palaemonetes*), and the heads of felines, probably jaguars in reclining positions along with elaborate designs with engraved lines.

Decorative elements such as necklaces and bracelets present in some zoomorphic pieces have been considered evidence of hominization or deification (Gutiérrez 2009). This allows us to interpret the symbolic use that society assigned to certain animals within its cosmology. For example, the representation of a shark wearing a collar between its head and body (Fig. 8.13) evokes the deep waters of the Humboldt Current that run along the SW Ecuadorian coast (Ontaneda and Valdez 2016) (Fig. 8.13).

Fig. 8.12 Chorrera whistle bottle. It has two resonance chambers that combine the body of a shrimp and the head of a feline

Fig. 8.13 Chorrera vessel representing a shark, which is wearing a collar between its head and body

Another interesting artifact has two jaguars arranged in opposition, each with their head on opposite sides of a ceramic pot with two wide holes. We interpret this object as representing a binary duality in their cosmology (MAAC: GA-7-2428A-82). A symbolic peculiarity is the half-open mouth, showing its fangs and hanging the tongue which, in the case of the jaguar, refers to the inner water of the mountains, as does squatting or crouching with legs bent and their tails up. Finally, frogs (Anuran amphibians of the *Ranidae* family) evoke terrestrial water (MAAC: GA-2-2752-84).

Summary and Conclusions

Our research and analysis of the MAAC collections highlight important and revealing insights into the introduction and consumption of various flora and fauna from a total of roughly 300 objects. Both plant and zoomorphic representations in the MAAC archaeological collections from coastal Ecuadorian cultures provided symbolic reference to pre-Hispanic social meanings of these food crops, fauna, domesticates, and their associated art and symbolism. The scientific inventory and records of the archaeological contexts were used in selecting the objects for analysis, taking into consideration stylistic, cultural, and chronological criteria. Thus, the plants and animals interpreted in this study cover more than 5000 years of prehistory from the Early Formative until the Integration period of almost the entire Ecuadorian coast. Morphological, iconographic, ethnohistorical, bibliographical, and archaeobotanical criteria were methodologically combined.

These plants and fauna have for the most part been documented at various sites by archaeologists and in the macro- and micro-botanical remains (organic remains, seeds, phytoliths, and granules of starch and pollen). The identification of the plants summarized here was based on interpretations made through pre-Hispanic ceramic art. The piece called "The Fruits of the Earth" (GA-1-2539-83) deserves special attention. Our research suggests the basic triad of Neotropical lowlands vegetal food is represented as cassava, maize and beans, all of which contain a high cultural symbolism (Veintimilla 2016: Fig. 17) (see Fig. 8.3).

A total of 27 plant species were classified as food, industrial, or psychotropic taxa. Among the different types of food, it is worth noting that maize, beans, squash, chilies, tubers, and roots such as cassava, sweet potato, and fruits such as custard apple, pineapple, sapote, and cacao played important roles in the cuisines, foods, and drink of ancient Ecuadorian cultures. However, maize, and tubers such as cassava and sweet potatoes, as well as various beans, squash, and peppers are still popular in Ecuadorian cuisine, particularly in the form of tortillas, soups, and seafood dishes such as *ceviche*, fermented intoxicants like maize beer or *chicha*, and Ecuadorian cacao for its fine aroma and flavor. These domesticates deserve particular mention with regard to consumption in various coastal cuisines. It is also evident that some specific fauna, such as camelids, deer, and various bird species were also central to the prehistoric diet along the coast.

Among the industrial plants, cotton was a very important crop whose early domestication appears to have been southwestern coastal Guayas Province and the Santa Elena Peninsula. The iconography provided a basis to identify psychotropic plants through symbolism and decorations on certain bottles that Gutiérrez (2011) links to special or sacred rites and ritual activities. It was possible that coca, ayahuasca, a Leguminosae possibly of the genus *Anadenanthera* (called *yopo* or *virola*), Chamico or Monte Hediondo (*Datura* ssp.) were also central to ancient ritual practice. The fauna and food crops represented in these ceramic vessels appear to be indicators of their alimentary importance, as well as their cultural, symbolic, mythical, or cosmological importance to pre-Columbian agricultural societies in Ecuador.

Notes

i "*Albarradas*" which are small reservoirs were constructed to capture rainwater and to recharge the aquifer. "*Camellones*" constitute small elevations constructed in flood areas to prevent crops from being destroyed (Marcos et al. 2004, Veintimilla 2004).

ii *Spondylus crassisquana* Lamark, sp. 1819: synonymous with the thorny oyster *Spondylus princeps* Broderip, sp. 1833. The Strombus species associated with southern coastal Ecuador was reclassified to *Lobatus galeatus*, sp.: It is a gastropod of the Strombidae family.

iii www.museos.gob.ec/fondosculturales/GA-1-2264-82.

References

Andrews J (1996) The Peripatetic Chili Pepper: Diffusion of the Domesticated Capsicums since Columbus. In: Foster N, Cordell L (eds) Chilies to Chocolate. Food the Americas gave the World. The University of Arizona Press, pp 81–93

Barceló JA (1989) Arqueología Lógica y Estadística: un análisis de las Estelas de la Edad del Bronce en la Península Ibérica. Unpublished doctoral thesis, Universitat Autónoma de Barcelona

Byrd K (1976) Changing animal utilization patterns and their implications: Southwest Ecuador (6500 B.P.-AD. 1400). Unpublished PhD dissertation. Department of Anthropology, University of Florida, Gainesville

Chase T (1988) Restos Fáunicos. In: Stothert K (ed) La Prehistoria Temprana de la Península de Santa Elena, Ecuador: Cultura Las Vegas. Guayaquil, Ecuador, Museo Antropológico, pp 171–178

Cope FW (1979) *Theobroma cacao* (Sterculiaceae). In: Simmonds NW (ed) Evolution of crop plants. Longman, London, pp 285–289

Demarest A (2004) Ancient Maya: the rise and fall of the rainforest civilization. Cambridge University Press, Cambridge

Damp JE, Pearsall DM (1994) Early cotton from coastal Ecuador. Econ Bot 48:163–165

Damp JE, Pearsall DM, Kaplan LT (1981) Beans for Valdivia. Science 212:811–812

Doebley JF (1990) Molecular evidence and the evolution of maize. Econ Bot 44:6–27

Estrella E (1990) El Pan de América: Etnohistoria de los alimentos aborígenes en el Ecuador. Colección 500 años. Ediciones Abya-Yala, Quito. (1995) La Flora Huayaquilensis de Juan Tafalla. Crónica e Iconografía de una Expedición Silenciada (1799–1808). Banco del Progreso, Jardín Botánico de Guayaquil

García-Caputi M (2008) Las Figurinas de Real Alto, Reflejos de los Modos de Vida Valdivia. Abya-Yala, Quito, Ecuador
García-Caputi M (2016) La Figurina Como Reflejo de un Modo de Vida Valdivia: Cronología y Uso Social de la Figurina a Través de un Método Comparativo entre Colecciones. Abya-Yala, Quito, Ecuador
Gentry AH (1993) A field guide to the families and genera of woody plants of Northwest South America (Colombia, Ecuador, Peru) with supplementary notes on Herbaceous Taxa. Conservation International, Washington D.C.
Gutiérrez A (2009) Dioses, símbolos y alimentos en los Andes. Interrelación hombre-fauna en el Ecuador prehispánico. Abya-Yala, Quito-Ecuador
Gutiérrez A (2011) El Eje del Universo: Chamanes, sacerdotes y religiosidad en la cultura Jama Coaque del Ecuador Prehispánico. Museo de América, Ministerio de Cultura de España
Harlan JR (1992) Crops and man, 2nd edn. American Society of Agronomy and Crop Science Society of America, Madison, WI
Jorgensen PM, León-Sánchez (1999) Catalogue of the vascular plants of ecuador. Mongr Syst Bot Missouri Botanical Garden 75:i–viii, 1–1182
Kaplan L, Kaplan LN (1996) Beans of the Americas. In: Foster N, Cordell L (eds) Chilies to chocolate: food the Americas gave the world. The University of Arizona Press, Tucson, pp 61–79
Lanaud C, Solórzano RL, Zarrillo S, Valdez F (2012) Origen de la domesticación del cacao **y** su uso temprano en el Ecuador. Revista Nuestro Patrimonio. Ecuador, pp 12–14. https://palanda.arqueo-ecuatoriana.ec/origen-de-la-domesticacion
Lathrap D, Collier D, Chandra H (1975) Ancient Ecuador: culture, clay, and creativity, 3000–300 B.C. Field Museum of Natural History, Chicago
Marcos JG (1986) Breve Prehistoria del Ecuador. In *Arqueología de la Costa Ecuatoriana, Nuevos Enfoques, e*dited by Jorge G. Marcos. Biblioteca Ecuatoriana de Arqueología, vol 1. Corporación Editora Nacional, Quito, pp 25–84
Marcos JG, Álvarez SA, Bazurco M (2004) Introducción General y Lineamientos Finales. In: Marcos JG (ed) Las Albarradas en la Costa del Ecuador: Rescate del conocimiento ancestral del manejo sostenible de la biodiversidad. Proyecto Albarradas, CEAA/ESPOL. Guayaquil, Ecuador, pp 11–30
Ontaneda S, Valdez F (2016) Chamanes & Divinités de l'Équateur Précolombien. Catalogue Musée du Quai Branly, París
Pearsall DM (1979) The application of ethnobotanical techniques to the problem of subsistence in the Ecuadorian formative. Unpublished PhD dissertation, Department of Anthropology, University of Illinois, Champaign-Urbana
Pearsall DM (1988a) La Producción de alimentos en Real Alto. Biblioteca Ecuatoriana de Arqueología, vol 2, Corporación Editora Nacional, Quito
Pearsall DM (1988b) An overview of formative subsistence in Ecuador. Paleoethnobotanical data and perspectives. In: Kennedy BV, LeMoine GM (eds) Diet and subsistence: current archaeological perspectives. Proceedings of the nineteen annual chancmool conference. The University of Calagary Archaeological Association, Calagary, Canada, pp 149-164
Pearsall DM (1996) Reconstructing subsistence in the Lowland Tropics: a case study from the archaeology from the River Jama, Manabi, Ecuador. In: Reitz EJ, Newsom L, Scudery SJ (eds) Case studies in environmental archaeology. Plenum Press, New York, pp 223–254
Pearsall DM (1997) Agricultural evolution and the emergence of Formative societies in Ecuador. In: Blacke M (ed) (in press) Development of agriculture and emergence of formative civilizations in Pacific Central and South America. Washington State Press Univ. Press, Seatle
Pearsall DM (2004) Plants and people in ancient Ecuador: the ethnobotany of the Jama River Valley. Case Studies in Archaeology. Jeffrey Quilter Series Editor. Wadsword/Thomson, Canada
Pearsall DM (2015) Paleoethnobotany: a handbook of procedures, 3rd edn. Left Coast Press, Walnut Creek California
Pearsall DM, Chandler-Ezell K, Zeidler JA (2004) Maize in ancient Ecuador: results of residue analysis of stone tools from the Real Alto Site. J Archaeol Sci 31:423-442

Pearsall DM, Duncan N, Jones J, Friedel D, Veintimilla C, Neff H (2016) Human-environment interactions during the early mid-Holocene in coastal Ecuador as revealed by mangrove coring in Santa Elena Province. Holocene J: 1–28

Pikersgill B (1984) Migrations of Chile peppers, *Capsicum* spp. in the Americas. In: Stone s (ed) South and Central America: pre-Columbian plant migration. Papers of the Peabody Museum of Archaeology and Ethnology. Harvard University Press, Cambridge, MA, pp 105–123

Piperno DR, Pearsall DM (1998) The origins of agriculture in the Lowland Neotropics. Academic Press, San Diego

Piperno DR, Stothert K (2003) Phytolith evidence for Early Holocene Cucurbita domestication in Southwest of Ecuador. Science 299:1054–1057

Plowman T (1984) The origins, evolution, and diffusion of coca, *Erythroxylum* spp. in South and Central America. In: Stone S (ed) Pre-Columbian plant migration. Papers of the Peabody Museum of Archaeology and Ethnology. Harvard University Press, Cambridge, MA, pp 124–163

Reichel-Dolmatoff G (1978) El Chamán y el Jaguar: Estudio de las drogas narcóticas entre los indios de Colombia. Siglo 21 editores, Colombia

Schultes RE (1972) An overview of Hallucinogens in the Western Hemisphere. In: Furst PT (ed) Flesh of the gods: the ritual use of Hallucinogens. Praeger Publishers, New York, pp 3–53

Stahl PW (1985) Native American cosmology in archaeological interpretation: tropical forest cosmology and the Early Valdivia Phase at Loma Alta. In: Thompson M, Garcia MT, Kense FJ (eds) Status, structure and stratification: current archaeological reconstructions. Proceedings of the sixteenth annual conference. The Archaeological Association of the University of Calgary, pp 31–37

Staller JE (2000) Figurinas Valdivia VII-VIII del sitio de San Lorenzo del Mate, provincia de Guayas, Ecuador y la transición Valdivia-Machalilla. Miscelánea Antropológica Ecuatoriana 9:99–133. Revista del Museo Antropológico del Banco Central del Ecuador, Guayaquil

Staller JE (2001) Documentando la transición Valdivia-Machalilla con figurinas Valdivia VII-VIII del San Lorenzo del Mate, provincia de Guayas, Ecuador. Cuadernos de Historia y Arqueología, Nos 51–53:93–179. Publicación de la Casa la Cultura Ecuatoriana. Benjamín Carrión, Núcleo del Guayas. Guayaquil

Staller JE (2006) The social, symbolic and economic significance of *Zea mays* L. in the Late Horizon Period. In: Staller JE, Tykot RH, Benz BF (eds) Histories of maize: multidisciplinary approaches to the prehistory, linguistics, biogeography, domestication, and evolution of maize. Taylor & Francis Group, NY, Routledge, pp 449–467

Staller JE (2010) Maize cobs and cultures: history of *Zea mays* L. Springer, Berlin, Heidelberg

Staller JE (2013) Ancient El Niño Events, human adaptation, and ecological transformations: early formative period (2400–1450 B.C.) Occupations in Southern Coastal Ecuador. Dialogo Andino, vol 41. Universidad de Tarapacá, Chile, pp 101–132

Staller JE (2014) Lightning and it manifestations: *Huacas* and *Ushnus*. In: Meddens F, McEwan C, Willis K, Branch N (eds) Inca sacred space: landscape, site and symbol in the Andes. Archetype Publications, London, pp 177–186

Staller JE (2015a) Maize. In: Metheny KB, Beaudry MC (eds) The archaeology of food: an encyclopedia. Alta Mira Press, New York. Imprint of Rowland & Littlefield, pp 296–298

Staller JE (2015b) Plant domestication. In: Metheny KB, Beaudry MC (eds) The archaeology of food: an encyclopedia. Alta Mira Press, New York. Imprint of Rowland & Littlefield, pp 407–409

Staller JE (2015c) Geological and archaeological evidence of El Niño Events along the Coast of El Oro Province Ecuador: Excavations at La Emerenciana, a Late Valdivia (ca. 2200 1450 B.C.) Ceremonial Center. Maskana 6(2):155–186. Universidad de Cuenca, Ecuador

Staller JE (2016) High altitude maize (*Zea mays* L.) cultivation and endemism in the Lake Titicaca Basin. J Bot Res 1(1):8–21

Stothert KE (1988) La Prehistoria Temprana de la Península de Santa Elena, Ecuador: Cultura Las Vegas. Miscelánea Antropológica Ecuatoriana, Serie Monográfica 10, Museo del Banco Central del Ecuador, Guayaquil

Stothert KE, Piperno DR, Andrés TC (2003) Terminal pleistocene/early holocene: human adaptation in coastal Ecuador: the Las Vegas evidence. Quatern Int 109(110):23–43

Valverde FDM (2004) Situación de la biodiversidad en las albarradas y sectores colindantes de la planicie costera de las Provincias de Guayas y Manabí. In: Marcos JG (ed) Las Albarradas en la Costa del Ecuador: Rescate del conocimiento ancestral del manejo sostenible de la biodiversidad. Guayaquil, Ecuador, CEAA/ESPOL, pp 127–192

Valdez F (2019) Evidencias arqueológicas del uso social del cacao en la Alta Amazonía. Revista de Historia, Patrimonio, Arqueología y Antropología Americana, Julio 1:117–134

Veintimilla Bustamante CI (2004) Plantas Silvestres y Cultivadas de las Albarradas de la Península de Santa Elena: Aproximación Ambiental y Cultural (Componente Paleo-Etnobotánico). In: Marcos JG (ed) Las Albarradas en la Costa del Ecuador: Rescate del conocimiento ancestral del manejo sostenible de la biodiversidad. CEAA/ESPOL, Guayaquil, Ecuador, pp 193–252

Veintimilla Bustamante CI (2013) Ruinas de una "ciudad" Prehispánica de los antiguos Manteños establecidos en los cerros Hojas – Jaboncillo, Manabí central: Análisis espacial y jerarquía social. *La Sociedad Prehispánica Manteña en los cerros Hojas – Jaboncillo*. Centro Cívico Ciudad Alfaro. Boletín Arqueológico, no. 1:53–157

Veintimilla Bustamante CI (2016) Las Plantas del Pasado a través del Arte Prehispánico. Research report, Museo Antropológico y Arte Contemporáneo (MAAC), Ministerio de Cultura y Patrimonio del Ecuador. www.museos.gob.ec/maac/investigación

Weinstein E (1996) The Serpent's children: the iconography of the Late Formative Ceramics of coastal Ecuador. Unpublished PhD dissertation, Department of Anthropology, University of Toronto, Canada

West JA (1996) A brief history and botany of Cacao. In: Foster N, Cordell LS (eds) Chilies to chocolate. Food the Americas gave the World. The University of Arizona Press, pp 105–121

Wing ES (1988) Dusicyon sechurae en contextos arqueológicos tempranos. In: Stothert K (ed) Cultura Las Vegas. Guayaquil, Ecuador. Miscelanea Antropológica Ecuatoriana Serie Monográfica vol 10. Museos del Banco Central del Ecuador, Guayaquil, pp 179–185

Yacovleff E, Herrera FL (1935) El Mundo Vegetal de los Antiguos Peruanos. Revista del Museo Nacional de Lima, Perú

Zeidler JA, Pearsall DM (1994) The Jama Valley Archaeological-Paleoethnobotanical project: an introduction. In: Zeidler JA, Pearsall DM (eds) Regional archaeology in Northern Manabí, Ecuador, vol 1. Environment, cultural chronology, and prehistoric subsistence in the Jama River Valley. University of Pittsburgh Memoirs in Latin American Archaeology, no 8. Pittsburgh, pp 1–11

Chapter 9
The *Achumera*: Gender, Status, and the San Pedro Cactus in Moche Ceramic Motifs and Iconography

Sarahh Scher

Abstract The Moche of north coastal Peru existed from approximately 200–900 as a series of culturally affiliated but politically divided areas. Their art is often quite naturalistic, representing a wide number of subjects. It does not, however, have a great deal of imagery relating directly to entheogenic plants. While there are definitely classes of artistic expression that refer to the experience of trance and spiritual transformation that is often the result of entheogen use explicit representation of entheogenic plants is rare. One type of representation that does seem to depict an entheogen is that of figures, called *achumeras*, depicted in ceramic vessels holding pieces of *Echinopsis pachanoi*, the San Pedro cactus. These women appear to occupy a space in society that considers them of medium status and of minimal gender, allowing these figures to comment not only on entheogen use among the Moche, but the complex role of gender and status in when and how the entheogen was used, how the plant itself was represented, and the representation of useful plants (as food and as spiritual catalyst) in Moche art.

Introduction

The Moche culture of north coastal Peru existed from approximately 200–900 C.E. as a series of culturally affiliated but politically separate areas. Moche iconography is often quite naturalistic, and represents a wide number of subjects, including humans, the natural world, and supernatural beings. The representation of the mescaline-bearing *Echinopsis pachanoi*[i] (and perhaps other species of the genus *Echinopsis*, see below) in Moche art indicates the plant was used as an entheogen, that is a medicinal or psychotropic plant, used for healing or obtaining altered states of consciousness to access the spiritual realms. The way in which the plant is represented often highlights social categories of gender and status that intersected with the ritual use of the plant. This research is specifically focused upon ceramic representations from the

S. Scher (✉)
Salem State University, Salem, USA
e-mail: sscher@salemstate.edu

J. E. Staller (ed.), *Andean Foodways*, The Latin American Studies Book Series,
https://doi.org/10.1007/978-3-030-51629-1_9

southern Moche area of female figures holding the cactus, apparently in a trance state induced by its psychotropic properties, in this case mescaline, as well its broader symbolic connections to healing and transformation. The figures are often referred to as *achumeras*, based on a contemporary folk name for the cactus, *achuma* or *huachuma*, although the plant is more widely known as the San Pedro cactus.

The overall depiction of plant use in Moche iconography will be considered, before examining the importance of *E. pachanoi* in ethnographic and ancient cultures of Peru. With this as a foregrounding, the gender, role, and status of the *achumeras* are examined relative to other depictions of status and role in Moche imagery. While not consumed as food, *E. pachanoi* was part of a wider Andean use of plants for both physical and spiritual sustenance.

Crops, Plants, and Entheogens in Moche Iconography

The repertoire of depicted plants in Moche iconography is limited, and most are used either to indicate location (such as *Tillandsia* sp. signifying the hilly landscape up-valley from the desert coast) or role in ritual activities such as feasting or divination (maize and beans are often depicted in these representations).[ii] The limited nature of these depictions which often exclude imagery, related to the tending of cultivated plants, is common to Moche iconography and ceramics. They are often focused on high-status individuals, objects, locations, and activities, framing representation in the guise of everyday life. Agriculture and cultivation (i.e., preparing the soil, planting, watering, weeding, etc.,) are not depicted in Moche motifs; instead a figure is portrayed who displays harvested plants, usually a stalk of maize in one hand and a bunch of tubers, probably yuca, (*Manihot esculenta* L.) in the other (Fig. 9.1). The result of agricultural success is shown in symbolic form, eliding the labor and the laborers required to produce the crops and instead focusing on the symbolic nature of the end products and the religious and political importance of agriculture.

The "harvester" figure is dressed in clothing common to higher status males: short, sleeveless tunics, frequently with broad necklaces overlapping them, and a kilt. Ear spools and an elaborate headdress complete the costume, with the headdress sometimes incorporating a crescent element that resembles a *tumi*, or ritual knife used in sacrificial ceremonies. Other headdresses reference the sea, with two projections similar to a V-shaped projection often seen on fishermen's headdresses.

A rare textile uncovered at the site of El Castillo in the Santa Valley (Chapdelaine and Pimentel 2003, Figures 37–38) depicts the yuca harvest, again featuring symbolic masculine figures in high-status clothing, including earspools, wrist ornaments, and animal-figure headdresses with avian and feline or vulpine attributes. Each of the main figures holds a stylized bunch of yuca tubers, and the central figure also holds a long object representing a digging stick. These examples emphasize the highly artificial way plants are represented in Moche motifs and iconography—not in terms of a lack of naturalistic representation, but how the motifs elide the manner in which

Fig. 9.1 Moche vessel depicting a "harvester" figure in relief (courtesy of the Museo Larco, Lima, Peru)

such plants were integrated into daily traditional food and the labor required to produce them. A similar elision occurs with animal husbandry.

Imagery of entheogenic plants in Moche ceramic depictions is equally bounded by convention and a reluctance to show mundane tasks such as plant collection. While there are definitely classes of artistic expression that refer to the experience of trance and of spiritual transformation that is often the result of entheogen use (Stone 2011: 193–203), the explicit representation of entire entheogenic plants is rare. Along with crowns of *Echinopsis* columns, the Moche also depict figures holding strings of *espingo* seeds (*Nectandra* spp.), which may have been combined with the leaves of jimsonweed (*Datura stramonium* L.) in order to induce narcosis and freedom from pain in sacrificial victims (Carod-Artal and Vázquez-Cabrera 2007: 48). The seeds are not shown in use, and rather are depicted as necklaces or on strands held by standing figures. These two plants appear to be the only ones with possible entheogenic use depicted in Moche art—*Datura stramonium* is not depicted,[iii] although its use is not established (Bourget 1994: 73). *Echinopsis* is the only one shown in use, and it is still in use as an entheogen in modern Peru, sometimes with *Datura*. *Nectandra* is used as a folk remedy for stomach ailments.

San Pedro Cactus (*Echinopsis* spp.) in Peru

Most discussions of entheogen use and representation in both the Andean past and present center on the San Pedro Cactus (*Echinopsis pachanoi*)—a tall, branching, ribbed columnar cactus that most often lacks spines, or on occasion has underdeveloped ones. It grows up to six meters in height and produces edible fruits. The species is likely a domesticate (Feldman Gracia 2011: 149). However, there are other members of the genus *Echinopsis* in the Andes. The wild *Echinopsis* all have spines, some of them quite large. *E. peruvianus*, *E. cuzcoensis*, *E. bridgesii*, and *E. terscheckii* are all called *achuma* or *wachuma* in contemporary Peru. Many of the *Echinopsis* species produce white flowers which bloom at night. This nocturnal flowering was seen as signifying the plant's connection to the world of night, darkness, and the supernatural (Schultes and Hoffman 1979: 169). All species of *Echinopsis* contain mescaline in the layer just inside its skin, in a concentration of approximately12%. However, *Echinopsis. peruvianus* has a particularly high concentration of mescaline (Feldman Gracia 2011: 147, 154). Colonial and contemporary practices indicate that the plant can be ingested raw or boiled into a thick, mucilaginous liquid. Both methods would yield wildly varying doses depending upon amounts ingested and method of preparation (Davis 1983: 375). In sufficient dosage, the mescaline acts as a hallucinogen, producing entheogenic visions. However, there are other important chemicals in the cactus that create a sedative effect that presages the full effect of the mescaline, which can include the sensation of flying, a heightening of the senses, and synesthesia (Feldman Gracia 2011: 154–156). Modern healers often includes other plants to the liquid mescaline as it is cooking down, intensifying its hallucinogenic effects (Schultes and Hoffman 1979: 168). It should be noted that another cactus, *Armatocereus laetus*, is a similarly tall, branching cactus that in modern practice can be used as a substitute for *Echinopsis*, however it is quite localized, and goes by the name *pishcol* rather than *achuma* (Davis 1983: 378).

Echinopsis sp. in Ancient Andean Imagery and Iconography

Evidence of the importance of the San Pedro Cactus to Andean culture has its origins long before the Early Intermediate Period and continues to the present. Guitarrero cave, which yielded the oldest textiles known in the Americas, also provided the earliest evidence for *E. peruvianus*, dated to between 8600 and 5600 B.C. However, the extent to which it was used as an entheogen is unknown (Lynch 1980: 101).

Many cultures from the Andean area depicted the *Echinopsis* cactus, sometimes in more detail and with more frequency than the Moche, as detailed below. However, cultures that depicted the cactus did not show the plant as ubiquitous. The nature of the cactus, and its association with a restricted area of human activity, meant that it was only represented within contexts related to shamanic visions and rituals of use. More quotidian plants, such as maize, beans, tubers, and fruits were more frequently represented. *Echinopsis pachanoi* appears in the highland Chavín (c. 900–200 B.C.) culture, associated with shamanic[iv] transformation into supernatural beings (Burger 1992: 134–35). The relief panels in the sunken circular court at Chavín de Huántar depict marching figures with human and animal features (Fig. 9.2). A tall section of cactus is held by a figure with both human and feline attributes, including protruding fangs and eyes with pendant irises. Other animal attributes include talons and feet that may be feline or avian, and hair represented as intertwining snakes. The figure's belt also terminates in snake heads, conveying a mysterious, powerful supernatural being. Beneath the panels depicting these cactus-bearing figures, other slabs of carved stone depict highly stylized jaguars crouching, with spiraling forms emerging from their bodies. The spirals may symbolize the swirling geometric shapes that presage the full effect of visual hallucinations induced by mescaline or the spiritual power emanating from the giant feline. A Chavín-style painted textile found in the south coast of Peru depicts the cactus, with serpent heads for roots, surrounded by hummingbirds and felines (Sharon and Donnan 1977: 378). The two animals are associated in modern shamanic practice with attributes and alter egos of the shaman. The snake roots are similar in style to the snake hair and belt on the Chavín de Huántar cactus bearers.

On the north coast of Peru, Cupisnique (c. 1400–500 B.C.) sculpture and ceramic vessels depict stands of *Echinopsis*, with a spotted feline (usually identified as a jaguar) crouching among them (Fig. 9.2). Often the cactus columns terminate in

Fig. 9.2 Cupisnique ceramic vessel depicting a serpent, feline, and *Echinopsis* (courtesy of Metropolitan Museum of Art (www.metmuseum.org) by OASC)

scrolled forms, similar to those depicted at Chavín de Huántar, which Sharon interprets as the whirling forms seen during shamanic visions (Sharon and Donnan 1977: 378–379). The cactus columns stand in for the hallucinatory experiences they bring, and the jaguar may be a representation of the transformed shaman who has used the cactus to travel into the spirit world (Cordy-Collins 1998: 167). Figure 9.2 depicts a crouching jaguar with a serpent emerging overhead, and to either side *Echinopsis* columns flanking a wave-and step form that symbolizes the mountains and the sea, while also echoing the independent swirling form seen in other representations. Jaguars and serpents are both common elements in contemporary shamanic visions (Stone 2011: 39–44), and indeed the pelage marks of the jaguar and the patterning of the serpent are rendered in the same circle-within-a-circle design, perhaps indicating this symbolic relationship. The wave-and-step is nearly ubiquitous in Andean artistic traditions, and frequently symbolizes the duality of northern and central Andean geography, which is precipitous near the coast. The coastal plain is extremely narrow, only 20 km in some regions, and widens to only 50 km in other coastal regions. Traveling from the coast over the cordillera to the eastern highlands and into the Amazonian jungle is only a distance of 200 km in some regions. In addition, rivers originating in the nearby mountains are the source of fresh water necessary to sustain agriculture on the coast. The worlds of low/high and dry/wet exist in an interrelated set of hydrological and ecological cycles. Andean shamans usually mediate the spiritual forces between these worlds and assuring their productive interaction. The *Echinopsis* columns flanking the wave-and-step visually function as a literal gateway to the symbolic power of the natural world. In this fashion, Cupisnique motifs and iconography represent the use and importance of *Echinopsis* and its importance to visionary religious practices within Moche culture.

The Nasca were roughly contemporary with the Moche. They were located on the southern coast of modern-day Peru, where the Andes retreat somewhat from the coast and the desert plain extends further to the east. Nasca iconography and ceramic motifs are often abstracted, and incorporate colors, patterning and repetition in a style that becomes increasingly "busy" throughout the culture's existence, this stands in contrast to Moche ceramic representations and imagery which is bi-or tri-chromatic and much more naturalistic. A Nasca ceramic vessel representing a mummy bundle has two columns of *Echinopsis* protruding from the shoulders, as well as star-like spine clusters decorating the arms and legs (Fig. 9.3). The prominent depiction of the spines may indicate that the cactus represented is not *E. pachanoi*, but rather *E. peruvianus*. The rest of the bundle is adorned with imagery related to supernatural creatures, ritual specialists, and the prominence of ritual or trophy heads in Nasca culture. Other vessels depict cacti that appear to be *Echinopsis* either by depicting the columnar forms or the ribbed structure. It has also been suggested (Proulx 2006: 167) that a star-like shape found in Nasca pottery decoration is a depiction of the cross-section of the plant.

The Lambayeque culture, which flourished in parts of the former Moche sphere from 700–1400, would depict a figure similar to the *achumeras* of Moche: seated female figures with closed eyes, holding a crown of *Echinopsis* (Fig. 9.4a). The right hand of the woman is held in a manner that may suggest consumption of a piece

Fig. 9.3 Nasca ceramic vessel depicting a mummy bundle. shoulders. Tello, *Paracas* LXXXI (1959) (drawing by Sarahh Scher)

Fig. 9.4 Left: Lambayeque vessel similar to Moche *achumera*. Right: Moche *achumera* vessel (courtesy of Museo Larco, Lima, Peru)

of the raw plant. This is one of many visual themes that would be retained by the cultures that followed the collapse of the Moche on the north coast.

Echinopsis and *Achumeras*

The most frequent Moche representations depicting *Echinopsis* are figures dressed as women, called *achumeras* after the modern folk name, represented in both three- and two-dimensional forms on ceramic vessels.[v] These women appear to occupy a space in society that considers them of medium status and of minimal gender, implying use of the cactus in rituals centered on healing individuals rather than in state-level rituals. Gender is generally regarded as the social identity built upon biological sex. That is, a set of ascribed qualities which are made visible through conventions of appearance, and repeatedly acted out throughout a lifetime.[vi] Gender in Moche ceramic representations is communicated through a system of costume which pair masculinity and maleness with loincloths and the accoutrements of warrior status. They are also depicted by a spectrum of head coverings ranging from simple kerchiefs to sculptural headdresses. Femininity and femaleness are depicted with long tunics which generally cover the body to the knees or further, an absence of a head coverings, and long hair. The more a figure strays from normative masculine and feminine expression, the more gender becomes indefinite, often exhibiting a blended male and female gender through their dress, or an ambiguity that moves away from sex and gender altogether (Scher 2010). While there seem to be more gender-complicated men than women (that is, individuals with a certain amount of gender ambiguity but who are somewhat marked with aspects of masculine identity) in Moche art, there are representations that appear to depict gender-complicated figures who are nominally women, seemingly associated with private rather than public and official spiritual practices.

Achumeras are depicted seated, wearing a long tunic. Their heads are usually covered by a mantle and they hold the crown of an *Echinopsis* cactus in an outstretched hand (Fig. 9.4b). As the cactus is not shown with any indications of spines, it is assumed the plant will be recognized as *E. pachanoi*. Most of the *achumeras* make up the full body of the vessel, while others are seated on top of larger rounded vessel bodies, and a few are in two-dimensional painted images. The fact that the depiction is usually of just the crown of the cactus can often make identification difficult. More detailed versions of this motif clarify the identification of the plant, while in other representations the crown it is rendered as a series of small bumps. Some scholars interpret these as representations of seeds or stones (Franco Jordán 2012: 22; Sánchez Davis 2012: 29). In contemporary folk practice, the crowns of the cactus are considered the most potent part.

In modern shamanic practice on the north coast of Peru, a four-ribbed cactus is believed to be particularly potent, perhaps because of its cosmological significance: the four ribs echoing the four quarters of the world, while its columnar form reaches to heaven above and the earth below as a kind of *axis mundi* (Sharon 2000: 13).

According to Alana Cordy-Collins, this belief exists despite no four-ribbed San Pedro ever being found, with five to seven ribs being most common (Cordy-Collins 1998: 168). While many of the representations of *achumeras* depict the object in their hand as having bumps similar to the crown of the cactus, the number is not always distinguishable. There is one piece from Huaca de la Luna, which clearly depicts four distinct bumps (Figs. 9.5 and 9.6). However, another from this site depicts three. It is possible that the number of ribs, and subsequently the number of bumps in the crown of the cactus, was not seen as visually important information. Rather, the crown itself seems to have been important to depict, as part of the larger designation of the figure as a user of the cactus.

Typically, *achumeras* have their eyes closed or covered, in an allusion to the trance induced by the San Pedro, as well as a possible reference to photophobia associated with mescaline usage, which has been documented in contemporary ethnomedical research by De Feo (2003: 251). Two ceramic vessels excavated from the Huaca de la Luna, a monumental site in the Moche heartland, (see Figs. 9.5 and 9.6), depict *achumeras* with a visor-like piece covering the eyes completely, an unusual addition

Fig. 9.5 Moche vessel excavated from Huaca de la Luna, depicting a type of *achumera,* with details of the *achumera* figures painted on the sides of the vessel (courtesy of the Phoebe A. Hearst Museum of Anthropology and the Regents of the University of California, photo by Sarahh Scher, 4–2945)

Fig. 9.6 *Achumera* vessel from Huaca de la Luna, with a similar eye covering to that seen in the vessel in Fig. 10 (courtesy of the Proyecto Arqueológico Huaca de la Luna, photo by Sarahh Scher)

to their costume not found elsewhere, but which may reinforce the notion of light avoidance in San Pedro users. The visor-like piece of Fig. 9.6 appears to be part of another accessory that is affixed under the mantle. The sculpted figure in Fig. 9.5 may indicate the same, it is also possible the dip in surface is meant to indicate where two differing materials meet and join.[vii] The two-dimensional versions on the sides of the vessel do not indicate such level changes, thus offering considerably less visual information. A notable ceramic vessel from Huaca Cao Viejo, a Moche site in the Chicama valley adjacent to the Moche Valley and in close interaction with the site of Huaca de la Luna, shows another female figure engulfed in the mantle that drapes her head (Fig. 9.7, left). The mantle is drawn far past the forehead and heavily shades her eyes, similar to the visors in the Huaca de la Luna pieces.

A representation similar to the *achumeras* from Huaca de la Luna (Fig. 9.7, center), is broken from a missing ceramic vessel, depicting a figure in a similar pose, and more importantly wearing a similar costume, although no San Pedro is shown. Its face is skeletal, with empty eye sockets, visible teeth, and a step-shaped form on the side of the face. This step may be a stylization of the zygomatic process of the skull as it crosses the ramus of the jaw. However, even if it is a stylization, the convention in Moche art was to reduce it to a step form, a symbol of power (Benson 1975: 119). While not exactly the same as the *achumeras*, the figure is similar enough to suggest a possible connection between the *achumeras* and the world of the dead, a spirit realm is reasonable to assume the *achumeras* would visit in their trances.

Fig. 9.7 Left: Moche *achumera* vessel excavated at Huaca Cao Viejo. Drawing by Sarahh Scher. Center: Moche vessel fragment depicting a skeletal figure similar to an *achumera* (courtesy of the Proyecto Arqueológico Huaca de la Luna, photo by Sarahh Scher). Right: Moche vessel depicting a standing male figure with loincloth, headdress, and a mantle over the head (courtesy of Museo Larco, Lima, Peru)

Achumera Costume, Gender, and Social Roles

The *achumera* figures appear to be nominally female as suggested by their long tunics. Contrary to more normative depictions of women, their hair is not exposed. The simple piece of fabric that *achumeras* wear as a head covering, however, is distinct from coverings associated with masculine gender. Rather it is a large, seemingly rectangular piece of fabric, usually called a mantle, draped rather than tied over the head, and not structured in any way. This mantle usually extends to the ground, covering the entire back and sometimes almost cocooning the body. A mantle draped over the head is a costume element that is closely associated with some types of ritual performance, but, significantly, not associated with polity-level public rituals.

The majority of female Moche depictions are not associated with head coverings, and visor-like elements have not been documented outside of the few identified from Huaca de la Luna. Rather, women's exposed hair, often braided, is seen as an important marker of feminine identity (Hocquenghem and Lyon 1980; Lyon 1978). Covering the hair and removing it from visual consideration may indicate this is a presentation of an older woman. Exposed, lustrous hair and braids are seen in contemporary Andean culture as a sign of youthful sexuality and an ability to bear children (Isbell 1997). The possible implications of the *achumeras* being older women are addressed below.

In a few depictions, such as Figs. 9.5 and 9.6 above, the *achumeras* wear earspools, which are a rare adornment for women, an indication of high status, and a complication of their gender that gives them masculine traits (Scher 2017). While it is possible that this is meant to be a male figure in women's dress, or simply wearing a long tunic, as seen elsewhere (e.g., Kutscher 1983: Figures 155-57, 306-7), it would be highly unusual for a man with such high-status jewelry to not only be wearing a woman's tunic, but to also be without an elaborate headdress, either a warrior's helmet or a sculptural headdress. Male figures in long tunics without such jewelry are associated with plainer head coverings, but none are like the draped mantle. Figures in masculine clothing, including a loincloths and short tunics, wearing typical male headdresses underneath a draped mantle, are well known (Fig. 9.7, right), but these versions retain the obvious sartorial markers of masculinity. If a man in a long tunic were intending to adopt a feminine appearance, the earspools would be such a contradiction of that identity that it is difficult to think that they would not be removed. However, it is possible such figures are meant to be deliberately ambiguous, covered with symbols of both masculine and feminine identities without strongly signifying either. Since masculinity in Moche art can be so strongly expressed through costume elements (in images of warriors, rulers, and important deities), the appearance of ambiguity should at the very least be regarded as a move away from the masculine side of the gender spectrum, and toward the feminine.

There is a further clue that *achumeras* are intended to be seen as nominally female. Moche ceramic vessels from the Santa valley (Fig. 9.8, left), depicts an *achumera* with a belt as well as *ticpis* (wedge-shaped garment pins) at the shoulders. The plain belt is often part of normal Moche female attire, but the *ticpis* indicate a clothing tradition associated with female costume in the highlands to the east of the Moche region (Scher 2011). This would seem to indicate that all or most *achumeras* have

Fig. 9.8 Left: Moche *achumera* vessel, possibly from the Santa Valley (courtesy Museo Larco, Lima, Peru). Right: Moche vessel from Huaca de la Luna, depicting a figure similar to the *achumeras* who may be male (courtesy of the Proyecto Arqueológico Huaca de la Luna, photo by Sarahh Scher)

feminine aspects, although some may be less explicitly feminine than others. It also seems to reinforce the idea that small-scale, private, or semi-private religious practices were the context in which these figures were engaging in trance work, a topic I will return to.

Some costume elements worn by the *achumeras* are not exclusively worn by these figures alone, and the fact that they are seen on other actors can be seen as part of the gender ambiguity of the type. There is another Huaca de la Luna piece (Fig. 9.8, right) which depicts a figure in similar garments to those of *achumeras*, but who appears to be intended as masculine. This figure also wears a head covering with a visor-like projection but holds a set of rattle-bells in the right hand instead of an *E. pachanoi* crown. The figure has a necklace of large round beads and earspools, and a long tunic with a complex step design. This figure is less associated with female traits than the *achumeras*. The rattle-bells instead of cactus in the figure's hand may indicate that it is associated with other male figures wearing loincloths holding the same kind of instrument. In addition, the structure of the head covering is different from that of the other visor-wearing figures. Rather than a shapeless, draped mantle with an affixed projection, it is a cap-like construction, with the "visor" being merely an extension of the flap that extends behind. In the back, a flap has clearly been denoted in slip, with a horizontal line demarcating the transition between cap and flap. In addition, there is a definite change in surface level between the figure's neck and the flap. In front of the figure, painted on the vessel body, a stirrup spout vessel with a wave or spiral design and a beaker-shaped vessel decorated with light and dark horizontal bands and a volute design (likely meant to indicate a *florero*, or flaring bowl). The vessels' presence may relate to the ceremonies being performed by the figure, but these vessel forms are not specifically associated with entheogen use and appear in many ritual contexts. The spiral and volute designs are echoed in the design on the visor. Lamentably, this design on the flap running down the back of the figure has eroded away. It is tempting to see the curled shapes on the visor as similar in meaning to the swirling forms associated with *E. pachanoi* in Cupisnique art, but this would only be speculation, especially since the more female figures do not have similar patterning.

Overall, the piece in Fig. 9.8, right seems to indicate someone who is nominally male, albeit wearing a long tunic, and a variation on the cap and flap head covering that has associations with maleness, with an extended shape that covers the eyes. However, covering of the eyes and the rattle-bells maintain an association with spiritual ritual. Rattles are essential for contemporary shamanic ritual practices on the north coast (Sharon and Donnan 1974: 53), and ethnohistoric documents state musical instruments were common to native ritual (Polía 1999: 97: 319, 415). Figures holding rattles and rattle-bells in Moche art are often depicted with their eyes closed, further associating these figures with trance ritual similar to that of the *achumeras* (Benson 1974: 30; Glass-Coffin 1999: 207), and connecting the visor-type coverings for the eyes in the Huaca de la Luna pieces with this wider iconographic depiction that signals seeing into the spirit world.

The costume elements and attributes exhibited by these pieces—heads covered with mantles or visor headdresses, earspools without further clear indication of masculine gender such as loincloths, and *E. pachanoi* crowns, rattle-bells, or references to the land of the dead—point to gender complication, and an elevated level of status from the baseline but well below the highest elites, and participation in rituals. The unusual incorporation of the visor in pieces from Huaca de la Luna may indicate a variation of *E. pachanoi* usage is unique to this site. The rituals in which these figures participate center around vision trances, as evidenced by the closed or covered eye imagery, and appear to be related to representations that explicitly depict healing, which is also associated with visionary experience. An example from El Brujo published by Mujica Barreda et al. (2007: 223) depicts a bareheaded woman holding a small child being touched by a figure in the same costume as the *achumeras*: a long tunic and a draped mantle. In this case, the mantle wraps around to the front of the figure, and her eyes as well as the eyes of both the mother and child are closed. A frequently published piece in the collection of the Peabody Museum of Archaeology and Ethnology (Fig. 9.9) depicts a healer transformed or transforming into an owl. She wears the familiar long tunic and mantle over her head and may be holding a crown of *E. pachanoi* in her left hand (the protrusion is damaged). Three-dimensional depictions show an open box with round objects inside, a *florero,* and a fanged supernatural figure lying underneath a blanket. Two-dimensional elements such as a warrior's costume, strings of *Nectandra* seeds, a bowl with conical objects, and a small wrapped rectangular object lie on the vessel's upper surface; on the sides a llama is tied to a stake, and an attendant sits on a platform, looking towards the llama and a set of four jars (e.g., Kutscher 1983: 305). All of the elements depicted indicate associations with transformation, ritual, and the land of the dead (the fanged

Fig. 9.9 Moche ceramic vessel depicting an owl shaman engaged in a ritual (courtesy of Peabody Museum of Archaeology and Ethnology, Gift of Friends of the Museum, 1916. © President and Fellows of Harvard College, PM 16-62-30/F728)

figure is often interpreted as either sick or deceased, and the llama and jars suggest either a sacrifice, a burial, or both).

Entheogens, Gender, and Status

In many cultures that practice trance states and vision healing through the use of entheogens, women do not participate fully in these activities until after menopause, as the stresses of the entheogens, and lengthy rituals in and of themselves, can be harmful to a pregnancy (Schultes and Hoffman 1979: 62–64). As women pass beyond the child-bearing years they become, as a social category, less differentiated from males through the cessation of menses (Brown 1982: 148). It is possible that this moving toward an undifferentiated center is what we are seeing with the *achumeras*, and would explain why braids, a marker of women in Moche sexual representations and an element of youth and sexual fitness in contemporary cultures (Isbell 1997; Hocquenghem and Lyon 1980) are not shown, and the hair is covered altogether.

Costume can also indicate status of *achumeras*. In general, Moche representations favor plain, undecorated textiles, but the depiction of textile decoration is associated with an elevation of status above the baseline. In addition, jewelry and other ornaments such as headdresses further elaborate the status and social role of an individual. *Achumeras'* mantles sometimes display textile patterns, which can be indicative, along with the earspools some wear, of elevated status. Among the mantles with decoration, a pattern of repeating pairs of horizontal lines is dominant. In the Andean tradition there seems to be an association of horizontally oriented textile designs with women and vertical designs with men (Conklin 1997: 127). Some of the *achumeras*' long tunics also have designs, although there appears to be no clear preference for a particular style. Most wear neck ornaments such as collars or necklaces, which indicate somewhat elevated status. While no individual *achumera* is heavily adorned, as a group they display elements of elevated status. The metal plating on the eye covering in Fig. 9.5 is normally seen on (male) figures of extremely high rank or status, but in much larger quantities, usually entire garments or even suites of garments. The presence of earspools, which is not only associated with high status, but also the masculinization of females, is otherwise associated only with women performing in a central ritual of Moche ideology (Scher 2017: 470–471).

The ritual, known as the Sacrifice Ceremony, involved the display and exsanguination of prisoners, whose blood was then consumed by priestly figures who may have been acting as representatives or avatars of Moche gods. The relationship between *E. pachanoi* and the public, polity-level nature of the Sacrifice Ceremony is not well understood. Véronique Wright has found that mural paintings in several large public spaces at the Huaca de la Luna contained *E. pachanoi* sap as a pigment binder (Wright 2008, 119, 121). It has been proposed that the women in small shelters seen in several two-dimensional depictions of sacrifice (e.g., Kutscher 1983: Figures 120, 123–124) are giving San Pedro to captives prior to their sacrifice, perhaps to make them more compliant as well as reinforcing the supernatural basis for the sacrifice

(Wright 2010: 312; Uceda and Tufinio 2003: 222). However, there is no iconographic proof of this hypothesis.

San Pedro is commonly cooked down into a thick, mucilaginous beverage for consumption in present day shamanic rituals. However, it is known that pieces of the cactus were also eaten to achieve similar effects. It is unknown which might have been the preferred method of ingestion. The appearance of the crown in the hands of the *achumeras* could suggest the cactus was eaten raw, but evidence for cactus sap as a binder for mural paint may mean the plant could be juiced (raw or cooked down, the latter is much more potent) as well. If *Echinopsis* was given to prisoners as part of their preparation for sacrifice, it is notable that the crowns are depicted in the hands of *achumera* figures but are not represented in the two-dimensional sacrifice scenes. It is possible these cacti were eaten raw, and voluntarily, by the *achumeras* as part of a ritual trance state, while prisoners may have been given the liquid version surreptitiously, in foods or beverages, or else they were compelled to ingest it, as part of their role in the sacrifices. This would work with the imagery of ceramic vessels seen in sacrificial iconography, however, the vessels could be holding so many other liquids (including intoxicants like corn beer or the *Nectandra-Datura* mixture noted above), that it is impossible to make a direct connection without further evidence.

The nature of the ritual being performed by the *achumeras* is not clear, as there are no representations of their context. It is possible that these are women who are preparing *E. pachanoi* for its use in the sacrificial rituals. This does not, however, explain why their eyes are closed or shielded, which could be a direct reference to their own ingestion, as well as the fact that there are many dissimilar elements between the *achumeras* and the female figures depicted in the small architectural structures. The closer visual associations with healing, trance, and the land of the dead seem to point to a more private ritual similar to modern healing practices than to the polity-level, highly-visible (in action or in representation) Sacrifice Ceremony.

The inclusion of earspools on some of these figures would seem to indicate that in at least some instances the ritual being performed was very important, enough so to confer this particular gender-complicating ornament. Since the only provenanced pieces with representations of earspools are from Huaca de la Luna, it is tempting to speculate that this particular elevation of gender-complicated women is a unique aspect of that site. However, it is equally possible that men rather than women at Huaca de la Luna performed San Pedro rites, and complicated gender with long tunics and substituting mantles for more masculine head coverings while retaining male earspools. It will take more analysis of archaeologically recovered pieces to determine which is the case. A tomb excavated in 1992 at Huaca de la Luna contained a male individual with braided hair (Montoya Vera and Uceda 1992: 43), which may indicate that such gender-crossing was possible at the site, but this aspect has not been studied further, and none of the objects listed in the tomb inventory indicates an association with an *achumera* identity. Lamentably, textiles in Moche graves are so poorly preserved that the weave of the textile remains are listed, but the garment shapes and relationship to gender identity are unknown.

The *achumera* representations definitely display aspects of an undifferentiated gender associated with both older men and women. This lack of differentiation can be seen as "both male and female" or "neither male nor female," as the aspects of human fertility have been shed. This complication makes them prime candidates for contact with spirit forces, as it facilitates communication the ambiguous realm of the supernatural to embody ambiguity oneself. How those forces were being approached, whether as a small-scale healing ritual or as part of a larger sacrificial ritual that was an important ideological basis for the polity's ruling class, is not yet clear, although it would seem to tend toward the *achumeras* not performing in public ritual. Most public rituals depicted in Moche art include the presence of figures dressed in the highest-status garments, elaborate versions of warrior costume that usually include metal plaques, sumptuous jewelry, and large multi-piece headdresses. These figures are also unambiguously male, with one notable exception (Scher 2011, 2017). The costume of *achumeras*, while somewhat higher status than the baseline, is far from the extravagant ensembles of these high-status figures. In addition, complex two-dimensional scenes that show the minor performers in these rituals do not show women dressed like the *achumeras*, or the San Pedro Cactus (*E. pachanoi*).

Conclusions

Like many foods in Moche art, the collection of the cactus is not depicted, and if it was processed into liquids, that too is not shown. Like the maize that would become beer (*chicha*), the symbolic aspects of the plant, rather than the mundane elements of collection/cultivation and subsequent preparation into ritually-consumed comestibles is highlighted. Maize beer cemented communal feasting that acted as a social binder of mutual obligations. The daily diet of the Moche was not the subject of its official art; rather, the foods of spiritual and political sustenance were emphasized. The *achumeras* and *Echinopsis*, like many aspects of Moche art, participate in a visual vocabulary that favors symbolic over descriptive representation. This has the effect of obscuring the place of the plant in Moche religious life, a kind of esoteric knowledge in and of itself, which participates in the power strategies of the elites who produced these representations (Bernier 2008; Chapdelaine, Kennedy, and Uceda 1995), and may have relied on the general shamanic tradition as a foundation for their own ceremonial power. Rather than celebrating their foodways, Moche artistic practices obscured them in favor of politico-religious symbolism.

Notes

i. Previously classified as *Trichocereus pachanoi* in earlier literature, however more current scholarship considers *Echinopsis* more correct.
ii. While agricultural products are usually quite recognizable, not all Moche plants are easily identifiable. A fruit called *ulluchu* appears prominently in scenes of sacrifice, and even has been recovered archaeologically, but has not been definitively identified botanically (McClelland 1977; Bussmann and Sharon 2009).

iii. Bourget (1994, 73), argues for the depiction of *Datura stramonium* through a representation of its spiky seed pod, but he only offers one example.
iv. The use of the term "shaman" to describe ritual specialists who use trance states and travel to supernatural realms in their practice is a complex issue. The original use of the term referred only to practitioners in some Siberian cultures, and there are scholars who argue that the term's use outside of that context is inaccurate. In general, the use of a local term for the person who performs such tasks is preferred, however in cultures where the native term is unknown, "shaman" has taken on a more generic use, describing the use of entheogens, drumming, dancing, and/or fasting to achieve a trance state and travel to the Other World for the benefit of individuals or groups. The term is often less clunky than "ritual specialist," and avoids over-specific terms such as "healer" or "priest." I try to use specific terms when they are warranted, but will use the generic shaman/shamanic where it is appropriate and to avoid unnecessary wordsmithing.
v. Another representation features a landscape scene with columns of *Echinopsis pachanoi* cactus, sometimes in mountainous terrain, accompanied by a feline figure, similar to those of Cupisnique (Cordy-Collins 1998).
vi. Judith Butler's *Gender Trouble* (1990) and *Bodies that Matter* (1995) lay out her performative theory of gender, which contends that gender is not a state of being but a continual performance which strives for but never achieves perfection. Sørensen (1997, 94) has studied the use of costume to communicate identity in Bronze Age Europe and has been an important voice in the study of gendered identity in archaeology.
vii. This figure's extended arm is missing, but the similarity to the two-dimensional depictions, and to Fig. 14, make the designation of *achumera* possible.

References

Benson EP (1974) Salesmen and sleeping warriors in Mochica art 41:26–34. Mexico
Benson EP (1975) Death-associated figures on Mochica pottery. In: Benson EP (ed) Death and the afterlife in pre-Columbian America: a conference at Dumbarton Oaks, 27 Oct 1973. Dumbarton Oaks Research Library and Collections, Trustees for Harvard University, Washington, pp 105–144
Bernier H (2008) Especialización artesanal en el sitio Huacas de Moche: contextos de producción y función sociopolítica. In: Castillo LJ, Bernier H, Lockard G, Yong JR (eds) Arqueología Mochica: Nuevos Enfoques (Actas del Primer Congreso Internacional de Jóvenes Investigadores de la Cultura Mochica, Lima 4 y 5 de agosto 2004). Fondo Editorial, Pontificia Universidad Católica del Perú, Lima, pp 33–52
Bourget S (1994) Bestiaire sacré et flore magique: Écologie rituelle de l'iconographie de la culture Mochica, côte nord du Pérou. Doctoral dissertation, University of Montreal
Brown JK (1982) Cross-cultural perspectives on middle-aged women [comments and replies]. Curr Anthropol 23(2):143–156
Burger RL (1992) Chavin and the origins of Andean civilization. Thames and Hudson, London
Bussmann R, Sharon D (2009) Traditional medicinal plant use in Northern Peru: tracking two thousand years of healing culture. J Ethnobiol Ethnomed 2(1):47

Carod-Artal FJ, Vázquez-Cabrera CB (2007) Semillas psicoactivas sagradas y sacrificios rituales en la cultura Moche. Revista de Neurología 44(1):43–50
Chapdelaine C, Kennedy G, Uceda S (1995) Activación neutrónica en el estudio de la producción local de la cerámica ritual en el sitio Moche, Perú. Bulletin de l'Institut Français d'Études Andines 24(2):183–212
Chapdelaine C, Pimentel V (2003) Un tejido único Moche III del sitio Castillo de Santa: Una escena de cosecha de yuca. Bulletin de l'Institut Français d'Études Andines 32(1):23–50
Conklin WJ (1997) Structure as meaning in Andean textiles. Chungará 29(1):109–131
Cordy-Collins A (1998) The jaguar of the backward glance. In: Saunders NJ (ed) Icons of power: feline symbolism in the Americas. Routledge, NY, pp 155–170
Davis EW (1983) Sacred plants of the San Pedro Cult. Bot Mus Leafl Harv Univ 29(4):367–386
De Feo V (2003) Ethnomedical field study in Northern Peruvian Andes with particular reference to divination practices. J Ethnopharmacol 85(2–3):243–256
Feldman Gracia L (2011) Coca y wachuma: sus prácticas y significados en la cultura andina y en Lima. PhD dissertation, Ciencias Sociales, Universidad Nacional Mayor de San Marcos. Programa Cybertesis PERU, Lima./z-wcorg/
Franco Jordán RG (2012) Oficiantes y curanderos Moche, una visión desde la arqueología. Pueblo Continente 23(1):18–26
Glass-Coffin B (1999) Engendering Peruvian Shamanism through time: insights from ethnohistory and ethnography. Ethnohistory 46(2):205–238
Hocquenghem AM, Lyon PJ (1980) A class of anthropomorphic supernatural females in Moche iconography Ñawpa Pacha 18:27–48
Isbell B (1997) From unripe to petrified. Cornell University eCommons, pp 1–37. Accessed 19 Oct 2008
Kutscher G (1983) Nordperuanische Gefäßmalerein des Moche-Stils. Verlag C.H. Beck, München
Lynch TF (1980) Guitarrero cave: early man in the Andes. Academic Press, New York
Lyon P (1978) Female supernaturals in ancient Peru. Ñawpa Pacha 16:95–140
McClelland D (1977) The Ulluchu: a Moche symbolic fruit. In: Cordy-Collins A, Stern J (eds) Pre-Columbian art history: selected readings. Peel Publications, Palo Alto, pp 435–452
Montoya Vera M, Uceda S (1992) Los Entierros. In: Proyecto Arqueológico Huaca de la Luna, Informe 1992. Proyecto Arqueológico Huaca de la Luna, Trujillo, Peru
Barreda M, Elías RF, Jordán CG, Quilter J, Cruz AM, Gamarra C, de la Cruz V, Cisneros HR, Alcalde SL, Verano JW, Merello MA (2007) El Brujo: Huaca Cao, centro ceremonial Moche en el Valle de Chicama, 1a edn. Fundación Wiese, Lima
Polía M (1999) La cosmovisión religiosa andina: en los documentos inéditos del Archivo Romano de la Compañia de Jesús 1581–1752. Pontificia Universidad Católica del Perú, Fondo Editorial, Lima
Proulx DA (2006) A sourcebook of Nasca ceramic iconography: reading a culture through its art. University of Iowa Press, Iowa City
Sánchez Davis ML (2012) Roles chamánicos en la sociedad moche en base al estudio de unas cerámicas del museo nacional de arqueología, antropología e historia del Perú, Lima. Pueblo Continente 23(1):27–33
Scher S (2010) Clothing power: hierarchies of gender difference and ambiguity in Moche ceramic representations of human dress, C.E. 1–850. PhD dissertation, Art History, Emory University, Atlanta
Scher S (2011) Clothing the body politic: gender, ethnicity, and status in Moche ceramic art. Paper presented at the College Art Association annual meeting, New York City
Scher S (2017) High-ranking women and masculine imagery in Moche art and burial ensembles. In: Scher SEM, Follensbee BJA (eds) Dressing the part: power, dress, gender, and representation in the pre-Columbian Americas, pp 450–481
Schultes RE, Hoffman A (1979) Plants of the Gods: their sacred healing and hallucinogenic powers. Healing Arts, Rochester, VT
Sharon D (2000) Shamanism and the sacred cactus. San Diego Museum of Man, San Diego

Sharon DG, Donnan C (1974) Shamanism in Moche iconography. Ethnoarchaeology 4:51–77
Sharon DG, Donnan C (1977) The magic cactus: ethnoarchaeological continuity in Peru. Archaeology 30(6):374–381
Sørensen MLS (1997) Reading dress: the construction of social categories and identities in bronze age Europe. J Eur Assoc Archaeol 5(1):93–114
Stone RR (2011) The jaguar within: Shamanic visions in ancient central and South American art. University of Texas Press, Austin
Uceda S, Tufinio M (2003) El Complejo Arquitectónico Religioso Moche de Huaca de la Luna: Una Aproximación a su Dinámica Ocupacional. In: Uceda S, Mujica E (eds) Moche: Hacia el Final del Milenio, vol 1. Universidad Nacional de Trujillo and Pontificia Universidad Católica del Perú, Lima, pp 179–228
Wright V (2010) Pigmentos y tecnología artística mochicas: una nueva aproximación en la comprensión de la organización social. Bulletin de 'Institut Francais d'Etudes Andines (Lima) 039(02):299–330
Wright V (2008) Étude de la polychromie des reliefs sur terre crue de la Huaca de la Luna Trujillo, Pérou. Archaeopress, Oxford, England

Chapter 10
The Symbolic Value of Food in Moche Iconography

Margaret A. Jackson

Abstract Moche art is well known for its apparent pictorial veracity in ceramic and architectural imagery. Yet, the visual record is not definitive about which images, if any, depict food. The importance of food is expressed through an assortment of indexical references to conceptions of deity and assertions of religious practice. Specific kinds of agricultural produce, particular fruit, and certain animals appear often, but whether they were selected for representation because of their nutritional values or symbolic qualities is unknown. Moche imagery rarely shows images connected with sumptuous feasting. The images on ceramic vessels do not contain images of bacchanalian events, no images of tables piled with platters of food at feasts or festivals. Rather than eating food, humans, zoomorphic and supernatural beings are shown handling an array of cups and vessels in various events and rituals. Such conflations seem to signal the specific kinds of associations central to the purpose of the ceramic assemblages in which they appear and the ideological foundations on which they were based. The questions raised by these disjunctions are the subject of the present research.

Introduction

Food is rarely depicted in Moche iconography, ceramic imagery, or textiles. Representations of prepared cuisine, bacchanalian revelries, or tables piled high with food are rarely if ever depicted. There are no images of roasting venison on a spit, butchering animals, chopping vegetables, or food in a market. Occasionally, participants in ritual processions are depicted holding a cup of maize beer or *chicha*, but there are no scenes depicting people eating or feasting. Nevertheless, archeological evidence clearly indicates that ritual feasting and community events were common throughout the annual cycle. Very few iconographic images depict food preparation, although there are rare examples of *chicha* (maize beer) production.[1]

M. A. Jackson (✉)
Department of Art: Studio, History, Education, University of New Mexico, Albuquerque, US
e-mail: mars@unm.edu

J. E. Staller (ed.), *Andean Foodways*, The Latin American Studies Book Series,
https://doi.org/10.1007/978-3-030-51629-1_10

Scholarly interpretations of Moche imagery frequently blur the distinction between "food" and images of plants and animals that are presented in artwork for symbolic value. While most plants and animals are potentially edible, the assumption that mimetic representation of plants and animals in Moche art is solely to evoke nutritional significance does not appear to be supported by the visual evidence. Well-known animal characters, such as the "Pato Guerrero," (a Muscovy duck dressed as a warrior with weapons), often reflect social, ideological, or militaristic associations, rather than importance as sources of food.[2] Likewise, images of trees, plants, and flowers carry a range of meanings apart from their nutritional connotations, and images of certain kinds of produce and key plant cultigens stand out as very significant (see Veintimilla and Caputi, Chapter 9 and Scher, Chapter 10 for additional insights into food in ceramic and iconographic depictions). Therefore, the focus of this research is to investigate the relationships between images of particular flora and fauna known to have been nutritionally significant to Moche culture that regularly appear on ceramics, and the larger symbolism of Moche art and iconography.

Images of food and food references in Moche culture may have been selected because of their relationship to the function of the ceramics or objects on which they appear. Finely made ceramic vessels were used for propitiation and celebration of supernatural forces, personified as mountain and ocean deities, as sources of water and environmental sustenance. Scores of examples show elaborate Moche bottles, jars, and flaring vases used in various ceremonial events. Thematic examples include the Presentation Scene, Mountain Sacrifice Scene, Copulation Ceremony, the Owl Curandera, and many others.[3] Such ceramic wares and figurines were not created solely for the purpose of inclusion in tombs, even though that is the context in which they are most often recovered by archeologists. The living were meant to use these ceramics. These were liturgical wares, crafted with ritual purposes in mind, and made by occupational specialists. (Russell and Jackson 2001). Various archeological examples cement the idea that particular ceramic assemblages were actively used in religious and sacrificial contexts.[4] By virtue of their connection to propitiation rituals, they would have been associated with agricultural abundance. The selection of imagery was tightly keyed to a well-understood visual vocabulary employed by master craftsmen, distilled and codified to the point of being pictographic in many instances.

In its most quotidian definition, food is comprised of basic staples required for nutrition—food and foodstuffs are the sustenance we cultivate and domesticate to stay alive. Most commonly, these include sources of protein, such as meat, fish, eggs, and mollusks, wild or cultivated fruits or vegetables, as well as starchy food sources, like quinoa, grain, maize, potatoes, or beans. In Moche imagery those depicted are visually mimetic—some have greater and lesser degrees of variability, realism, and abstraction. Leaving aside interpretations of depictions featuring live animals in naturalistic contexts—those not shown as prepared, cooked, or served on plates—and images of humans hunting, fishing, and gathering plants or animals, this research investigates the variability of cultigens in Moche ceramic art, articulating the ways in which certain plants are given symbolic eminence and hypothesizing their connection to larger societal concerns about sustenance and crop harvest.

The evidence suggests the connection between food and the imagery found on fineware vessels most often rests in the conjunction of religious practice and liturgy, as it intersects with notions of divine beneficence and agricultural fertility. Here, "fineware" refers to a recognizable ceramic type, differentiated by temper, construction technique, and context (Russell et al. 1998; Jackson 2000; Quilter 2010). Specialized fineware ceramics, often elaborately modeled and painted, were used by humans to mediate between natural forces and the desires or petitions of those who employed them. Thus, images of deities merged with cultigens represent a particular amalgam of concepts specifically attuned to agricultural cycles, rather than to the mundane necessity of eating. Various instances of how that nexus is made visible in the artwork are presented here.

"Food" in Moche Culture

Coastal people had a range of foodstuff available; including squash, maize, manioc, peanuts, potatoes, and common beans, as well as guava, avocado, *ají* peppers, and protein from venison, as well as marine resources such as fish and shellfish. The specific proportions of food resources changed significantly over time and according to proximity. Pozorski (1976) notes the earliest occupations in the Moche Valley, during the Cotton Preceramic (2500–1800 B.C.) show heavy dependence on cotton cord and twining probably for manufacturing fishing nets, along with protein supplied almost exclusively by marine resources. The earliest cultigens were cotton, gourds, squash, *ají* peppers, and lima beans. At this point in time, two of the three main species under cultivation were industrial plants: cotton and gourds—plants that could serve to provide raw materials for clothing and netting (Pozorski 1979: 165–172). Initial Period and Early Horizon (1800–200 B.C.) occupations show increasing evidence for the development of irrigation agriculture, corresponding to inland intensification of agriculture, an increase in variability of food cultigens, and significantly larger seed size for plants such as gourds and squash. Maize was domesticated in Mesoamerica c. 6000 B.C. (Staller 2010), and became an important cultigen along the coast of Peru (including the Moche Valley) by 1200 B.C. (Hastorf et al. 2006: 429). For communities near the coast, marine sources represented nearly 100% of protein; for inland sites, meats included deer, large shellfish, marine resources, local fauna, and later, llamas, which appear for the first time in the archeological record (Pozorski 1979: 174).

The evidence suggests that by the Early Intermediate Period (EIP) ca. 200 B.C., Moche social complexity and stratification were accompanied by food specialization. The data show a dramatic uptick in use of domesticated llamas for meat and fiber, and also guinea pig and a de-emphasis on marine resources. At the Huacas de Moche, the protein source was almost entirely llama meat (Pozorski 1979: 175). Llama herds were being controlled and maintained for the purpose of supplying meat and possibly labor and fiber. Plant food production increased dramatically, as did canal irrigation. Cotton, gourd, squash, corn, peanuts, and common beans emerged as major crop plants, probably because they could be produced in great quantity on large irrigated fields, meanwhile fruits seem to have decreased in importance. "All of the dominant cultigens except squash could easily be reduced to forms suitable for storage, thereby facilitating an accumulation of surplus agricultural products" (Pozorski 1979: 176). She observes that "agricultural and animal protein production increased in efficiency and narrowed in scope until virtually all meat came from a single source, llamas, and field crop cultigens, especially maize, dominated the plant species inventory" (Pozorski 1979: 181). While Pozorski posits a causal link between food surplus and increased social stratification, she eschews further interpretation of sudden adoption of particular foods.

It is noteworthy that widespread maize consumption does not appear on the coast until almost 1000 years later than its initial introduction to the Andes (Hastorf and Johannessen 1994: 430). Maize remained an incidental crop for a long period of time, and its importance was primarily related to consumption as a fermented intoxicant, i.e., *chicha.* Maize was adopted into already existing agricultural systems, and initially cultivated in small plots alongside more traditional crops. Its later dietary prominence seems to have occurred relatively quickly, coinciding with rapid changes in the Andean political economy and associated with increased social and political complexity. Hastorf and Johannessen (1994) propose social or ideological reasons for the increased societal emphasis on the cultivation of maize. They note that throughout the Americas, maize was a remarkably widespread and powerful bearer of cultural meaning (Hastorf and Johannessen 1994: 428), suggesting it might have served as an agent of change in its own right in certain instances (Staller 2006).

The question of the sudden transformation of maize into a dietary staple is not satisfactorily answered in purely economic terms. Maize is not necessarily superior to other staples; various other cultigens yield higher nutritional value per unit of water and labor (potatoes and beans being ready examples). Originally domesticated in Mesoamerica and brought to the Andes in the context of long-distance interaction, in certain environments, maize is actually a precarious and difficult crop to grow. Hastorf and Johannessen (1994: 443) note that in many cultures, the relationship between humans and cultivated maize is very close, requiring considerable human intervention to succeed, "corn in a sense appears to belong conceptually with people rather than with other plants – in the realm of culture rather than nature." They propose an ideological impetus as being at least somewhat responsible for the significant jump in maize cultivation in Early Intermediate Period Peru. Maize beer or *chicha* was commonly consumed in various rituals and rites and left as offerings to deities or sacred places in the landscape (see Staller 2006, 2010).

The assertions by Hastorf and Johannessen (1994), Staller (2006, 2010), and others underscore what the imagery suggests. The Moche clearly had a cultural emphasis on maize as a crop closely associated with rituals and rites and by extension, elite status. Ideological elevation of particular food crops also coincides with the dramatic rise in political complexity of the Moche, with an increased need for dependable staple resources.

Iconography Associated with Food

Food cultigens are among the most striking images in Moche iconographic repertoire precisely because they show "naturalism" to great advantage. Ceramicists employed sleek, volumetric forms to convey key attributes of subjects, in most cases making particular species easily recognizable. The ripened produce of key crops such as maize, squash, and manioc are frequently depicted (Fig. 10.1). Less common fruits and legumes are also shown, including beans, *ají* peppers, peanuts, and potatoes. In most, but not all cases, the ripened products are shown alone—entire plants, stalks, and leaves are rarely depicted. This speaks to Moche artists' tendency to employ *pars pro toto* or abbreviated visual vocabulary, where a single definitive element stands for an entire subject or larger concept.

These depictions did not serve as a visual registry of Moche's entire botanical inventory. Indeed, only a small fraction of coastal flora is represented in the artistic record, among hundreds that grow on the north coast. Larco Hoyle (2011 [1938–1939], 1:46–52, 263) identifies only 30 different plant varieties and 20 various trees and cacti in the artistic record. Iconographic and symbolic focus on such a limited

Fig. 10.1 Naturalistic image of Manioc plant (Courtesy of Museo Larco ML006641, Lima, Peru)

range of species speaks to the function of the ceramics themselves, as representatives of a specialized cohort of liturgical vessels, rather than to any social urge to document nature. The vessels were intended to mediate proper religious observance and their motifs articulate particular relationships and abstract concepts. Complex compositions featuring multiple characters and non-naturalistic juxtapositions serve to illustrate the ways in which specialized visual vocabulary functioned in Moche cultures. Depictions of maize, manioc, potato, peanut, frogs, beans, and stacked plates of food demonstrate this point.

Moche imagery is distinguished by its use of compound iconographies, where clusters of symbolic elements combine to generate complex ideas. A naturalistic maize ear may refer to ripeness and bounty; or signify a good harvest, a particular time of year, or a tasty meal. A ceramic vessel modeling realistic ears of maize, but animated through the addition of a fanged supernatural face (Fig. 10.2) compounds these ideas, amalgamating them with an entirely different cluster of concepts. In this case, the animated maize-character is known as Ai Apaec, very familiar in Moche art, originally so-named by Larco Hoyle (2001 [1938–1939], 2: 273), based on his understanding of ancient Muchic language, and widely used in various publications (Gölte 2015; Jackson 2008). Ai Apaec is alternatively glossed as the Fanged God (Benson 1972), the Mountain Deity (Morales 2003), the Decapitator (Alva 1994; Uceda 2001), the Supernatural Warrior with Serpent Belt (Donnan 1978), and various others. All refer to the same godlike Moche supernatural being associated with blood sacrifice, agricultural bounty and water. The disagreement and confusion over the identity

Fig. 10.2 Ai Apaec as ripened maize, flanked by two servants (Courtesy of Museo Larco, ML003296, Lima, Peru)

figure appear to arise from the fact that this fanged entity appears in a multitude of pictorial contexts—leading some scholars to posit multiple complementary deities (Makowski 2005), or the same character as a culture hero protagonist in variable scenes (Rucabado 2015). Benson notes that an image of maize marked with an Ai Apaec deity face is surely not a representation of a separately identifiable maize god, per se, but a demonstration of the sacred nature of the plant (Benson 1997: 126).

Various authors point to widespread ethnographic and ethnohistoric accounts suggesting maize was a gift from the gods, seen as somehow kindred to humans and strongly associated with ethnic identity—its cultivation requires human care and its upright stalks are humanlike with flowing hair and outstretched arms (Hastorf 1994: 398). Such characteristics provide reasons why maize became the center of belief and symbolism throughout the Americas, and may shed light on its semi-deified appearance in Moche imagery. On the coast of Peru, seventeenth century priest, Antonio de la Calancha (1974 [1638]), recounts an indigenous creation myth that emphasizes the divine nature of maize and other cultigens. In tones that clearly echo the Biblical, he writes that Pachacamac,[5] the son of the solar deity, created a man and woman, but the two were dying from hunger. When the woman begged for food, the Sun impregnated her with his rays to create a son that would help her. Yet, Pachacamac was jealous of his new half brother and killed the infant. Clearly, the jealousy of Cain and Abel underlies Calancha's telling, but the story bears important differences, for Pachacamac subsequently dismembers the infant and buries the body parts, from which plants sprout. Maize sprouted from the teeth, manioc came from ribs and bones, as did other kinds of edible roots, and from the flesh grew squash, *pacay,* and other fruits (Calancha 1974 [1638] II, cap. 19: 931–932). This reincarnation suggests maize, manioc, and particular fruits were sacred.

Maize was conflated not only with deity, but also ritual practice, as indicated in various depictions of deified maize. As seen in Fig. 10.2, Ai Apaec-Maize is sometimes flanked by two maize-servants. Here, the composition makes clear reference to a body of Moche art, a group of three, featuring a central, supernatural being flanked on either side by attendants, made identifiable as temple servitors by the presence of hooded garments or bird heads.[6]

Such servitors populate the backdrop of many complex scenes. One well-known example is the Arraignment of Prisoners, where, behind a vivid depiction of prisoners racing up the front of a gigantic step-platform pyramid, a series of sacrificial events are conducted by bird-headed creatures; the servitors appear in small vignettes on all sides of the painted vessel.7 Based on the many variants of maize compositions, it seems apparent that multi-figured Ai Apaec-Maize representations refer in some manner, to religious events that likely occurred at one of the Moche platform pyramid temples (e.g., Huaca de la Luna). The paralleling of the multi-figured, deified maize cobs with the larger thematic body of artwork draws attention to the presumed ritual purpose of the ceramic vessel on which the imagery occurs.

Manioc (yuca or cassava root), was also associated with divine beneficence, and a frequent theme on Moche ceramics. Indeed, manioc and maize are the two main food cultigens that are consistently depicted as deified, or closely associated with the spiritual realm; manioc is often shown conflated with Ai Apaec attributes (Fig. 10.3a

Fig. 10.3 Ai'Apaec as manioc plant, ceramic bottle (Courtesy of Museo Larco ML003266)

and b). Images of manioc typically include the juncture of the plant stalk, with its characteristic "eyes," and numerous bulbous tubers (or capsules) that grow below ground. Pearsall (2008: 106) describes manioc cultivation as "productive and undemanding," noting that the plant produces more carbohydrates per edible portion than other lowland roots and tubers. Various early chroniclers mention manioc cultivation in their accounts (see Ugent et al. 1986, 1982). At the time of the Spanish conquest, it is clear manioc was widely cultivated along the coast and eastern flanks of the Andes with a wide variety of crops, and that it represented considerable social and economic value (see e.g., Acosta [1590] 1954; Cieza de Leon [1553] 1924; Cobo [1653] 1956; Fornee [1586] 1965; Oviedo [1535] 1959; Palomino [1549] 1965).

Organic remains pertaining to manioc have been documented archeologically on the north coast as early as 1800 B.C. (Ugent et al. 1986; Pearsall 2008: 114), yet the archeological data suggest an unrealistically low reflection of what was probably a fairly high prevalence of this foodstuff. The data suggesting a low incidence may be related to the manner in which manioc was processed for consumption. Several varieties of the tubers cannot be eaten directly, which are referred to as 'bitter cassavas' must be processed to remove cyanogenic glycoside toxins. The so-called "sweet" varieties of manioc have no such toxins, and when cooked, readily transform into soft edible chunks that have a nutritious flour from which a starchy paste can be made. Indeed, Ugent et al. (1986) assert that only the "sweet" manioc varieties were cultivated in western Peru (and western Mexico). The toxic "bitter" varieties were grown in Amazonia, although both varieties occur in the intermediate Andean territories (Ugent et al. 1986: 98). Circumstantial evidence for manioc processing is found in the ceramic record, in the form of various kinds of grinding bowls, whose interior surfaces are gouged and roughened to shred the tough tubers. This type of ceramic bowl was definitely manufactured in the Chicama Valley at Cerro Mayal (Russell and Jackson, 2001: 165). Evidence from early coastal sites indicates that in many cases, the roots were cooked by simply placing them directly into the fire (Ugent et al. 1986: 96). This may account for the scarcity of archeological remains.

Evidence that manioc was conceptually allied with maize is implied by the frequency with which they are depicted together. It is telling that complex depictions

Fig. 10.4 Supernatural being, holding manioc and maize plants, in 5-point mountain peak (Courtesy of Museo Larco, ML003352, Lima, Peru)

often show the cultigens in the hands of a standing supernatural being, a seeming referent to Ai Apaec, against a 5-point mountain backdrop (Fig. 10.4). Such scenarios have a clear reference to the Moche Mountain Sacrifice scene, which most scholars interpret as directly tied to sacrificial petitions, rituals, and rites for water and fertility from the mountain deity. Once again, these images are conceptually linked to propitiation of a deity for the purpose of a rich agricultural bounty, and logically tied to the larger purpose of the ceramic vessels themselves. This is because the ceramics in a self-referential way have a symbolic link to the mountain deity, which is vitally important to the cultivation of staple crops such as manioc and maize. The roles of such vessels in the observance of religious veneration is evident by their motifs and vessel function.

Surprisingly, the potato, an important cultigen, does not appear in deified form or ever depicted as Ai Apaec. Moche depictions of corn became part of the symbolism associated with ethnic identity and kingship, manioc also appears in the spiritual realm, but one of the most important food crops in the Andes, potatoes, does not appear to have such spiritual or cultural associations. Potatoes appear along the Peruvian coast as early as 1200–2000 B.C. (Ugent, Pozorski and Pozorski 1982); however, its cultivation is most common in western highland valleys and higher altitudes. When depicted naturalistically, the humble tuber is presented in quietly modest forms—realistic, bulbous shapes, characterized by the presence of small, arched, indented "eyes." The potato's propensity for protuberant, compound growths seems to have fascinated the Moche, as they are rarely depicted as singular subjects.

They are most commonly depicted as a part of composite imagery, with faces and heads of humans or animals "growing" from their bodies.

Potatoes appear to be symbolically linked to the inner realm of the earth, what Quechua speakers in the highlands refer to as *uku* pacha. This is the *mundo interior*, source of formation and beginnings, dwelling place of all things past and future. (Staller 2008; Urton 1981) *Uku pacha* has metaphorical connotations to the womb. Millones (2001: 55) shows imagery of living creatures emerging from potato eyes as analogous to primordial emergence from *pacarinas* (i.e., ancestral places of origins or emergence into the existing creation cycle). Yet, corpses eventually return there. *Uku pacha* stands in contrast to *kay pacha*, the present, terrestrial realm. Thus, potatoes, as denizens of the *mundo interior*, may have been regarded with some amount of trepidation. However, this association may be literal as it also relates to the fact they are root crops. Potatoes with strange shapes, or those fused together like conjoined twins were considered to be especially powerful omens for a bountiful harvest, or even magical talismans. The strangely twisted, compounded images of potatoes and distorted heads, faces, and animals evoking strong ideas and feelings, forming a pictorial cohort, what Larco Hoyle (1938/39: 200]) referred to as the *mundo hororroso*. Yet, there is tremendous variability among compound imagery, and it is possible there are references to a multitude of narrative themes. The kinds of creatures that might grow from a potato "eye" include snakes and jaguars, or seals and sea birds. (Fig. 10.5a). The surrealistic quality of the compositions suggests hallucinogenic visions or altered states of consciousness. Often, potatoes are conflated with embryonic or malformed human shapes, which some authors see as representations of people with specific illnesses or physically deformities (Fig. 10.5b), and in particular, individuals with disfiguring tropical diseases like tumor-causing verruga and lesion-causing leishmaniosis *(uta)* (Liddell 2000; Urteaga-Ballon 1991). Benson (2012: 23) noted that the frequency of the incidence of leishmaniosis appears to correspond to geographic locations associated with the cultivation of potatoes, observing annual cultivation

Fig. 10.5 Composite potato imagery often features animated potato figures who appear to have *"uta"* lesions at nose and mouth (Courtesy of Museo Larco, ML007289, Lima, Peru)

of potatoes occurs at roughly the same time as maximum infectiousness of insect-carried diseases. Guardians of potato fields were known to commonly suffer from these afflictions, and indeed, according to some sources, having one of the diseases was a requisite for the job.

In some cases, images of leishmaniosis sufferers with missing noses or lips include facial markings or tattoos on cheeks or other body parts. A Moche vessel at the Museum of Göteborg from the Chicama Valley, depicts frogs on the face of an *uta* disfigured individual (cat. 30.28.3,p shown in Wassen 1934a: Fig. 25). Did such tattoos indicate body markings or other symbolism? The cultural perception that frogs as have symbolic associations to agricultural fertility, seems appropriate to the possible role of these individuals. An evocation of crop guardians on a liturgical vessel whose function was propitiatory seems to justify their presence, although the larger meanings of iconographic components found in surrealistic potato/animal composites remain unclear in most cases.

Peanuts, like potatoes, grow underground, however, they are not part of the frightening cohort of *mundo hororroso* images. Initially domesticated c. 7600 B.C. in the highlands, the peanut is commonly rotated for cultivation with maize, due to its capacity to fix nitrogen and add nutrients to soil (Dillehay 2007). Spectacular, naturalistic gold and silver peanuts are featured in one of the necklaces from the tombs of Sipán, and peanuts often appear in simplified form on painted vessels. Yet, their iconographic meanings are not readily apparent. When shown as anthropomorphized, peanut representations generally include the telltale markings of fibrous, segmented shells, along with the addition of a human head at one end and sometimes with arms (Fig. 10.6). Peanut-men are never deified—peanuts are not shown as Ai Apaec, with fangs, or other markers of associations to deities. They do, however, appear as flute players, suggesting a symbolic relationship to successful planting or harvesting.

Larco Hoyle describes twentieth-century coastal traditions related to planting that involved celebration, mock combat, music, and dancing (Larco Hoyle 2011 [1938–1939], 1:256). These, in turn, seem somehow relevant to various scenes in Moche fine line painting that include active protagonists accompanied by small groups of

Fig. 10.6 Anthropomorphized peanut musician. (Courtesy of Museo Larco, ML002214, Lima, Peru)

Fig. 10.7 "Botanical Frog" composite frog-feline-plant creature; the manioc stalk-as-spine is clearly visible, as are bean pods and maize stalks (Courtesy of the Museum of Archeology, Anthropology and History, National University of Trujillo UNT-185; photo by author)

musicians. The Ribbon Dance scene, for example, has been paralleled with the Inca *baile con soga*, conducted to celebrate the beginning of the planting season and end of *Camay*, a ritual form of combat (Hocquenghem 1987: 122–123; Molina 1959, 82–83). Among the highlanders, such events are associated with lunar cycles, generally the first new moon after the December solstice (summer). Besides flute music, the events are associated with a range of ritual behaviors, lasting roughly three to four weeks. Thus, while it is impossible to know if peanut-men played flutes to celebrate the planting season, their performance probably is related to ritual behavior associated with the annual cycle.

A pictorial theme much more clearly tied to agriculture and food cultigens is what McClelland (2011: 35) calls the Botanical Frog**.** This is a creature that stands out among Moche ceramic motifs because of its complex, food-related symbolism. Originally called the *"sapo-jaguar"* by Larco Hoyle (1948: 44), the Botanical Frog combines aspects of frogs and felines, with images of food plants, like corn, *ají*, manioc, and beans (Fig. 10.7). The overall frog-like shape is at times offset by feline aspects including straight front legs (sometimes striped like a feline), rounded feline ears, feline nose, and in a few rare examples, teeth, or fangs. Its single most consistent feature is the manioc tubers (manioc capsules) on its backside connected to a "spine" showing the characteristic "eyes" of a manioc stalk or trunk. While other elements might be more or less variable, all pieces in McClelland's (2011) sample had these tubers as a diagnostic feature. The notched manioc stalk forming the spine in some examples turns into manioc "horns" on the head. The creative morphing of realistic plant knowledge into an abstract concept is among the more compelling features of the depiction. Given that manioc propagates from cuttings planted horizontally, its sprouts emerge from one end, and clusters of tubers grow down from the other. McClelland (2011: 35) correctly notes that in this configuration, manioc and the body of Botanical Frog resemble each other and are deliberately conflated.

McClelland identifies the animal's basic form as most akin to a "Smoky Jungle Frog" (*Leptodactylus pentadactylus*)**,** an aggressive species, with tiny fingers that resemble claws. Although this species is found on the north coast, it is more commonly distributed in the upper Amazon along the eastern slopes of the cordillera.

Several unique characteristics such as the "ear flaps" and spirals at the corner of the eyes, the stripes on front legs and spotted "pelage" markings suggest this represents the species *Leptodactylus pentadactylus*, which screams like a feline when grabbed (*ibid.*). Her identification rejects *Bufo marinus* as a possible representation even though they are common on the north coast. *Bufo marinus* is culturally significant for its hallucinogenic qualities, which would imply a direct reference to shamanic practice; yet, because the *Bufo marinus* has no recognizable traits in the Botanical Frog, the visual argument is lacking. Instead, numerous factors point to fertility and water associations. Reproductive success among frog species is generally tied to temperature, humidity, and the availability of water—factors that directly parallel successful agricultural cultivation (McClelland 2011: 32). They prefer moist environments, which on the arid north coast only exist in rivers, freshwater marshes, and aquaculture ponds such as *wachaques*. Along the coastal valleys, their proliferation would be associated with a specific temporal season.

That the Moche prominently featured nonlocal species such as *Leptodactylus pentadactylus* is perhaps surprising, however, the headwaters of major Amazonian tributaries are in some places only a four-day hike from the coast. Amazonian lore includes a long-standing tradition of frog imagery that was likely familiar to *costeños*. Wassen (1934b) carried out a comparative ethnography from various different indigenous communities and published detailed work to investigate broad commonalities on this subject. Frog imagery is ubiquitous, and frog motifs have wide distribution and significant time depth. Examples of small frog images among Aymara speakers around the Lake Titicaca involve placing them on top of the hills as a means of bringing down rain (Wassen 1934b: 344), Bandelier (1910) reported carved frogs on Koati island. Other specific examples include the Tucuna of the upper Marañon, where girls carry sticks with carved images of frogs, a sign of fertility, as part of feasts associated with rites of passage (Wassen 1934b: 333). Ethnographic sources from Guyana documented the *Hyla venulos* frog, which emits a call similar to a cow when rain is expected the next day. It is perhaps for this reason the frog was considered an infallible weather prophet (Wassen 1934b: 332).

While ethnographic examples bolster the Botanical Frog motif as having probable associations with agricultural fertility concepts, the assertion is significantly strengthened by its representation in pictorial narratives, as articulated in the sample by McClelland (2011). In one complex low relief composition, multiple characters engage in a Harvest Scene (Fig. 10.8). Divided into two registers, supernatural beings gesture toward clumps of produce, apparently maize, with a large Botanical Frog among them. Maize stalks, manioc capsules, and other plants sprout from its body. Below, a procession of anthropomorphic monkeys carry large bags of produce toward a higher-status overseer. McClelland (2011) concludes the Botanical Frog likely embodies the Moche concept of agriculture. From the perspective of visual culture the composite frog creature was never intended to represent a real animal; it's a deliberate ideographic evocation of a larger concept.

In almost every major Moche artistic theme involving food cultigens, complex imagery, and iconographic elements are linked to individuals, deities, rituals, and activities surrounding agricultural production. Images of common beans and large

Fig. 10.8 Harvest Scene with deity and Botanical Frog, rollout drawing of press relief imagery on ceramic bottle (Mint Museum of Art, after McClelland 2011: Fig. 17b)

lima beans (*Phaseolus. vulgaris* and *P. lunatus* L.), present a most remarkable exception. Beans are highly nutritious and rich in protein, legumes, among the most longstanding Andean food staples. Indeed, beans were some of the earliest cultigens in the New World. Samples from Guitarerro Cave date back to 8000 B.C. (Smith 1980: 111). Yet, along the north coast, the consumption of beans and their associations with particular social classes has changed over time. Archeologists working on the north coast have documented samples from various sites in the Moche and Chicama Valley. The evidence suggests that the local diet included chili, squash, lucuma, avocado, common beans, and maize (Ryser 2008: 398). The absence of lima beans during the Early Intermediate Period is surprising because data indicate a much higher presence during the earlier Initial Period Chavín culture and subsequent Chimú culture in the Late Intermediate Period. Ryser (2008: 397) suggests that subsistence changes with regard to lima beans among the Moche are related to changes in the political economy, asserting that non-nutritional factors provide evidence for a socially constructed "non-subsistence classification" for beans in Moche culture (2008: 398), this is readily supported by Moche imagery indicating that more common bean varieties were a staple in the Moche diet.

Lima beans (*pallares*) have distinct symbolic associations among plant representations in the Moche imagery. They are the only plant truly anthropomorphized, complete with heads, arms, and legs (Benson 1972: 82). Conspicuously associated with elite status, beans are depicted as warriors, messengers, and even appear to be counting beans, all of which give tantalizing clues about social roles and activities (Donnan 1978). Beans are very rarely shown as fully naturalistic. Modeled and painted images of high status bean warriors exhibit typical warrior characteristics, including crescent headdress, large earspools, war club, and shield; the backside of the body, however, has the form of large beans (Fig. 10.9). Other than their warrior status, little else gives direct clues about the connection between beans and warriors. In some versions, the bean warrior confronts a deer warrior (see McClelland and Donnan 2009: Figs. 32 and 109, for examples), and based on iconographic interchangeability of deer and sacrificial captives found elsewhere in Moche art (Donnan 1997), one suspects that the bean warrior may be the eventual victor in the scene.

Beans are also related to scenes of running messengers, not as protagonists, but because the messengers appear to carry them in small bundles. The messenger theme was initially identified by Larco Hoyle (2011 [1938–1939]) who noted that messengers run with dynamic striding pose and wear an ornate headdress, ear spools, and

Fig. 10.9 Painted bottle with seated Bean Warriors (VMFA 82.193, Virginia Museum of Fine Arts, Richmond). (Gift of Mr. and Mrs. Sandford G. Etherington. Photo: Katherine Wetzel © Virginia Museum of Fine Arts)

wrist cuffs. Their distinctive, disk-shaped headdress serves as a key identifier, and copper or copper–gold alloy disks of this kind are known archeologically (Donnan 1978: 76). The headdress disk typically has a feline face in its center, with rounded bosses or raised spheres surrounding the edges. It would have been worn attached to a wrapped textile headband, itself ornamented with a feline head. Zoomorphic versions convey an impression that only the swiftest of creatures would be messengers (e.g., foxes, falcons, owls, centipedes, and hummingbirds). Expanded versions of the scene generally show multiple runners traversing hilly territory holding small bundles.

Rafael Larco Hoyle (2001 [1938–1939]) had an unshakable conviction that Moche ceramics incorporated script elements analogous to Maya hieroglyphs. Numerous painted scenes show seated figures—not all of whom are necessarily human, facing each other manipulating groups of marked beans. He interpreted the lines and spots on these images of beans as evidence of writing, and spent considerable effort seeking to discover the underlying meaning. He found archeological evidence of beans with inscriptions, and asserted that the bundles carried by the messengers were very similar to *llama* skin bundles, also retrieved in archeological excavations (2001 [1938–1939]: Figs. 188, 189, 196a, b). Ultimately, he was unsuccessful in discovering an internal system of notation in the marked beans, although his initial intuition remains compelling. The marked beans continue to be a source of interest, as they seem surely to have been carriers of some type of glyphic, mnemonic, or numerical meaning.

Ryser (2008: 400) observed that 77% of the vessels with primary anthropomorphized lima bean warrior imagery are associated with the Moche Phase IV style. Their sudden prominence in the imagery coincides with an overall emphasis on captives, combat, status, and blood sacrifice during Phase IV. Anthropomorphized bean characters are marked with a particular social status best described as semi-elite. They do not wear high elite regalia or have divine attributes, such as those associated with corn and manioc, but rather are usually associated visually with human concerns.

It may be as Larco Hoyle suggested (2001 [1938–1939]), that in Moche pictographic lexicon, beans signified a social activity and office so widely established and well understood, they did not function as signifiers for a food source, but instead as an ideograph for high status messengers or agents of elites. One might even speculate the nature of the messages being transmitted—urgent affairs of state, important results of prognostication effected through use of marked beans, perhaps even numbers or tallies of accounting signified by beans. All these would be good hypotheses, yet lacking additional data, impossible to prove. Equally speculative, therefore, would be any hypothesis about how they figure into the larger liturgical purpose of the ceramics themselves. In any case, it seems true that for Moche ceramicists, lima beans did not reside in the same conceptual category as other food cultigens.

Despite Moche's general symbolic insistence on depictions of ripe produce, rather than on images of prepared food or feasting, there is one persistent theme featuring images of foodstuffs prepared and served–the Stacked Plates motif. Consistently shown as pairs of large open bowls, brimming with food, the theme appears in both painted and sculpturally modeled varieties. A significant number of surviving examples feature one bowl with peanuts over another with *ají* peppers (although the elongated pods might also indicate manioc capsules) (Figs. 10.10a). Foods depicted include maize, potato, pacay, fruits, and occasionally fish or crayfish (*camaron del rio*). That the plates are intended to function as pairs is evident by the painted or sculpted addition of a cord joining the plates.

Fig. 10.10 **a** Modeled bottle depicting stacked pairs of plates containing peanuts and aji peppers (Courtesy of Museo Larco, ML006428, Lima, Peru); **b** Rollout drawing of painted bottle with Presentation of Stacked Plates scene (Donnan and McClelland Fig. 5.69; after Larco 1939:lam. 31)

(b)

Fig. 10.10 (continued)

Various painted Moche bottles convey an idea of how paired plates were used, illustrating the self-referential nature of liturgical bottles. In one example, a lord richly arrayed is seated inside a structure atop a stepped pyramid (Fig. 10.10b). This lord gestures toward the funerary vessels immediately surrounding him. A standing figure faces him, and they appear engaged in an interaction, presumably involving the presentation of the many rows of paired gourd bowls behind him. The Moche bowls are noteworthy because they illustrate a characteristic feature of the imagery—there are no insignificant figures—no nondescript servants carrying these bowls, only the bowls themselves merit inclusion in the illustration. They are shown having feet, walking themselves toward the pyramid. On the opposite side of the same vessel is another stepped pyramid with several additional figures sit facing the lord. One is dressed almost identically to the seated lord but lacks a feline headdress. Given that these two clusters of characters appear on opposite sides of the same bottle, it is unclear whether these figures represent the same lord in a narrative sequence or a different person altogether. They all seem to be chanting or singing. Rows of tall-necked jars with convex lids stand to the right. Like the paired offering plates, the jars are also notably anthropomorphic through the addition of feet and ropes around their "necks," conceptually equating them to prisoners. Sacrificial prisoners were usually paraded to the pyramid to have their blood let. These "captured" vessels walk themselves forward to pour out their liquid contents into shallow offering bowls. The addition of feet serves as an artistic convention to indicate movement, visually reiterating the central action and message of the scene: offerings are being made. Although these bowls contain food offerings, their presence is in no way related to eating. Their significance derives from their status as offerings, rather than their alimentary value.

The exact events that triggered the presentation of paired-plate offerings are unknown. It's likely, however, that in some cases they were offered as part of high status funerary events. This may not have been the only context in which they were used, but the well-known Moche Burial scene, as articulated by Donnan and McClelland (1979: 7, Figs. 2122) makes the connection clear. Complex fine line painted images show a very high status burial that featured many, many pairs of plates, shown stacked inside the burial chamber of the deceased person. It may be that the presentation of plates, illustrated in the previous figure, represents the ritual precursor to the inclusion of the food into tombs as funerary offerings. Gumerman (1994: 395–410) notes the presence of many such large, open bowls with food offerings in high status tombs at Pacatnamu—over half of which contained ears of maize. He also observes that proportionally, the contents of funerary food offerings were significantly different from normal dietary patterns, and that maize cobs with unusual numbers of rows were being selected; which suggests that food for the dead carried increased "symbolic" value.

When compared alongside painted versions, modeled pairs of stacked plates become more comprehensible, their value becomes clear. Depictions of paired bowls of food are almost certainly direct references to making offerings. Because the sculpted compositions contain no narrative content or indications of action events, they stand alone as metonymic references to the activity of making offerings. The activity of making offerings—of these particular offerings—may have been associated with specific rituals, which themselves were triggered by certain conditions, but deriving meaning would have been a constructive effort on the part of the viewer. The meaning depended on knowledge of the social context in which the object was used and by the person using it. The ceramic vessel and the human agent who used the vessel provided mediation for the transmission of meaning.

Ultimately, the same holds true for all the vessels discussed here. In every case, the vessel's imagery carries multiple meanings that require human agents to make conceptual connections, and these connections depended on cultural knowledge to supply the catalyst.

Summary and Conclusions

Moche art cannot be regarded in any way as if it were a vernacular snapshot into their world. It is, and always has been, first and foremost a symbolic language developed in response to very specific needs and functions. It was never intended to be a catalog of the Moche natural world. Because we are obliged to look back upon the Moche through a tiny keyhole of archeological time, it's appropriate to investigate the internal structure of visual communication through alternately contextual and hermeneutic lenses. Scholars are disadvantaged by a lack of cultural knowledge regarding referents to such symbolism. Nonetheless, every investigation of imagery produced by Moche culture must always return to the basic intersection of place, agency, and vehicle.

Although described elsewhere, what I refer to is a fundamental matrix where meaning is constructed to such imagery (Jackson 2011, 2012). Architecture, costume, and liturgical arts, as the three primary contexts in which Moche's specialized iconography most commonly appear, together generated a critical societal nexus where meaning resided. Over the years, excavations of royal tombs and other sites have made it clear that much of the iconography on ceramics refers to social roles that were acted out in real life. As "actors" in the events, the participants held status and gained status from social enactments related to veneration and propitiation of deity, celebration of key events, such as planting and harvest, and other religiously tuned actions, including blood sacrifice and funerals. Moche monuments were clearly a primary venue for assertion and affirmation of ideological constructs, as such they became the conceptual markers for place, loci for social cohesion, self-identity, and physical manifestation of ideological beliefs. As staging areas for the performance of ritual events, the monuments provided ideal places for enactment of liturgy and other social narratives.

Costume and accouterments jointly supported the performance aspects of agency—they were both part of, and props for, the compelling sequences of events to which the iconography refers. Fineware ceramics, such as those discussed here, created for the observance of religious liturgy, constitute a third area of cultural production. They were actors in the performance of ritual, they mediated between human participants and intangible forces, and physically confirmed that performance had occurred. Ultimately, they became primary vehicles for the commemoration and dissemination of social narratives. Thus, Moche fineware bottles and jars do not depict food as something to be consumed or savored, but rather, as something in a larger ideological construct surrounding the veneration and celebration of the continuance of agricultural beneficence.

This brief assessment of Moche food imagery reveals an iconographic emphasis on certain kinds of food as especially meaningful in an overall liturgical process dedicated to agricultural propitiation. Various factors readily support the association of deity in symbolic representations of maize and manioc. Hastorf (2003) has written in detail about the social and economic reasons that certain foods, maize, in particular, enjoyed elite status. Gumerman (1993) makes clear that Moche elites carefully selected "special" ears of corn to be placed in burials. Manioc as major agricultural staple, which heretofore has received relatively little attention in the Andean literature, also appears regularly in combination with Ai Apaec imagery and conflated with the Botanical Frog motif. As a plant whose very lineage was said to directly descend from the sun god, a staple whose characteristics allowed for long-term nutritional viability and flexible time of harvest (especially important during off-peak seasons or times of hardship), it seems perplexing that manioc does not appear with greater frequency in the archeological record. The artistic record suggests that it should be more solidly represented among those data—a *lacunae* that deserves further investigation.

Notes

1. See Museo Larco ML002277, for example.
2. See Uceda et al. (1994), Lam. 12, pp. 296, for image; for more on zoomorphic symbolism, see Jackson (2016).
3. Narrative themes are discussed at length by various authors, as for example, the Presentation Scene, also known as the Sacrifice Scene (Donnan 1978), Mountain Sacrifice Scene (Zighelboim 1995), Copulation Ceremony (Bourget 2006), the Owl Curandera (Benson 1972)
4. Images of stacked plates of food, as seen in the Moche Burial Scene (Donnan and McClelland 1979), for example, correspond with food offerings in high status burials at Sipán and elsewhere. Other examples include San Jose de Moro, where a Moche Priestess was interred with a ceremonial goblet associated with blood

sacrifice (Donnan and Castillo 1994), and Huaca de La Luna, where bodies of sacrificial victims, apparently captive warriors, were accompanied by modeled prisoner effigies and portraits that had been deliberately smashed (Bourget 2006).
5. In addition to a supernatural being, Pachacamac refers to a sacred place and oracle venerated throughout coastal prehistory. Located south of present-day Lima, it was the site of monumental architecture and a temple where a sculpted "idol" was venerated. It's power was derived in part by its ability to predict earthquakes and therefore has an indirect association with Illapa or lightning; (Calancha ch.19: 918–940; and, Staller, ed. note).
6. For image, see Lowe Art Museum, LAM94.0021.02; or Museo Larco ML 002,968.
7. American Museum of Natural History, no.1/907; see drawing by Donna McClelland, Jackson 2008: Fig. 2.5.

References

Acosta PJ de (1954 [1590]) Historia natural y moral de las Indias. In: Obras del P. Jose de Acosta, Mateos PF (ed.), Biblioteca de autores espanoles. Ediciones Atlas, Real Academia Espanola, Madrid
Alva W (1994) Sipán. Colección del Cultura y Artes de Perú, Lima
Benson EP (1972) The Mochica. Praeger Publishers, New York
Benson EP (1997) Moche art: myth, history and rite. In: Berrin K (ed.), The spirit of Ancient Peru. Thames and Hudson and Fine Arts Museum, San Francisco
Benson EP (2012) The worlds of the moche on the north coast of Peru. University of Texas Press, Austin
Bourget S (1990) Des tubercules Pour la Mort. Bulletin de l'Institut Français d'Études Andines 19(1):45–85
Bourget S (2001) Rituals of sacrifice: its practice at the Huaca de La Luna and its representation in moche iconography. In: Pillsbury J (ed.), Moche art and archaeology in ancient Peru. Center for Advanced Study of the Visual Arts, National Gallery of Art, Washington DC
Bourget, Steve (2006) *Sex, Death and Sacrifice in Moche Religion and Visual Culture.* University of Texas Press, Austin.
Burger R, van der Merwe N (1990) Maize and the origin of highland chavin civilization: an isotopic perspective. Am Anthropol 92(1):85–95
Campana C, Morales R (1997) Historia de una Deidad Mochica. A and B, SA., Lima
Chambi N, Chambi W (1995) Ayllu y papas: cosmovisión, religiosidad y agricultura en Conima, Puno. Asociación Chuyma de Apoyo Rural "Chuyma Aru", Puno [Peru]
Cieza de Le6n, Pedro de (1924 [1553]) La Cr6nica General del Peru, Urteaga HH, vol 7. Historiadores Clasicos del Peru, Lima
Cobo BP (1956 [1653]) Historia del Nuevo Mundo, Mateos PF, Obras del P. Bernabe Cobo. Biblioteca de Autores Espanoles. Ediciones Atlas, Real Academia Espanola, Madrid
Deza MC (1995) Estudio de la dieta y recursos alimenticios de los antiguos habitantes del complejo arqueologico, Huacas de Moche. Informe de Practicas Pre-Profesionales de Arqueologia, Facultad de Ciencias Sociales, Universidad Nacional de Trujillo, Trujillo, Peru
Dillehay TD, Rossen J et al. (2007) Preceramic adoption of peanut, squash, and cotton in northern Peru. Science 29, 316(5833):1890–1893 (June)
Dobkin de Rios M (1977) Plant hallucinogens and the religion of the Mochica—an ancient Peruvian people. Economic Botany 31(2):189–203 (April)

Donnan CB (1978) Moche art of Peru: Pre-Columbian symbolic communication. University of California Press, Los Angeles, Museum of Cultural History
Donnan CB (1997) Deer hunting and combat. In: Berrin K (ed.), The spirit of ancient PeruFine Arts Museum of San Francisco. Thames and Hudson, San Francisco
Donnan CB, Castillo LJ (1994) Excavaciones de tumbas de sacerdotisas Moche en San Jose de Moro, Jequetepeque. In: Uceda S, Mujica E (eds.), Moche: propuestas y perspectivas. Universidad Nacional de La Libertad, Lima
Donnan CB, McClelland D (1979) The burial theme in moche iconography. In: Studies in Precolumbian art and archaeology, no. 21. Dumbarton Oaks Research Library and Collection, Washington DC
Donnan CB, McClelland D (1999) Moche fineline painting: its evolution and its artists. University of California—Los Angeles Fowler Museum of Cultural History, Los Angeles
Dunn ME (1979) Ceramic depictions of maize: a basis for classification of prehistoric races. American Antiquity, 44(4):757–774 (Oct.)
Fornee N de (1965 [1586]) Descripcion de la tierra del corregimiento de Abancay, de que es Corregidor Niculoso de Fornee. In: de la Espada DMJ (ed.), Relaciones geograficas de Indias-Peru, vol. 2. Biblioteca de Autores Espanoles. Ediciones Atlas, Real Academia Espanola, Madrid
Foster N, Cordell LS (1992) Chilies to chocolate: food the Americas gave the world. University of Arizona Press, Tucson
Graves C (2001) La papa: tesoro de los Andes: de la agricultura a la cultura/The Potato. Treasure of the Andes, Centro Internacional de la Papa/International Potato Center, Lima
Golte J (2015) La cerámica Mochica como sistema de communicación. In: Castillo LJ (ed.), El Arte Mochica del Antiguo Perú. Museo Larco, Lima; Obra Social "La Caixa", Barcelona
Guaman Poma de A (1993) Nueva corónica y buen gobierno, vols 1–3, Pease F (ed). Fondo de Cultura Económica, Lima
Gumerman G IV (1994) Corn for the dead: the significance of zea mays in moche burial offerings. In" Johannessen and Hastorf (eds.), Corn and culture in the prehistoric New World, ch. 22. Westview Press, Boulder (Colorado)
Hastorf CA (2003) Andean luxury foods: special food for the ancestors, deities and the elite. Antiquity 77:545–554 (Sept.)
Hastorf CA, Whitehead WT, Bruno MC, Wright M (2006) The movements of maize into middle horizon Tiwanaku, Bolivia. In: Staller JE, Tykot RH, Benz BF (eds.), Histories of Maize, multidisciplinary approaches to the prehistory, linguistics, biogeography, domestication, and evolution of Maize. Elsevier Academic Press, San Diego
Hastorf CA, Johannessen S (1994) Becoming corn-eaters in prehistoric America. In: Johannessen and Hastorf (eds.), Corn and culture in the prehistoric new world, ch. 24. Westview Press, Boulder (Colorado)
Hastorf CA, Johannessen S (1993) Prehispanic political change and the role of maize in the central andes of Peru. Am Anthropol 95(1):115–138
Jackson MA (2008) Moche art and visual culture. University of New Mexico Press, Albuquerque
Jackson MA (2011) Moche as visual notation: semasiographic elements in moche ceramic imagery. In: Boone E, Urton G (eds.), Their way of writing. Dumbarton Oaks Research Library and Collection, Washington DC
Jackson, MA (2012) The mediated image: reflections on semasiographic notation in the Ancient Americas. In: Joshua E (ed.), Agency in Ancient Writing. University Press of Colorado, Denver
Johannessen S, Hastorf CA (eds) (1994) Corn and culture in the prehistoric new world. Westview Press, Boulder (Colorado)
Larco Hoyle R (2001 [1938, 1939]) Los Mochicas, vol 1 & 2. Museo Arqueológico Rafael Larco Herrera, Lima
Larco Hoyle R (1948) Cronología arquología del norte del Perú. Sociedad Geográfica Americana, Buenos Aires
Liddell K, Liddell LJ (2000) Cutaneous leishmaniasis manifested in the Moche pottery of ancient Peru. British J Dermatol 142(57):86

Makowski K (2005) Hacia la reconstrucción del panteón Moche. In El Mundo Sobrenatural Mochica, Universidad de Varsovia, Polonia and PUCP, Fondo Editorial, Lima
McClelland DM (2011) The Moche botanical frog. Arqueología. Iberoamericana 10:30–42
Melka TS (2010) The moche Lima beans recording system, Revisited. Folklore 45
Morales Gamarra, R (2003) Iconografía litúrgica y contexto arquitectónica en Huaca de la Luna. In: Uceda and Mujica (ed.), Moche Hacia el Final del Milenio. PUCP/UNT, Lima
Oviedo GF de (1959 [1535]) Historia general y naturalde las Indias. In: de Tudela Bueso JP (ed.), Gonzalo fernandez de Oviedo. Biblioteca de Autores Espanoles. Ediciones Atlas, Real Academia Espanola, Madrid
Palomino D (1965 [1549]) Letter to the king of Spain. In: de la Espada DMJ (ed.), Relaciones geograficas de Indias-Peru, vol. 3. Biblioteca de Autores Espanoles. Ediciones Atlas, Real Academia Espanola, Madrid
Pearsall DM (1994) Issues in the analysis and interpretation of archaeological maize. In: Johannessen and Hastorf (eds.) Corn and culture in the prehistoric new world, ch. 15. Westview Press, Boulder (Colorado)
Pearsall DM (1978) Early movement of maize between Mesoamerica and South America. J Steward Anthropol Soc 9:41–75
Pearsall DM (2008) Plant domestication and the shift to agriculture in the Andes. In: Helaine S, William I (eds.), *Handbook of South American Archaeology*, ch.7. Springer, New York
Pozorski S (1976) Prehistoric subsistence patterns and site economics in the moche valley, Peru. Ph.D. Dissertation, University of Texas, Austin
Pozorski S (1979) Prehistoric diet and subsistence of the moche valley, Peru. World Archaeol 11(2):163–184 (Oct.)
Quilter J (2010) The moche of ancient Peru: media and messages. Harvard University, Peabody Museum Press, Cambridge MA
Rucabado J (2015) Ai Apaec. Héroe Mochica. In El Arte Mochica del Antiguo Perú, Museo Larco, Lima
Russell G, Margaret AJ (2001) Patronage and political economy at cerro mayal, Peru. In: Pillsbury J (ed.), moche art and archaeology in ancient Peru. Center for Advanced Study of the Visual Arts, National Gallery of Art, Washington DC
Russell G, Banks LL, Jesus BR (1998) The Cerro mayal workshop: addressing issues of craft specialization in moche society. In: Shimada I (ed.), Andean ceramics: technology, organization, and approaches, MASCA Research Papers in Science and Archaeology, Supplement to Vol. 15. Museum Applied Science Center for Archaeology, University of Pennsylvania Museum of Archaeology and Anthropology, Philadelphia
Ryser GL (2008) Moche bean warriors and the paleobotanic record: why privilege beans? In Castillo LJ (ed.), Arqueología mochica: nuevos enfoques. Fondo Editorial, Pontificia Universidad Católica del Perú, Lima, Perú
Smith AF (2011) Potato: a global history. Reaktion Books, London
Smith E Jr. (1980) Plant remains from Guirarrero Cave. In: Lynch JF (ed.), Guitarrero cave, early man in the Andes. Academic Press, New York
Staller JE (2006) The social, symbolic, and economic significance of *Zea mays* L. in the late horizon period. In: Staller JE, Tykot RH, Benz BF (eds.), Histories of Maize, multidisciplinary approaches to the prehistory, linguistics, biogeography, domestication, and evolution of Maize. Boston: Elsevier Academic Press, San Diego
Staller JE (2008) Dimensions of place: the significance of centers to the development of andean civilization: an exploration of the *Ushnu* Concept. In Staller JE (ed.), *Pre-Columbian landscapes of creation and origin,* Ch. 9. Springer, New York
Tate J (1998) Maize variability in the moche valley Peru. Ph.D. Dissertation, Anthropology, Northern Arizona University, Flagstaff. Accessed https://cline.lib.nau.edu/search/X?SEARCH=Maize%20Variability%20in%20the%20Moche%20Valley%20Peru%2C
Towle MA (1961) The ethnobotany of Pre-Columbian Peru. Aldine Publishing Company, Chicago

Uceda S (2001) Investigations at Huaca de la Luna, Moche Valley: an example of moche religious architecture. In: Pillsbury J (ed.), Moche art and archaeology in ancient Peru. Center for Advanced Study of the Visual Arts, National Gallery of Art, Washington DC

Uceda C., Santiago, Ricardo Morales G., Jose Canziani A. and Marie Montoya V. (1994) Investigaciones sobre la arquitectura y relieves polícromos en la Huaca de La Luna, valle de Moche. In *Moche: Propuestas y Perspectivas.* edited by Santiago Uceda and Elias Mujica, Universidad Nacional de La Libertad, Lima.

Urteaga-Ballon O (1991) Medical ceramic representation of nasal leishmaniasis and surgical amputation in ancient Peruvian civilization. In: Aufderheide AC, Ortner D (eds.), Human paleopathology: current syntheses and future options, pp 95–101, Zagreb Paleopathology Symposium 1988, Smithsonian Institution Press, Washington, DC

Urton G (1981) At the crossroads of the earth and the sky. University of Texas Press, Austin

Ugent D, Sheila P, Thomas P (1986) Archaeological manioc *(Manihot)* from coastal Peru. Economic Botany 40(1):78–102 (January)

Ugent D, Sheila P, Thomas P (1982) Archaeological potato tuber remains from the casma valley of Peru. Economic Botany 36(2):182–192 (Apr.–Jun.)

Vasquez Sanchez V, Teresa Rosales T (2011) Origen del Maíz Andino: Arqueobotánica y Genética Molecular. Sian Revista Arqueológica, Año 16, Edición 22, (Dec.). Trujillo

Wassén H (1934a) The frog-motive among the south American Indians. Ornamental Studies (part 1). Anthropos, Bd. 29, H. 3./4., pp. 319–370 (May.–Aug.)

Wassén H (1934b) The frog in Indian mythology and imaginative world (part 2). Anthropos, Bd. 29, H. 5./6., pp. 613–658 (Sep.–Dec.)

Weiss P (1961) La asociación de la uta y veruga peruana en los mitos de la papa, figurados en la cerámica mochica y chimu. Revista de Museo Nacional (Lima) 30:65–77

Yacovleff E, Herrera F (1934–35) El mundo vegetal de los antiguos peruanos. Revista del Museo Nacional (Lima), tomo 3:241–322

Zighelboim A (1995) Mountain scenes of human sacrifice in moche ceramic iconography. In: Zighelboim A, Barnes C (eds.), Current research in andean antiquity. J Steward Anthropol Soc 23(1–2):153–88. Urbana (1995).

Part II
Andean Foodways, Indigenous Customs, and Transformations among Colonial and Contemporary Andean Cultures

Chapter 11
Maize in Andean Food and Culture: Interdisciplinary Approaches

John E. Staller

Abstract Interdisciplinary evidence on Andean maize (*Zea mays* L.) consumption indicates that it was primarily consumed as a fermented intoxicant or maize beer (*chicha*). Its consumption with other food crops and meat was that it was commonly added to stews or as a side dish with meat and fish, unlike Mesoamerica maize was not processed in the Andes. Processing maize in an alkali solution, such as lime water [$CaOH_2$], or wood ash and then cooking or steeping it before grinding it into flour (*masa*) with *manos* and *metates* and making it into tamales, tortillas etc., a process referred to as *nixtamal*, incrementally increased its nutritional value, and economic importance; however, this was not the case in the Andes. Nevertheless, maize played a central role in the rise of Andean civilizations and complexity primarily because of its roles in pre-Columbian Andean rituals and rites. Ethnohistoric accounts and ethnographic evidence are presented indicating that despite the fact the Spaniards preferred maize to other food staples, maize beer consumption was rejected because of its association and importance to traditional rituals, rites, and ethnic identity among indigenous Andeans. This research on maize also indicates the changes in preparation and consumption and associations with cuisines, rites, rituals, and rituals.

Keywords Maize · Andes · Food science · Archaeology · Ethnohistory

Introduction

Interdisciplinary evidence on Andean maize (*Zea mays* L.) consumption indicates that it was primarily consumed as a fermented intoxicant or maize beer (*chicha*). Its consumption with other food crops and meat was that it was commonly added to stews or as a side dish with meat and fish. Unlike Mesoamerica, maize was not processed in the Andes. Processing maize in an alkali solution such as lime water usually consisting of lime (calcium hydroxide), [$CaOH_2$] or wood ash then cooking or steeping the kernels to remove the transparent skin or pericarp before grinding it

J. E. Staller (✉)
Botanical Research Institute of Texas (BRIT), Fort Worth, USA
e-mail: jstaller@earthlink.net

J. E. Staller (ed.), *Andean Foodways*, The Latin American Studies Book Series,
https://doi.org/10.1007/978-3-030-51629-1_11

into flour (*masa*) with grinding stones, that is, *manos* and *metate*, which incrementally increased its nutritional value and economic importance; however, this was not the case in the Andes. Nevertheless, maize played a central role in the rise of Andean civilizations and complexity primarily because of its roles in pre-Columbian Andean rituals and rites. Ethnohistoric accounts and ethnographic evidence are presented indicating that despite the fact the Spaniards preferred maize to other food staples, maize beer consumption was rejected because of its association and importance to traditional rituals and ethnic identity among indigenous Andeans. This research on maize also indicates significant changes in the preparations, consumption, and associations with cuisines, rites, and rituals in different regions of the Andes.

There has been growing acknowledgment in the literature that specific Andean foods and cuisines are highly marked culturally and ethnically. Foodways convey complex social meanings as well as provide critical information about ancient political economies, regional interaction, social complexity, and acculturation among and between cultures and regions. The Andes are characterized by complex ecologies and diverse environments dominated by the second-highest mountain range in the world (Fig. 11.1). The extreme geography results in food crops that are specifically adapted to different altitudes of the cordillera and dense tropical forests of the Upper Amazon. Once such crop is maize (*Zea mays* L.), one of the most important foods in the Andes. In Andean cuisines, people consumed maize as a vegetable to accompany other foods rather than as a grain or as the focal point of meals (Valcárcel 1946: 477–478; Sauer 1950: 494; Nicholson 1960). Additionally, Andean people consumed maize as the fermented intoxicant maize beer or *chicha* (Nicholson 1960; Coe 1994). The brewing of *chicha*, its consumption, and social meaning had great significance for the Inca State and earlier civilizations; however, its role as a food staple is debatable.

The general perception in the anthropological sciences is that maize played an important role in the origins of agriculture and early subsistence in many areas of the New World (Ford 1969; Pearsall 1999, 2003; Pozorski and Pozorski 1987; Pearsall and Piperno 1990). Anthropologists, and particularly archaeologists, have long documented that maize played a central role in the rise of complex sociocultural development and pre-Columbian Andean political economies (Murra 1972, 1976, 1980). Some scholars assert that maize was an economic food staple that had a primary role in the origins and expansion of Andean civilizations (Pozorski and Pozorski 1987; Piperno and Pearsall 1998).

In addition to archaeological evidence, early chroniclers describe how people consumed maize giving the impression that it was the principal foodstuff. For example, various accounts state that the main food along the coast was "*maize and chile and greens,*" along with meat protein derived from maritime or aquatic resources such as fish, crab, shrimp, mollusks; or with terrestrial mammals such as brocket deer; or domesticated mammals such as camelids, particularly llamas and alpacas, and guinea pigs (Cieza de León 1998 [1553]; Molina 1989 [ca. 1557]; Guamán Poma 1980 [1583–1615]; Garcilaso de la Vega 1966 [1609]). Early colonial accounts from the coast may have understated the role of highland foods, such as

Fig. 11.1 Topographic Map of South America. The Andes has considerable variability with regard to altitude, topography, and ecological and environmental diversities. The cordillera spans the entire length of western portion of South America and represents the second highest mountain range on earth. The Amazon Basin has the largest contiguous area of the tropical rain forest and the greatest species biodiversity in the world. (Courtesy of Google Earth)

potatoes and other native tubers and grains, as well as squashes, peanuts, and other plant foods in the Andean diet.

Spanish conquest of the New World transformed Andean foodways and habitats through the introduction of exotic plants and animals from other parts of the world, particularly Western Europe and Asia; however, during the sixteenth century, Spaniards commonly relied on Andean foodstuffs for sustenance. Ethnohistoric accounts emphasize that early European explorers preferred maize, along with beans and squash, over other more exotic New World plant foods such as cassava or bitter manioc (*Manihot esculenta* Crantz) (Sauer 1969: 53–55). Although the early Spanish conquistadores preferred maize as a vegetable or side dish with other foods, meat and fish, they were not particularly inclined to consume it as beer, and subsequently the role of chicha changed significantly during the colonial period. *Chicha* remained popular among indigenous communities for their feasts and ritual celebrations; however, Spaniards viewed indigenous Andean people as having inferior status. Another factor that influenced the negative attitudes toward *chicha* by Spaniards was related to the Catholic Church. The Church condemned *chicha* because it was both an intoxicant and Catholic officials perceived of the associated rituals as idolatrous and sinful (Arriaga 1968 [1621]; Cobo 1990 [1653]). Even though Spaniards ascribed negative connotations to *chicha* as a beverage associated with individuals of low status, the social value of *chicha* remained strong for many indigenous people during the colonial era and is still important today for many inhabitants of the Andes, especially in traditional communities in the highlands.

Archaeological research from various pre-Columbian time periods and historical accounts from the colonial period have greatly influenced the perceptions and assertions that scholars in the social and biological sciences have made with regard to the economic importance of maize and its role in the origins of Andean civilizations and complex social organization (Lanning 1967; Damp 1984; Pearsall 1992, 1993, 1994; Piperno and Pearsall 1998). Although some scholars advocate for the primacy of maize in the development of Andean cultures, archaeological evidence indicates that maize was often one of many food crops cultivated to create varied cuisines in the Andes. The research presented here addresses how perceptions regarding the primacy of maize were formed, assumed, and asserted as well as the evidence presented in archaeological and ethnohistorical literature showing the social and culinary roles of maize.

Maize in Andean Food and Culture: Processing and Nutrition

Maize has a long history of cultivation in the Central Andean region. Maize was domesticated in Mesoamerica at c. 6000 B.C. (Matsuoka et al. 2002; Vigouroux et al. 2003; Benz and Staller 2006, 2010). Chronological evidence indicates that maize initially was introduced from Mesoamerica to the Andean coast at around

3500 B.C. (Pearsall 1990, 1999; Pozorski and Pozorski 1987; Pearsall and Piperno 1990). The earliest ^{14}C dates derive from the Valdivia site of Loma Alta (Norton 1982; Pearsall 2003: 223); however, the earliest directly AMS dated kernels are c. 2490 ± 40 B.C. (Zarrillo et al. 2008: 5006; Staller 2010a: Table 3.1). Recent multidisciplinary research from food science, ethnobotany, and stable carbon isotope analysis of ancient skeletons indicate that maize did not have a primary role in the Andean diet until long after its introduction c. 5500 B.P. (see e.g., Burger and van der Merwe 1990; van der Merwe et al. 1993; Gil 2003; Gil et al. 2006; Tykot and Staller 2002; Tykot et al. 2006; Staller and Carrasco 2010).

Analysis of ethnohistorical accounts indicates that various people in the New World referred to maize by the Taíno-Arawakan word, *mahiz*, or "life-giver" (Weatherwax 1954; Sauer 1969). It is possible that such references to the cultigen as the "life-giver" may have predisposed early European accounts to emphasize the economic importance of maize in the rise of New World civilizations (Coe 1994). Maize was clearly the primary subsistence staple in pre-Columbian Mesoamerica. Its economic importance is evident in the cuisines of various regions and documented by paleodietary evidence from stable carbon isotopes signatures from ancient skeletons (e.g., Christenson 2010; Friewald 2010; Cheetham 2010; Misra et al. 1992). In Mesoamerica, processed maize flour was consumed and formed into tamales, tortillas, enchiladas, tacos, etc. as the main component of meals (see Coe 1994; Wacher 2003; Cheetham 2010; Katz et al. 1974; Misra et al. 1992). The patterns of maize consumption in the Andes differ significantly from those in Mesoamerica.

Recent research in food science has documented that the preparation and processing of maize in Mesoamerica called *nixtamal* incrementally increase its nutritional benefits. Maize processing, that is, soaking or cooking kernels in an alkaline solution usually consisting of lime (calcium hydroxide), and wood ash (potassium hydroxide) removes the transparent skin or pericarp from the kernels (Staller 2018). This makes grinding the corn kernels into flour or *masa* easier, and maize processing into flour and this is what increases its nutritional benefits (Katz et al.1974; Misra et al. 1992; Coe 1994: 14; Wacher 2003: 738–739; Cheetham 2010; Staller 2018). The nutritional benefits of nixtamalization arise from alkaline processing convert corn-bound niacin to free niacin, so that niacin and vitamin B3 can be absorbed into the body during consumption (Katz et al. 1974; Cheetham 2010; Staller 2018). Vitamin B3 is also derived from the amino acid tryptophan, which is an essential micronutrient (Katz et al. 1976; Haas and Levin 2006). *Nixtamal* initially referred to any food processing involving removal of the pericarp from any grain by soaking in an alkali process (Haas and Levin 2006). However, in recent years, it has come to refer almost exclusively to maize processing, particularly in Mexico and Central America (Staller 2018).

Various lines of evidence indicate that in contrast to Mesoamerica where maize was a primary dietary staple, this was not necessarily the case in the Andes. This is in part related to the fact that in the Andes, maize was not processed through nixtamalization (Katz et al. 1974; Wacher 2003; Staller 2018). It was also related to the fact that numerous native Andean food staples (e.g., potatoes, various other tubers, quinoa, peanuts, squashes) had and continue to have much larger and more important roles in different regional Andean cuisines (Staller 2010a). When Andean people did consume maize with different meals, it was often served as a side dish, rather than ground into flour or *masa* as in Mesoamerica (Sauer 1969; Staller 2018). Furthermore, when maize was ground to make Andean foods, such as *humitas* or *locro*, maize was not soaked or processed in an alkaline fluid as was done in Mesoamerica.

Maize is highly mutagenic, and this is reflected by the great diversity of varieties or landraces in different regions of the Andes (Anderson 1947; Breiger 1958; Grobman et al. 1961; Timothy and Harry 1963; Timothy et al. 1961; Rivera 2006; Staller 2016). The phenotypic diversity of maize varieties is the product of human influence or deliberate conscious selection, or artificial selection (unconscious selection), for certain favorable traits (Timothy et al. 1961; Smith 2001; Staller 2010a, 2015a, b). This multitude of maize varieties has close historic meaning and importance to traditional ethnic identity and worldview. Maize varieties play a particularly central role in indigenous ethnic identities that link cultures to particular regions and to sacred places in the surrounding landscape (Dietler and Hayden 2001; Jennings 2004; Jennings and Bowser 2009; Staller 2006, 2008b).

Despite the fact that maize in the Andes was not processed as it was in Mesoamerica and maize was not a food staple around which indigenous societies built meals, it was and continues to be highly significant to rites and rituals and has great cultural and political importance (Staller 2010a). Maize was directly related to the spread of agricultural economies and sociocultural complexity as evident from the agricultural terraces and built landscapes throughout the central Andes dating back to the formative periods, during pre-Columbian times such terraces were primarily used for the cultivation of maize (see Murra 1973, 1975, 1980; Morris 1979, 1993; Chávez 1989, 2012; Chávez and Thompson 2006).

Andean Worldview and the Natural World: A Sacred Landscape

The ecological and environmental complexity of the Andes also played critical roles in the traditional worldview and particularly the importance of maize to the rise and spread of Andean civilizations. Andean religious ideology is a veneration of the natural world and a cultural perception of a sacred landscape (Zuidema 1982; Sullivan 1987, 1988; Staller 2006, 2008a, b, 2010a, 2014; Staller and Stross 2013). The natural landscape, its cycles, and rhythms are perceived as interrelated and dynamic expressions of mythology, history, economics, and intrinsically related to ethnic identity and various culinary traditions (Zuidema 1973, 1982, 2002; Sullivan 1988; Ramírez 2005).

Andean worldview assumes a mutual interdependence and an intrinsic interrelationship between culture and nature. Thus, the well-being of *huacas* depends upon a constant flow of ritual offerings from the living to deceased ancestors and the spiritual domain as a form of reciprocity. *Huaca* is a Quechua term for "sacred or extraordinary" had direct reference to certain places in the natural world (Sullivan 1988; Staller 2014). It is an all-encompassing term that refers to places in the landscape, mountains, certain locations associated with myth and legend, certain objects that were out of the ordinary or unusual in some ways, even venerated ancestors were also referred to by this term Garcilaso de la Vega (1966 [1609]: 73, 76–77; Staller 2008b: 269–270, 20). *Huacas* were believed to contain supernatural power in the form of a life force or animating essence *enqa* and were perceived as essential to the survival and well-being of nearby cultures (Sullivan 1988; Zuidema 2002). Spiritual energy or animating essence can be both hostile and friendly to humans (Allen 1988: 44–54, 65–66). This spiritual energy or *enqa* residing in the landscape, on mountain peaks, natural springs, caves, rivers, lakes, etc., and are ranked according to their presumed power (Zuidema 1982; Urton 1981: 39–48, 1985: 258–259; Rostworowski de Diez Canseco 1986; Ramírez 2005). Ritual and sacrificial offerings are essential to maintaining cosmological equilibrium and facilitating the exchange or circulation of animating essences among these spiritual entities in cultural and geographic spaces (Buechler and Buechler 1971: 9; Allen 1988: 207–208; Classen 1993: 14; Staller 2006: 453; z 2008b).

The ethnohistoric accounts clearly emphasize maize played a variety of roles and functions involved with Andean feasting, ceremonial activities, curing rituals, and religious rites of passage (Classen 1993: 80; Staller 2008b: Fig. 9.3). Among the Inca, maize played not only important social roles, but also involved a symbolic reference to elite status as well as an ethnic marker central to rituals surrounding the Cult of the Sun and the Dead as well as ancestor veneration (Betanzos 1987 [1551]; Cobo 1990 [1653]). There are numerous examples from various ancient Andean cultures that indicate maize was associated with various religious cults and consumed as *chicha* in ritual ceremonies long before the arrival of the Spaniards (Fig. 11.2). Spiritual forms and their symbolic connotations are therefore conditioned in part by the geography and also extended to the associated biota. Natural features and the plants and animal species adapted to such places were also perceived as sacred and temporally couched with reference to sacred rather than secular time (Turner 1985; Tilley and Bennett 2001; Staller 2014). Such places are associated with ritual performance, human sacrifice, and burials, all of which are sanctified and propitiated through ritual feasting and offerings (Cobo 1990 [1653]: 115–119; Rostworowski de Diez Canseco 1986: 31, 62; Tilley and Bennett 2001: 336). Therefore, Andean periodicities have particular references to sacred places on the horizon and to celestial bodies associated with them, particularly with the coming of climatic or seasonal events (Zuidema 1982, 1997, 2002, 2008). Thus, *chicha* was essential to feasts and feasting and particularly to ritual acts of reciprocity that reaffirm the social order and provide a communal means of venerating the surrounding landscape. For example, in central Peru, the Inca would plant maize in August, during the annual festival of *Hailli*, the point in the annual cycle in which the earth was "opened" (Fig. 11.3).

Fig. 11.2 The ruler or *Sapa* Inca is drinking *chicha* with his right hand and toasting the Sun during the Festival of the Sun or *Inti Rami*, which occurred each June. A woman sitting to the right is pouring *chicha* from a jar (*aríbalo*) into cups (*keros*) (From Guamán Poma 1980: 220 *fol.* 246 [1583–1615])

Fig. 11.3 The Inca planted maize each August during the festival of *Hailli*, the time to "open the earth." The woman on the right holds a cup (*kero*) of *chicha* in her right hand. The *chicha* or maize beer was often spilled during cultivation and planting as an offering to the earth to make the fields "fertile" (From Guamán Poma 1980: 1050 *fol.* 437 [1583–1615])

Maize Beer or Chicha in Andean Political Economy and Rituals

Early Spanish accounts indicate that the inhabitants of the Andes frequently consumed maize beer or *chicha*, but they did not eat much maize directly (Cieza de León 1977 [1551]; Molina 1989 [1575]; Garcilaso de le Vega 1966 [1609]; Arriaga 1968 [1621]). *Chicha* had a central role in the rise of Andean civilizations because it was of great importance to numerous rituals and rites during which it was often given as a ritual offering to *huacas* or sacred places in the surrounding landscape and to *Pachamama* the mother earth (Murra 1972, 1980; Otero 1951; Zuidema 1986, 1989; Staller 2008b, 2010a, b).

The social-cultural significance of maize to indigenous cultures is reaffirmed by the importance it continues to have in their lives and ethnic identities (Morris 1979, 1993). Indigenous brewers took great care not to spill a single drop and when it is spilled accidentally they explain it by saying, "the earth is thirsty" (Cutler and Cardenas 1947). In some regions, young indigenous girls customarily spill *chicha* on the earth during planting to make it more "fertile" (Cutler 1946; Coe 1994). Such customs and beliefs go back to the conquest (see Fig. 11.3). Cutler and Cardenas (1947: 33) suggest that mildly alcoholic brews were so common in some regions of the Andes, particularly Bolivia and Peru, that they may have contributed substantially to the ancient diet. However, ethnobotanical and isotopic evidence appear to indicate that this was not the case, at least not until late in the initial period and in other regions of the cordillera until much later (Tykot et al. 2006; Gil et al. 2006).

The early chronicler Fernandéz de Oviedo described maize processing in the preparation of *chicha*; stating that the indigenous cultures soaked the maize kernels in water until they spouted (*joras*), then the *joras* were boiled in water containing certain herbs for a considerable amount of time (Oviedo and Valdez 1969 [1535]: 136). Chroniclers make no mention of soaking the maize in an alkali solution nor is there any mention of the addition of saliva or chewed maize added to the fermenting mixture in the ethnohistoric accounts (e.g., Cieza de León 1977 [1551]; Arriaga 1968 [1621]). The *joras* and herbs were allowed to cool, kernels were left to sit for 3 or 4 days in a fermented mixture of liquid and then consumed as *chicha* or maize beer (Oviedo and Valdez 1969 [1535]: 136). In the Andean highlands and along the lowlands and coast *chicha* has been traditionally prepared by allowing the kernels to spout in the shade until the spouts were about an inch in length. The spouted kernels or *joras* chewed traditionally by women initiate the fermentation process and formed into small balls that are then flattened and laid out to dry. Naturally occurring enzymes in the saliva catalyzes the breakdown of starch in the kernels into maltose. The mashed fermented kernels were then strained and boiled and stirred for many hours with lime (Staller 2006: Fig. 32.1a). There is no mention of the addition of saliva or chewed maize along the coast, solely in the highlands.

Maize beer has a pale straw color, a slightly milky appearance, and a slightly sour aftertaste, a flavor reminiscent of hard apple cider. It is drunk either young and sweet or mature and strong. Maize beer usually contains only a small amount of alcohol, about 1–3% (La Barre 1938; Morris 1979). Allowing the maize kernels to spout makes it easier to grind them. Significantly, there is an association between the introduction of maize into the Andes and the appearance of grinding stones, manos and metates, and ceramics at archaeological sites (Damp and Vargas 1995).

In certain regions of the Andes, such as southern highland Ecuador, central and southern Peru, and highland Bolivia *chicha* also refer to fermented intoxicants made from *molle,* the Peruvian pepper tree (*Schinus molle* L.). The fruit of *Schinus molle* has a bright pink color and are often sold as "pink peppercorns." However, *S. molle* is not related to the true pepper (*Piper nigrum* L.), which is Asian, and native to southern India (Goldstein and Coleman 2004).

Andean cultures in the cordillera traditionally have maize beer or *chicha* manufactured by women (*chicherías*) in the cordillera, while along the pre-Columbian coast it is men (*chicheros*), (Arriaga 1968 [1621]: 12; Murra 1975 [1972]; Moore 1989; Rostworowski de Diez Canseco 1999; Cummins 2002). The women who prepared the *chicha* in the highlands would often spit or put saliva in the brew to augment the fermentation process. This may be related to the fact that in the highlands the Andeans often chewed coca with lime because of the altitude (Coe 1994).

Inca administrative centers in the lowlands and along the coast commonly prepared *chicha* from maize were cultivated in the surrounding landscape. Maize cultivation was solely for the purpose of ritual feasting sponsored by the Inca state for its subject populations (Arriaga 1968 [1620]: 34; Morris 1993; Morris and Thompson 1985). The Inca put considerable technological effort into transforming the mountainous landscape through the construction of vast irrigation terraces. Territorial expansion was based in part upon large-scale maize production on such terraces in both in the highlands and along the coast (Murra 1972, 1980). The transformation of mountainsides into garden terraces has pre-Inca origins, but they took highland terrace construction to an unprecedented scale (Morris 1993). The mountainside terraces near Pisac in the Urubamba Valley were particularly important because the maize variety of this region was manufactured into beer or *chicha* at various Inca administrative centers (Staller 2006: 463–464; Morris and Thompson 1985). Its importance among the Inca in Cuzco related to the perception that the valley of Urubamba was sacred, and only the maize landrace cultivated there could be used to prepare *chicha* and cuisines that were consumed by Inca nobility (Fig. 11.4).

Chroniclers state that maize beer or *chicha* was also central to ancestor veneration and rites associated with agricultural fertility (Arriaga 1968 [1621]: 56; Staller 2006: 454–456). Arriaga (1968 [1621]: 41) stated that maize beer or *chicha* was the principal offering, "*the best and most important part*…" of Inca rituals. Such associations emphasize the importance of drinking and toasting *chicha* as a recognition of ethnic status as well as hierarchy and rank (MacCormack 2004: 107; Staller 2008b: 456–458, 2016). However, the large-scale consumption of fermented intoxicants, particularly beer, was the critical component of almost every Andean social, political, or religious transactions (Murra 1972; Staller 2006: 454).

Maize beer played a fundamental role in all fertility rites and rituals and its consumption was what linked humans to their gods through the fecundity of the earth (Otero 1951). The ancient cultural and symbolic association between *chicha*

Fig. 11.4 Urubamba, the sacred valley of Machu Pichu located near Cuzco, the Inca capital. Colonial accounts referred to it as Valley of Yucay and considered it the most important region of the Inca Empire for maize production. The Urubamba River is fed by several rivers, and the valley was revered because of its geography and climate. *Chicha* consumed by the ruler and highest-ranked lineages in Cuzco was required to have been cultivated in Urubamba, and these maize varieties were spread throughout their empire. (Photo by Harold Hietala)

and mother earth (*Pachamama*) continue to be reflected in Andean customs. Carl Sauer (1950: 494) states, "The Spanish annalists give the impression that more of it was drunk and less eaten….". The term "*chicha*" is suggested to be taken from the Arawakan word derived from "*chichal*" or "*chichiatl*" (Sauer 1950; Nicholson 1960). *Chichia* and *atl* mean "to fertilize" and "water," respectively, while "*chi*" means "with," and "*chal*" means "saliva" or "with saliva," or "to spit" (Nicholson 1960). The etymology of the term *chicha* is also believed to pertain to a Mesoamerican term, and not a South American Arawakan word; the Nahuatl "*chichiya,*" (Stross, personal communication). According to the chronicler Molina (1989 [1575]), *chicha* means, "to become sour, bitter," referring in this case to the beer after the fermentation process (Staller and Stross 2013). Moreover, indigenous Maya populations in various regions of Guatemala still use "*chicha*" to refer to maize-based fermented intoxicants (Staller 2006: 449–450). The term was adapted among pre-Columbian indigenous cultures and contemporary Andeans to generally refer to maize beer as *chicha.* The Quechua term for maize beer is *aqha*, the Aymara is *kufa*; however, *chicha* with reference to beer is now commonly used throughout the Andes (Nicholson 1960).

Fig. 11.5 Depiction of the *Sapa* Inca addressing a circle of stone idols. These idols on the mountain summit represent mountain spirits, referred to as *apu* or *wamani*. The Inca rulers were divine and therefore believed to have the power to make *huacas* "speak". (From Guamán Poma 1980: 235 *fol.* 261 [1583–1615])

Ethnohistoric accounts indicate that coastal and highland Andean societies were prohibited by the Inca state from sowing their own fields before they had sown the terraces and their fields for making maize beer *chicha* (Arriaga 1968 [1620]: 42). The Andean people would also offer or sacrifice round masses or balls of corn meal or porridge mixed with salt called *yanhuar sanco* or *sancu* to *huacas* (Arriaga 1968 [1620]: 47). The porridge or *yahuar sanco* was also used in purification rituals during the festival called *Citua* (Cobo 1990 [1653]: 166). The rite of *Citua* was held just before the rainy season and the planting of crops. This was also the time of the year when Andeans become most susceptible to illness and disease (Rösing 1995). During purification rituals held during the *Citua* festival, chroniclers indicate that maize was used to purify living space and to affect healing among the sick. These rituals and medicinal uses were considered idolatrous by the Church (Cobo 1990 [1653]: 166).

Ritual purification also extends to what the Spanish chroniclers like Arriaga (1968 [1620]: 47–50) interpreted as confession while in the indigenous tradition it related primarily to physical ailments, spiritual cleansing, and uncovering social disorder. Before the cleansing rituals performed during *Citua*, all non-Inca people who had physical defects or sickness were ordered out of Cuzco (Cobo 1990 [1653]). Once they left the imperial capital, the Inca inhabitants would bathe in the springs and rivers, which were located along one of the 41 sight lines called *ceques* that radiated from the Temple of the Sun (*Coricancha*) to the various *huacas* in the surrounding valley (Zuidema 1964, 1997, 2008). Underground canals channeled *chicha* and water from the *ushnu* or well in the central plaza and may also have directed the flow of water in and out of Cuzco, the Inca capital (Zuidema 1964, 1980: 324–331, 1986: 182, 1989: 414, 2008; Classen 1993: 68).

It is through the consumption and offering of *chicha* that Andean people initiated the religious festivals to *huacas*. Inca rulers would toast the Sun (*Inti*) during the festival of the Sun as well as *huacas* as a form of symbolic alliance and reciprocity. The Inca believed that only their rulers could make a *huacas* "speak" (Garcilaso de la Vega 1966 [1609]: 223; Guamán Poma 1980 [1583–1615]: 220 *fol.* 246, 235, *fol.* 261) (Fig. 11.5). Maize was also associated with the Cult of the Sun and veneration of the Inca dynasty (Staller 2006: 464).

According to Spanish chroniclers, during the early colonial period, the significance of maize was closely tied to the Inca Solar deity (*Inti*) and to veneration of former rulers and nobles of high status and rank (Staller 2008b). Its significance to the ruling class and to dynastic imperatives is evident by its presence in the form of gold maize stalks in the Garden of the Sun located in a courtyard of the Temple of the Sun or Golden Enclosure (*Coricancha*), the most sacred temple in Cuzco (Garcilaso de la Vega 1966 [1609]; Cobo 1990 [1653]: 48–50). The Inca created a split economy based upon local food crops, potatoes, tubers of various kinds, and raising camelids, at the local level versus large-scale state-sponsored cultivation (Murra 1973, 1980; Morris 1979, 1993; Wing 1986).

The early sixteenth-century Spanish accounts also emphasis maize as a prestige crop because of its importance to fostering social and political alliances in the context of long-distance interaction, the redistribution of maize, and as a form of reciprocity in exchange for corveé labor to the Inca state (Murra 1980; Morris 1979, 1993). All state-sponsored cultivation involved maize, which was consumed in various ways and generally redistributed as a form of reciprocity to subject communities in exchange for corvée labor or *mit'a* provided to the state for services rendered or commodities provided for redistribution to subject communities during periods of economic stress in the annual cycle (Murra 1980, 1982). Maize produced on state-owned fields was also often redistributed to subject communities as beer or *chicha* (La Barre 1938: 224). Pre-Columbian and colonial imagery often depicted llamas carrying beverages particularly maize beer and various food staples for long-distance interaction between the highlands and the coast (Guamán Poma 1980 [1583–1615]; Donnan 1978, 1996; Donnan and McClelland 1999; Cummins 2002).

Inca storehouses (*collqas*) for foodstuffs and other products were distributed throughout their empire (Murra 1972, 1980; Morris 1979). For the most part, in the cordillera, they stored maize, however, in the *puna* and *Altiplano* environmental zones, important foods for storing included freeze-dried ducks and tubers, particularly sweet potatoes, yams, manioc, potatoes, and oca were cultivated (Cieza de León 1998 [1553]; Betanzos 1996 [1551]; Molina 1989 [ca. 1557]; Murra 1975, 1980, 1982). Ethnohistoric accounts reveal a great deal about how food crops, subsistence, and utilitarian resources were managed, stored, and redistributed in the Andes as well as their role in local economies (Betanzos 1996 [1557]; Cieza de León 1977 [1551]; Cobo 1983 [1553]; Garcilaso de la Vega 1966 [1609]; Guamán Poma 1980 [1583–1615]).

John Murra (1980) asserted that pre-Columbian Andean cultures sustained larger population densities throughout the cordillera than in the colonial periods after the Spanish conquest. Larger population densities were sustained through an economy based upon traditional forms of reciprocity or complementarity that clearly predates Inca expansion (Murra 1975 [1972]; Morris 1979; Morris and Thompson 1985; Staller 2006, 2008b).

Maize in Andean Foodways

Maize serves various culinary roles in contemporary Andean food and culture that extend beyond *chicha* production. Regional Andean cuisines and food continue to be highly diverse, with maize being one of many food crops. Its importance in different regions also varies, particularly with regard to the highlands, lowlands, and along the coast. Ground maize is common in a popular contemporary Andean dish called *humitas*. *Humitas* are often filled with meat or made into stews with various ingredients (Figs. 11.6 and 11.7).

Many of these dishes combine native Andean and Eurasian introduced plant and animal species. Among contemporary Andean cultures, maize is generally consumed with calorie-rich domesticates specifically root crops, tubers, and rhizomes, most notably manioc, potatoes, and sweet potatoes in stews such as *Locro*. For example, *Locro* is a national dish in many Andean countries including Argentina, Peru, Bolivia, and Ecuador and is made with a host of ingredients including potatoes, beans, squash

Fig. 11.6 *Humitas* are a popular dish in Ecuador and Peru. They have the appearance of Mexican tamales and are made from finely ground corn, generally filled with ground beer or pork, and often wrapped in corn husks. (Courtesy of Google Images)

Fig. 11.7 *Humitas en olla* is an Andean stew made up of ground corn and kernels and popular throughout the Andes. It is prepared with milk, maize, onions, beans, and tomatoes and has a creamy smooth texture. (Courtesy of Google Images)

Fig. 11.8 Ecuadorian *Locro de papas* are made from ground corn with diced potatoes, onions, celery, avocados, and cheese in a creamy sauce. Along the coast shrimp and cheese are sometimes added to this stew. (Courtesy of Google Images El Diario)

as well as maize (Coe 1994: 20) (Fig. 11.8). In some regions of the Andes, particularly Argentina and eastern Bolivia, ground maize is made into tamales that are filled with meat and wrapped in corn husks (Fig. 11.9). Andean maize is also parched (*chojllo*), popped (*rosetas*), boiled (*chocohoka*), or roasted (*kamcha* or *hamka*) (Valcárcel 1946: 477–482; Sauer 1950: 494).

Andean food crops that appear to have played a much larger economic role than maize include manioc, sweet potatoes, *oca*, achira, and particularly grains such as quinoa, amaranth (*Amarenthus caudatus* L) also known in Quechua as *kiwicha* (van der Merwe et al. 1993; Tykot et al. 2006; León 2017). It is becoming increasingly clear that native domesticated plant species played a much larger role in the pre-Columbian Andean diet than has been generally suggested in the literature (Benz and Staller 2006; Staller and Stross 2013). Root crops such as potatoes (*Solanum tuberosum* L.) formed a primary ingredient in many Andean cuisines and dishes. Botanists have identified over 3,000 varieties of this tuber in the Andes. Potatoes were a primary food crop dating back to c. 5000 BP (León 2017: 37, 68, 75; Chandler-Ezell et al. 2006). Some of the more notable dishes include boiled potato as a base or with *ají*-based sauces (*ibid*.). French-fried potatoes are a typical ingredient in Peruvian stir-fries, including the classic dish, *lomo saltado*. In Ecuador, it is featured in *locro de papas* a hearty stew of potato, squash, and cheese.

When gourds were introduced into Europe in the sixteenth century, they were generally associated with Old World species of cucumbers and melons. Many New

The conquistador Pedro Cieza de León had already traveled through other regions of the Americas. Spanish accounts generally infer a clear food preference for maize over other indigenous food crops such as quinoa or the potato. Moreover, the widespread maize cultivation along coastal streams and on artificial terraces along the western cordillera further reinforced the impression that in the Andes, as in New Spain (Mesoamerica) maize was a primary food staple, a grain, analogous to wheat and barley in the Old World (Staller 2010a, b).

When Pizarro and his army arrived in northernmost coastal Peru, Cieza de León (1998 [1553]: 49) states, "*they found nothing but some maize and those roots that they eat* [manioc or sweet potato]." In reference to the coastal valleys of Peru, Cieza de León (1998 [1553]: 57) observed that villages were dispersed and that Andeans, "*cut the forest on the slopes and sow their maize and other food stuffs.*"

Early Spanish conquistadores were amazed at the desolation along the desert coastlines and the nearby mountainous terrains, at the same time, they were amazed by the complex ecological and cultural diversity of the high sierra. How could such coastal societies have developed urban centers and constructed such enormous pyramids? Many of them stated that it was no doubt related to maize cultivation (Cieza de León 1993 [1554]; Molina 1989 [ca. 1557]; Garcilaso de la Vega 1966 [1609]; Guamán Poma 1980 [1583–1615]). Maize, cotton, and other indigenous root crops such as the potato (*Solanum tuberosum* L.) as well as legumes such as the peanut (*Arachis hypogaea* L.) were grown along the coastal streams and seemingly every place where water could be channeled by the vast irrigation networks constructed by many indigenous cultures.

The Andes have numerous indigenous cultigens and plants that played a major role in ancient, colonial, and contemporary cultures and their related cuisines. Many foods and meals were complemented with the ancient domestic of chili peppers (*Capsicum* spp.) or *aji*. *Aji* was and continues to be the domesticated pepper of choice in Bolivia, Ecuador, Peru, and Chile due to their tolerance for cold and poor soils (Staller 2010a). Freeze-dried potatoes, or *chuño*, continue to be widely traded and exchanged, a practice facilitated by their resistance to spoilage and lightweight. Such characteristics and adaptive characteristics continue to be important to indigenous communities in highland Peru and Bolivia (Sauer 1950). However, maize is the most extensively researched and is perhaps one of the most important of all cultigens, not because of its role as a dietary staple, but because of its roles in traditional Andean rituals and religious ideology, and particularly with ethnic identity (Rostworowski de Diez Canseco 1983; Piperno and Pearsall 1998; Staller 2010a, 2016; Staller and Stross 2013, Staller et al. 2006).

Discussion and Conclusions

The archaeological, anthropological, and ethnohistorical evidence clearly indicate that *chicha* or maize beer was central to Andean political economy. Although widespread, pre-Columbian Andean civilizations generally emphasized maize as a prestige crop, rather than as an economic food staple, instead maize and *chicha* or maize beer were used in fostering social alliances in the context of long-distance interaction (Staller 2009b). Three factors that are particularly striking about the role of maize in indigenous ritual and religion in the Andes is the importance it held for ritual offerings, calendric rites, and the scale of cultivation. Early explorers mention indigenous Andeans of the Sierra in Ancash prepared corn by soaking ears in water sometimes for as long as 2 or 3 months and cooking it (León 2017). However, maize was not processed, that is, soaked nor cooked in an alkali solution; nevertheless, maize kernels or *jora* were soaked and ground to make *chicha* (Jennings 2004; Jennings and Bowser 2009). The differences between Mesoamerica and the Andes are the way in which cobs and kernels were prepared and processed. While maize beer largely played this role in the development of Andean civilization and ritual practice, religious cults, and ethnic identity, it was not nixtamalized or ground into tortillas, tamales, *pozolé* (enchiladas) etc., as it was in Mesoamerica.

The multidisciplinary research presented here indicates that in the Andes maize was primarily consumed as a fermented intoxicant, *chicha* or maize beer related to feasting, rituals, and rites associated with traditional religion and culture. When consumed with various cuisines, it was generally eaten as a side dish or as boiled cobs not as the focal point or primary food in meals. When it was prepared it was generally by parching, popping, boiling, or roasting (Weatherwax 1953; Valcárcel 1946; Sauer 1950: 494). Among Andean cultures, the main role of maize was as a side dish or as *chicha*. It was its consumption as a fermented intoxicant that made maize such an important food crop to the rise and expansion of various pre-Columbian Andean civilizations.

Although the introduction of Andean plants, food crops, and domesticates appear to have played profound roles upon cuisines worldwide particularly in Western Europe, the introduction of European plants and animals also played important roles in indigenous Andean cuisines. Moreover, the introduction of Indo-European species into the cordillera has to varying degrees, transformed the environment, ecology, and natural landscape of the Andean cordillera from Colombia to Tierra del Fuego (Staller 2010a).

Among indigenous cultures, maize and *chicha* continue to play central roles in traditional rituals and rites regarding ritual feasting and the marking of the annual cycles (Coe 1994; Rösing 1995). Overall, the interdisciplinary evidence suggests that native-domesticated food crops such as quinoa, potato peppers, squash, played primary roles to Andean diets, but it was maize that had the greatest impact to pre-Columbian political economies, to the rise and expansion of Andean civilizations, and to the ancient political economies. The recent research surrounding foodways are presenting a strikingly different perception of the economic role of maize in Andean prehistory that to a large degree challenges our longstanding theoretical perceptions as well as our general understanding of its role as a primary food staple and association to agricultural origins.

References

Allen CJ (1988) The hold life has: Coca and cultural identity in an Andean Ayllu. Smithsonian Institution Press, Washington D.C.

Anderson E (1947) Corn before Columbus. Pioneer Hi-Bred Corn Co. Bandelier, Des Moines; Adolf F (1910) Islands of Titicaca and Koati. The Hispanic Society of America, New York

Arriaga Fr PJ de (1968) [1621] The extirpation of idolatry in Peru (Translated and edited by Keating C). University of Kentucky Press, Lexington

Austin DF (1988) The taxonomy, evolution and genetic diversity of sweet potatoes and related wild species. In: Gregory P (ed) Exploration, maintenance, and utilization of sweet potato genetic resources. CIP, Lima, pp 27–60

Benz BF, Staller JE (2006) The antiquity, biogeography, and culture history of Maize in the Americas. In: Staller JE, Tykot RH, Benz BF (eds) Histories of Maize: multidisciplinary approaches to the prehistory, linguistics, biogeography, domestication, and evolution of Maize. Routledge, Taylor & Francis Group, New York, pp 647–673

Benz BF, Staller JE (2010) The antiquity, biogeography, and culture history of Maize in Mesoamerica. In: Staller JE, Tykot RH, Benz BF (eds) Histories of Maize in Mesoamerica: multidisciplinary approaches. Routledge, Taylor & Francis Group, New York, pp 267–275

Betanzos J de (1968) [1551] Suma y Narración de Los Incas. Biblioteca de Autores Españoles, vol 209. Atlas, Madrid.

Betanzos J de (1987) [1551] Suma y Narración de Los Incas. Rubio CM (ed). Madrid.

Betanzos J de (1996) [1557] Narrative of the Incas (Translated by Hamilton R and edited by Buchanan D). University of Texas Press, Austin

Brieger FG (1958) Races of Maize in Brazil and other Eastern South American countries. National Academy of Sciences, National Research Council, Washington

Bruman HJ (2000) Alcohol in Ancient Mexico (Forward by Furst PT). University of Utah Press, Salt Lake City

Buechler H, Buechler J (1971) The Bolivian Aymara. Holt, Rinehart and Winston, New York

Burger RL, van der Merwe NJ (1990) Maize and the origin of highland Chavín Civilization. Am Anthropol 92:85–95

Camporesi P (1993) The magic harvest: food, folklore, and society. Blackwell Publishers Ltd., Cambridge, UK

Ceruti C (2004) Human bodies as objects of dedication at Inca Mountain Shrines (North-Western Argentina). World Archaeol 36(1):103–122

Chandler-Ezell K, Pearsall DM, Zeidler JA (2006) Root and tuber phytoliths and starch grains document manioc (*Manihot esculenta*), arrowroot (*Maranta arundinacea*), and llerén (*Calathea* sp.) at Real Alto site, Ecuador. Econ Bot 60:103–120

Chávez KL (1989) The significance of Chiripa in Lake Titicaca Basin developments. Expedition 30:17–26

Chávez S (2012) Agricultural terraces as monumental architecture in the Titicaca Basin. In: Burger RL, Rosenswig RM (eds) Early new world monumentality. University Press of Florida, Gainesville, pp 431–453

Chávez SJ, Thompson RG (2006) Early Maize on the Copacabana Peninsula: implications for the archaeology of Lake Titicaca Basin. In: Staller JE, Tykot RH, Benz BF (eds) Histories of Maize: multidisciplinary approaches to the prehistory, linguistics, biogeography, domestication, and evolution of Maize. Routledge, Taylor & Francis Grp., New York, pp 415–428

Cheetham D (2010) Corn, colanders, and cooking: early Maize processing in the Maya lowlands and its implications. In: Staller JE, Carrasco M (eds) Pre-Columbian foodways: interdisciplinary approaches to food, culture and markets in ancient Mesoamerica. Springer, New York, pp 23–69

Christenson AJ (2010) Maize was their flesh: ritual feasting in the Maya highlands. In: Staller JE, Carrasco M (eds) Pre-Columbian foodways: interdisciplinary approaches to food, culture and markets in ancient Mesoamerica. Springer, New York, pp 577–600

Cieza de León P (1998) [1553] The discovery and conquest of Peru. Chronicles of the New World Encounter (Translated and edited and by Cook AP, Cook ND). Duke University Press, Durham and London

Cieza de León P (1977) [1551] El Señorio de los Incas, 2nd edn. Universo, Lima

Classen C (1993) Inca cosmology and the human body. University of Utah Press, Salt Lake City

Cobo B Fr (1983) [1653] History of the Inca Empire; an account of the Indian's customs and their origin together with a treatise on Inca legends, history, and social institutions (Translated and edited by Hamilton R). University of Texas Press, Austin

Cobo B Fr (1990) [1653] Inca religion and customs (Translated by Hamilton R and forward by Rowe JH). University of Texas Press, Austin.

Coe SD (1994) America's first cuisines. University of Texas Press, Austin

Cutler HC (1946) Maize in South America. Botanical Museum Leaflets, Harvard University 12(8):257–291

Cutler HC, Cardenas MH (1947) Chicha, a native South American Beer. Bot Mus Leafl (Harvard University) 13(3):33–60

Cummins TBF (2002) Toasts with the Inca: Andean abstraction and colonial images on Quero vessels. University of Michigan Press, Ann Arbor

Damp JE (1984) Environmental variability, agriculture, and settlement processes of coastal Ecuador (3,300–1,500 B.C.). Current Anthropol 25:106–111

Damp JE, Vargas SP (1995) The many contexts of early valdivia ceramics. In: Barnett WK, Hoopes JW (eds) The emergence of pottery. Smithsonian Institution Press, Washington D.C., pp 157–168

Dietler M, Hayden B (2001) Feasts: archaeological and ethnographic perspectives on food, politics, and power. Smithsonian Institution Press, Washington D.C.

Donnan CB (1978) Moche art of Peru: pre-Columbian symbolic communication. Museum of cultural history, University of California, Los Angeles, CA

Donnan CB (1996) Moche. In: Boone E (ed) Andean art at Dumbarton Oaks. Dumbarton Oaks Research Library and Collection, Washington D.C.

Donnan CB, McClelland D (1999) Moche fineline painting its evolution and its artists. UCLA Fowler Museum of Cultural History, Los Angeles

Ford JA (1969) A comparison of formative cultures in the Americas. Smithsonian contributions in anthropology, vol II. Smithsonian Institution Press, Washington D.C.

Friewald CR (2010) Dietary diversity in the upper Belize river valley a zoological and isotopic perspective. In: Staller JE, Carrasco M (eds) Pre-Columbian foodways: interdisciplinary approaches to food, culture and markets in ancient Mesoamerica. Springer, New York, pp 399–420

Garcilaso de la Vega, "El Inca" (1966) [1609] Royal commentaries of the Incas and general history of Peru. Parts I–II (Translated by Livermore HV and forward by Toynbee AJ). University of Texas Press, Austin

Gerard J (1975) [1633] The herbal or general history of plants. The complete 1633 ed. revised by Johnson T. Facsimile reprint of the ed. printed by Islip A, Norton J, Whitakers R. London, under title: The herball, or generall historie of plantes. Dover Publications, New York.

Gil AF (2003) *Zea mays* on the South American periphery: chronology and dietary importance. Current Anthropol 44(2):295–300

Gil AF, Tykot RH, Neme G, Shelnu NR (2006) Maize on the Frontier; isotopic and macrobotanical data from central-western Argentina. In: Staller JE, Tykot RH, Benz BF (eds) Histories of Maize: multidisciplinary approaches to the prehistory, linguistics, biogeography, domestication, and evolution of Maize. Routledge, Taylor & Francis Grp., New York, pp 199–214

Goldstein DJ, Coleman RC (2004) *Schinus molle* L. (Anacardiaceae) *Chicha* production in the Central Andes. Econ Bot 58(4):523–529

Guamán Poma de Ayala (Waman Puma), Felipe (1980) [1583–1615] Nueva Cronica y Buen Gobierno. Murra JV, Adorno R, Jorge, Urioste L (eds). 3 vols. América Nuestra. Siglo XXI. México D.F.

Grobman A, Salhuana W, Sevilla R (1961) Races of Maize in Peru, their origins, evolution and classification [by] and in collaboration with P.C. Mangelsdorf. National Academy of Sciences-National Research Council, Washington D.C.
Hastorf CA, Whitehead WT, Bruno MC, Wright M (2006) The movements of Maize into the middle horizon Tiwanaku, Bolivia. In: Staller JE, Tykot RH, Benz BF (eds) Histories of Maize: multidisciplinary approaches to the prehistory, linguistics, biogeography, domestication, and evolution of Maize. Elsevier/Academic Press, San Diego, pp 429–448
Haas EM, Levin B (2006) Staying healthy with nutrition the complete guide to diet and nutritional medicine. Celestial Arts, Berkeley, CA
Harlan JR, de Wet JMJ (1973) On the quality of evidence for the origin and dispersal of cultivated plants. Current Anthropol 14(1–2):51–62
Hyslop J (1990) Inca settlement planning. University of Texas Press, Austin
Isbell WH, Groleau A (2010) The Wari Brewer woman: feasting, gender, offerings, and memory. In: Klarich EA (ed) Ancient kitchens: new directions in the study of daily meals and feasts. University Press of Colorado. Colorado
Jennings J (2004) La Chichera y el Patrón: Chicha and the energetics of feasting in the prehistoric Andes. Archeol Papers Am Anthropol Assoc 14(1):241–259
Jennings J, Bowser BJ (2009) Drink, power, and society: an introduction. In: Jennings J, Bowser BJ (eds) Drink, power, and society in the Andes. University Press of Florida, Gainesville, pp 1–27
Katz SL, Hediger ML, Valleroy LA (1974) Traditional Maize processing in the new world. Science 184(4138):765–773
Lanning EP (1967) Peru before the Incas. Prentice-Hall, Englewood Cliffs
La Barre W (1938) Native American Beers. Am Anthropol 40:224–234
León E (2017) 14,000 years of food in Peru. Fondo Editorial de la Universidad San Martin de Porres, Lima
MacCormack S (2004) Religion and society in Inca and Spanish Peru. In: Phipps E, Hecht J, Martín CE (eds) The Colonial Andes: tapestries and silverwork 1530–1830. The metropolitan museum of art, New York. Yale University, New Haven and London, pp 101–112
Matsuoka Y, Vigouroux Y, Goodman MM, Sanchez J, Buckler E, Doebley J (2002) A single domestication for Maize shown by multilocus microsatellite genotyping. Proc Natl Acad Sci USA 99:6080–6084
Misra MK, Ragland KW, Baker AJ (1992) Wood ash composition as a function of furnace temperature. Biomass Bioenerg 4(2):103–116
Molina C de (El Cuzqueño) (1989) [ca. 1575] Relación de los ritos y fábulas de los Ingas. In: Urbano H, Duvoils P (eds) Fábulas y mitos de los Incas. Crónicas de América series. Historia 16, Madrid, pp 47–134
Moore JD (1989) Pre-Hispanic Beer in coastal Peru: technology and social context of prehistoric production. Am Anthropol 91:682–695
Morris C (1979) Maize Beer in the economics, politics, and religion of the Inca State. In: Gastineau C, Darby W, Turner T (eds) The role of fermented beverages in nutrition. Academic Press, NY, pp 21–34
Morris C (1993) The wealth of a native American State. Value, investment, and mobilization in Inka economy. In: Henderson J, Netherly PJ (eds) Configurations of power. Cornell University Press, Ithaca, pp 36–50
Morris C, Thompson DE (1985) Huánuco Pampa an Inca city and its hinterlands. Thames and Hudson, New York and London
Murra JV (1972) El control vertical de un maximo de pisos ecológicos en la economia de las sociedades Andinas. In: Murra JV (ed) Visita de la Provincia de León de Huanuco en 1562, Tomo II. Universidad Nacional Hermilio Valdizan, Huanuco, pp 429–476
Murra JV (1973) Rite and crop in the Inca State. In: Gross DR (ed) Peoples and cultures of South America. Natural History Press, Garden City, New Jersey, Doubleday, pp 377–389
Murra JV (1975) Maiz, tuberculos y ritos agricolas. In: Murra JV (ed) Formaciones económicas y politicas del mundo Andino. Lima, Instituto de Estudios Peruanos, pp 45–57

Murra JV (1980) Economic organization of the Inca State. JAI Press, Greenwich, Connecticut

Murra JV (1982) The *mit'a* obligation of ethnic groups to the Inca State. In: Collier G, Rosaldo R, Wirth J (eds) The Inca and Aztec States, 1400–1800. Academic Press, NY, pp 237–264

Moseley ME (1975) The maritime foundations of Andean civilization. Cummings Pub. Co., Menlo Park, California

Nicholson GE (1960) *Chicha* Maize Types and *Chicha* Manufacturing in Peru. Econ Bot 14:290–299

Norton P (1982) Preliminary observations on Loma Alta, an early Valdivia Midden in Guayas Province, Ecuador. In: Menendez CZ, Norton P (eds) Primer Simposio de Correlaciones Antropológicas Andino-Mesoamericano. Salinas, Ecuador. Junio, 1971. ESPOL, Guayaquil, Ecuador, pp 101–120

Norton P, Lunnis R, Nailing N (1983) Excavaciones en Salango, Provincia de Manabí, Ecuador. Miscelanea Antropologica Ecuatoriana 3(3):9–72. Museo Antropológico del Banco Central del Ecuador, Guayaquil.

Otero GA (1951) La piedra mágica. Instituto Indiginista Americano Mexico D.F.

Oviedo F de y GV (1969) [1535] De La Natural Hystoria de Las Indias. A Facsimile Edition issued in Honor of Sterling A. Stoudemire, University of North Carolina Studies in the Romance Languages and Literature, No. 85, Chapel Hill: University of North Carolina Press.

Pearsall DM (1992) The origins of plant cultivation in South America. In: Wesley Cowan C, Watson PJ (eds) With the assistance of Benco, N. L. The origins of agriculture: an international perspective. Smithsonian Institution Press, Washington D.C. and London, pp 173–205

Pearsall DM (1993) Contributions of Phytolith analysis for reconstructing subsistence: examples from research in coastal Ecuador. In: Pearsall DM, Piperno DR (eds) Current research in Phytolith analysis: applications in archaeology and paleoecology. MASCA, University of Pennsylvania Museum, pp 109–122

Pearsall DM (1994) Issues in the analysis and interpretation of archaeological Maize in South America. In: Johannesson S, Hastorf CA (eds) Corn and culture in the prehistoric new world. Westview Press, Boulder, pp 245–272

Pearsall DM (1999) The impact of Maize on subsistence systems in South America: an example from the Jama River Valley, Coastal Ecuador. In: Gosden C, Hather J (eds) The prehistory of food: appetites for change. Routledge, London, pp 419–437

Pearsall DM (2003) Plant food resources of the ecuadorian formative: an overview and comparison to the central Andes. In: Scott RJ, Burger RL (eds) Archaeology of formative Ecuador. From the 7 and 8 October 1995 symposium at Dumbarton Oaks, Dumbarton Oaks Research Library and Collection, Washington D.C., pp 213–257

Pearsall DM, Piperno DR (1990) Antiquity of Maize cultivation in Ecuador: summary and reevaluation of the evidence. Am Antiq 55:324–337

Piperno DR, Pearsall DM (1998) The origins of agriculture in the lowland neotropics. Academic Press, San Diego

Pozorski S, Pozorski T (1987) Early settlement and subsistence in the Casma Valley, Peru. Iowa Press, Iowa City, Iowa

Ramírez SE (2005) To feed and be fed: the cosmological basis of authority and identity in the Andes. Stanford University Press, Stanford, CA

Rivera M (2006) Prehistoric Maize from Northern Chile: an evaluation of the evidence. In: Staller JE, Tykot RH, Benz BF (eds) Histories of Maize: multidisciplinary approaches to the prehistory, linguistics, biogeography, domestication, and evolution of Maize. Routledge, Taylor & Francis Grp., New York, pp 449–467

Rösing I (1995) Parman Purina: going for rain. Anthropos 90:69–88

Rostworowski de Diez Canseco M (1983) Estructuras Andinas de Poder. Ideología religiosa y politica. Institutos Estudios Peruanos. Lima

Rostworowski de Diez Canseco M (1986) Estructuras Andinas del Poder. Ideología Religiosa y Política, 2nd ed. Instituto de Estudios Peruanos, Lima

Rostworowski de Diez Canseco M (1999) History of the Inca Realm (Translated by Iceland HB). Cambridge University Press, Cambridge, UK

Sauer CO (1950) Cultivated plants of South and Central America. In: Steward JH (ed) Handbook of South American Indians. Physical anthropology, linguistics and cultural geography of South American Indians, vol 6. Smithsonian Institution Bureau of American Ethnology Bulletin 143. U.S. Government Printing Office, Washington D.C., pp 487–543

Sauer CO (1952) Agricultural origins and dispersals. American Geographic Society, New York

Sauer CO (1969) The early Spanish main. The University of California Press, Berkeley

Smith CE Jr (1968) The new world centers of origin of cultivated plants and the archaeological evidence. Econ Bot 22:253–266

Smith BD (1998) The emergence of agriculture, 2nd edn. Scientific American Library, New York

Smith BD (2001) Documenting plant domestication: the consilience of biological and archaeological approaches. Proc Natl Acad Sci USA 98:1324–1326

Smith BD (2005) Reassessing Coxcatlan cave and the early history of domesticated plants in Mesoamerica. Proc Natl Acad Sci USA 102:9438–9944

Stahl PW (2008) South American animal domesticates. In: Silverman H, Isbell WH (eds) Handbook of South American archaeology. Springer, New York, pp 121–130

Staller JE (2006) The social, symbolic, and economic significance of *Zea mays* L. in the late horizon period. In: Staller JE, Tykot RH, Benz BF (eds) Histories of Maize: multidisciplinary approaches to the prehistory, linguistics, biogeography, domestication, and evolution of Maize. Routledge, Taylor & Francis Grp., New York, pp 449–467

Staller JE (2008a) An introduction to Pre-Columbian landscapes of creation and origin. In: Staller JE (ed) Pre-Columbian landscapes of creation and origin. Springer, New York, pp 1–9

Staller JE (2008b) Dimensions of place: the significance of centers to the development of Andean civilization: an examination of the *Ushnu* concept. In: Staller JE (ed) Pre-Columbian landscapes of creation and origin. Springer, New York, pp 269–313

Staller JE (2010a) Maize cobs and cultures: history of Zea mays L. Springer, Berlin, Heidelberg

Staller JE (2010b) Ethnohistoric sources on foodways, feasts and festivals in Mesoamerica. In: Staller JE, Carrasco M (eds) Pre-Columbian foodways: interdisciplinary approaches to food, culture and markets in ancient Mesoamerica. Springer, New York, pp 23–69

Staller JE (2014) Lightning and its manifestations: *Huacas* and *Ushnus*. In: Meddens F, McEwan C, Willis K, Branch N (eds) Inca sacred space: landscape, site and symbol in the Andes. Archetype Publication, London, UK, pp 177–186

Staller JE (2015a) Maize. In: Metheny KB, Beaudry MC (eds) The archaeology of food: an encyclopedia. Alta Mira Press, New York. Imprint of Rowland & Littlefield, pp 296–298

Staller JE (2015b) Plant domestication. In: Metheny KB, Beaudry MC (eds) The archaeology of food: an encyclopedia. Alta Mira Press, New York. Imprint of Rowland & Littlefield, pp 407–409

Staller JE (2016) High altitude Maize (*Zea mays* L.) cultivation and endemism in the Lake Titicaca Basin. J Bot Res 1(1):8–21

Staller JE (2018) Origin, evolution, biogeography, and cultivation of Maize: the role of nixtamal to nutrition and economic importance in ancient and contemporary food and culture. In: Sanz N (ed) The origin and evolution of food production and its impact on consumption patterns. Centre for Research and Advanced Studies (Cinvestav), México D.F., pp 179–195

Staller JE, Carrasco MD (2010) Pre-Columbian foodways in Mesoamerica. In: Staller JE, Carrasco M (eds) Pre-Columbian foodways: interdisciplinary approaches to food, culture and markets in ancient Mesoamerica. Springer, New York, pp 1–20

Staller JE, Stross B (2013) A lightning in the Andes and Mesoamerica: Pre-Columbian, colonial, and contemporary perspectives. Oxford University Press, New York and Oxford

Sullivan LE (1984) Sacred music and sacred time. World Music 26(3):33–52. Berlin

Sullivan LE (1987) Nature: the worship of nature. In: Eliade M (ed) The encyclopedia of religion. Macmillan Publishing Co. NY, pp 324–328

Sullivan LE (1988) Icanchu's drum. An orientation to meaning of South American religions. MacMillan Publishing Co., New York

Sullivan W (1996) The secret of the Incas. Basic Books, NY

Tilley C, Bennet W (2001) An archaeology of supernatural places: the case of west Penwith. J Roy Anthropol Inst 7(2):355–362

Timothy DH, Harry D (1963) Races of Maize in Ecuador. National Academy of Sciences-National Research Council, Washington D.C.

Timothy D, Bertulfo Pena V, Ricardo Ramirez E (1961) Races of Maize in Chile, in collaboration with William L. Brown and Edgar Anderson. National Academy of Sciences-National Research Council, Washington D.C.

Turner T (1985) Animal symbolism, totemism, and the structure of Myth. In: Urton G (ed) Animal Myths and Metaphors in South America. University of Utah Press, Salt Lake City, pp 49–106

Tykot RH, Staller JE (2002) On the importance of early Maize agriculture in coastal Ecuador: new data from the late Valdivia phase site of La Emerenciana. Current Anthropol 43(4):666–677

Tykot RH, van der Merwe NJ, Burger RL (2006) The importance of Maize and marine foods to initial period/early horizon subsistence in highland and coastal Peru. In: Staller JE, Tykot RH, Benz BF (eds) Histories of Maize: multidisciplinary approaches to the prehistory, linguistics, biogeography, domestication, and evolution of Maize. Routledge, Taylor & Francis Grp., New York, pp 187–197

van der Merwe N, Lee-Thorp JA, Scott Raymond J (1993) Light, stable isotopes and the subsistence base of formative cultures at Valdivia, Ecuador. In: Lambert JB, Grup G (eds) Prehistoric human bone archaeology at the molecular level. Springer-Verlag, New York, pp 63–97

Urton G (1981) At the crossroads of the earth and the sky: an Andean cosmology. University of Texas Press, Austin

Urton G (1985) Animal metaphors and the life cycles in an Andean community. In: Urton G (ed) Animal Myths and Metaphors in South America. University of Utah Press, Salt Lake City, pp 251–284

Urton G (1999) Inca Myths. University of Texas Press, Austin

Valera B (1968) [1594] Relación de las Costumbres Antiguas del Peru. Introduction by F. Loayza, Lima

Valcárcel LE (1946) Indian markets and fairs in Peru. In: Steward JH (ed) Handbook of South American Indians. The Andean civilizations, vol 2. Smithsonian Institution Bureau of American Ethnology Bulletin 143. U.S. Government Printing Office, Washington D.C., pp 477–482

Vigouroux Y, Matsuoka Y, Doebley J (2003) Directional evolution for microsatellite size in Maize. Mol Biol Evol 20(9):1480–1483

Wacher C (2003) Nixtamalization, a Mesoamerican technology to process Maize at a small-scale with great potential for improving the nutritional quality of Maize based foods. In: 2nd international workshop: food based approaches for a healthy nutrition, vol 23–28, pp 735–744

Wing ES (1986) Domestication of the Andean Mammals. In: Vuilleunier F, Monasterio M (eds) High altitude tropical biogeography. Oxford University Press, New York, pp 246–264

Weatherwax P (1954) Indian corn in Old America. The MacMillan Company, New York

Zarrillo S, Pearsall DM, Raymond JS, Tisdale MA, Quon DJ (2008) Directly dated starch residues document early formative Maize (*Zea mays* L.) in tropical Ecuador. Proc Natl Acad Sci USA 105(13):5006–5011

Zuidema RT (1964) The Ceque system of Cuzco: the social organization of the capital of the Inca. International archives of ethnography, supplement to volume 50. E.J. Brill, Leiden

Zuidema RT (1973) Kinship and ancestor cult in three Peruvian communities: Hernández Príncipe's account in 1622. Bulletin, Institut français d'études andines 2(1):16–23. Lima.

Zuidema RT (1982) Myth and history in ancient Peru. In: Rossi I (ed) The logic of culture. Advances in structural theory and methods. J. F. Bergin Publishers Inc., South Hadley Massachusetts, pp 150–175

Zuidema RT (1986) Inka Dynasty and irrigation: another look at Andean concepts of history. In: Murra JV, Wachtel N, Revel J (eds) Anthropological history of Andean Politics. Cambridge University Press, Cambridge, pp 177–200

Zuidema RT (1989) Reyes y guerreros: Ensayos de cultura Andina. Fomciencias, Lima, Peru

Zuidema RT (1997) Cosmovisión Inca y astronomía en el Cuzco: Nuevo Año Agrícola y Sucesión Real. In: Aranda AG (ed) Pensar America Comovisión Mesoamerica y Andina. Actas de las VI Jornadas del Inca Garcilaso celebrated in Montilla on the 11 to 13 September 1996. Córdoba, pp 251–270

Zuidema RT (2002) Inca religion: its foundations in the central Andean Context. In: Sullivan LE (ed) Native religions and cultures of central and south America. Continuum, London & NY, pp 236–253

Zuidema RT (2008) The astronomical significance of ritual movements in the calendar of Cuzco. In: Staller JE (ed) Pre-Columbian landscapes of creation and origin. Springer, New York, pp 249–268

Chapter 12
Imperial Appetites and Altered States: The Spanish Transformation of the Inca Heartland

R. Alan Covey

Abstract In 1532, agricultural production in the region surrounding Cuzco, the Inca capital, focused on producing surpluses of staple goods and channeling them to the capital for use in imperial feasts and ceremonies. By 1600, the region was producing both Andean (coca leaf, maize) and Eurasian (wheat, sheep) commodities for the growing colonial centers, particularly the mining city of Potosí. In the intervening decades, the former Inca heartland experienced a transformation in land use and the consumption of food and drink. This chapter explores the changes in Inca foodways in the Cuzco region, the introduction of Eurasian crops and animals, and the concomitant spread of commodity production of staple products. The chapter concludes by discussing how the establishment of non-Andean foods buttressed the colonial divide between "Indian" and "Spaniard," and how important Inca commensal practices—especially the consumption of maize beer and coca leaf—were reshaped. Colonial administrators described the public consumption of maize beer as a threat to morality and public health and punished Andean elites who sponsored drinking events. At the same time, officials made coca leaf cheap and available for all indigenous people, even though they acknowledged coca chewing was an impediment to the spread of Christianity.

Introduction

The men of the Pizarro expedition did not enter the Inca world prepared to farm or herd. Most of the animals that accompanied them at the vanguard of the European invasion were auxiliaries of war—horses accustomed to the sensory assault of battle, and fierce dogs trained to kill—and the men bore no seeds or tools for planting into the Andean highlands. The Spaniards appropriated their food from indigenous populations, and by the time they arrived at their fateful encounter at Cajamarca, they were long accustomed to surviving on native foods. With the

R. A. Covey (✉)
The University of Texas at Austin, Austin, TX, USA
e-mail: r.alan.covey@austin.utexas.edu

J. E. Staller (ed.), *Andean Foodways*, The Latin American Studies Book Series,
https://doi.org/10.1007/978-3-030-51629-1_12

Spanish occupation of the central Andes, the imperial transformation of highland foodways took a new course. In Cuzco, the Inca capital, the intersection of Inca and Spanish foodways unfolded through a decades-long process in which new crops and animals were introduced, the means of production were reallocated, and indigenous labor was forced into unsustainable commodity production practices. This chapter outlines these processes, also describing the ways that food and drink helped to accentuate the divide between different colonial identities (e.g., Christian/Gentile, Spaniard/Andean, European/Indian) (see also Martinez, Chap. 2 for insights on colonial beverage consumption among different social groups and deFrance, Chap. 3 regarding Spanish perceptions of appropriate food animals).

Agriculture and Diet in Inca Cuzco

Eyewitness accounts of the Spanish invasion of the Inca empire make it clear that Europeans ate and drank from Inca storehouses, frequently receiving hospitality and local entertainment at Inca way-stations and administrative centers (e.g., Xérez 1985[1534]). Given the accounts of starvation that accompanied early Spanish expeditions in Panama and northern South America, it is not surprising that the invaders were more apt to comment on the abundance of Inca food than on its foreignness.[1] The Spaniards knew maize already as a staple in the Caribbean, which they consumed as a food and as a beer called *chicha.*[2] As they traveled into the highlands, Spaniards ate the meat of Andean camelids for the first time, as well as fresh and dried potatoes and other tubers. Inca accounts of Spanish food consumption by the Pizarro expedition (e.g., Betanzos 1999 [1550s Part 2, Chap. 25) note an indigenous conviction that the invaders did not know how to eat properly, and that they preferred to eat meat (duck and camelid) and maize bread and to drink maize beer. As Spaniards learned more about newly encountered Andean staples, they compared them to known foods: llamas were "sheep" (*ovejas*), and guinea pigs were like domesticated rabbits; potatoes were like bland truffles or turnips (*turmas, nabos*), and quinoa was similar to rice (e.g., Sancho de la Hoz 1968 [1534]: 327–328; Valverde 1959 [1539]: 314) (Fig. 12.1).

Spanish writers frequently mention the flavors of Andean foods, suggesting that they ate them at least occasionally. Nevertheless, it appears that the colonizing Spaniards also attempted from early on to establish some semblance of their own dietary customs as Christians. For example, the 1534 chronicle of Pedro Sancho de

[1]For example, Cieza de León (1984 [1553, Chap. 16]) comments on the tenuous food supply on his expedition with Juan de Vadillo through Anserma, on the way to Popayán in Colombia: "in some parts we did not find maize, or any other thing to eat; and as for meat, it was more than a year that we didn't eat it, unless it was from horses and dogs that died; and we didn't even have salt, such was the misery that we endured." The chronicler continues with an account of Spaniards who, being in a desperate state of hunger, ate an indigenous stew containing human flesh.

[2]The Caribbean names were more commonly used than the Quechua words for maize (*sara*) and beer (*aqha*).

Fig. 12.1 Woodcut of Andean camelids from Pedro de Cieza de León's 1553 chronicle (*folio* 124v.). Spaniards referred to the llama and alpaca using European names, such as sheep, in the early years of the conquest

la Hoz (1968: 311 [Chap. 11]) recounts that when the Spaniards first met the prince Manco Inca, he told them that he was trying to acquire fish for them to eat, because he understood that the following day was a meatless one for Christians.[3] Whereas Spaniards noted the lack of healthful food as a reason for their numerous deaths in the Americas, the emphasis on bread and wine, and on herding and gardening, also held

[3]Manco Inca, the erstwhile ally of the Spaniards in the early 1530s, should not be confused with the founding Inca ancestor, Manco Capac. See Earle (2010) on the articulation of Christian/European diet since the earliest days of Spanish colonization in the Caribbean. The 1534 chronicle of Francisco de Xérez describes the men of the Pizarro expedition as "Christians" nearly 100 times, and his prolog describes how Spaniards in the Americas have conquered great territories, "maintaining themselves with the beastly supplies of those [Indians] *who have no knowledge of bread or wine*, suffering with herbs and roots and fruits…" (1985: 60: *manteniéndose con los mantenimientos bestiales de aquellos que no tenían noticias de pan ni vino, sufriéndose con yerbas y raíces y frutas…* emphasis in translation added).

significance for the religious ideology they attempted to impose upon indigenous Andean cultures (e.g., Cañizares-Esguerra 2006; Earle 2012).

Even before the Pizarro expedition reached Cuzco, its members had observed Inca food practices. Pedro Pizarro (1986 [1571]) recalled years later that the Inca lords ate a morning meal, drank maize beer throughout the day, and had a light meal in the evening. At the capital, it was not uncommon for the ruling elite to host public feasts with the residents of the city and with the mummified remains of former Inca lords and ladies (Pedro Pizarro 1986 [1571]; Biblioteca Nacional, Ms. 20193). Food and drink at these events were prepared by the Inca *mamakuna*, women who possessed the skills and attributes to make food and drink for sacred objects and royal mummies and to provide those entities with appropriate sustenance (see also Staller, Chap. 5 for insights on chicha production and ancestor veneration). There were also frequent festivities in which the ruling Inca dispensed vast quantities of maize beer to crowds that came from the city and from provincial regions (e.g., Betanzos 1999 [1550s]). Royal largesse depended on intensive farming and herding practices in the imperial heartland, as well as a system of storehouses throughout the region that held a wide range of foodstuffs (Covey et al. 2016). Inca nobility had maintained control over the best agricultural resources and stored food through estates belonging to the founding pairs of royal lineages or *panaqas*, although the cult of the Sun also possessed farmland and herds (Fig. 12.2).

Archeological evidence for Inca foodways is limited and tends to make rough inferences based on the distribution of production infrastructure, rather than the contents of domestic middens or storehouses. The impressive valley-bottom terraces in the Cuzco Basin and nearby Sacred Valley (Vilcanota-Urubamba River) encourage interpretations of elite amplification and control of maize agriculture (e.g., Farrington 1983; Bauer 2004), especially when these material remains are juxtaposed with ethnohistoric descriptions of Inca ideology and ritual, which focused on maize agriculture (e.g., Bauer 1996). Although the contents of excavated storehouses in the Cuzco region have been found to contain maize, many held multiple crops, including fruits, dried potatoes or *chuño*, chenopods, and other taxa (see Covey et al. 2016 for a discussion of storage in the Cuzco region). Excavated middens from Inca-era households in the Cuzco region demonstrate a range of food remains, including animal bone from camelids, guinea pig, bird, and deer (e.g., Covey 2015; Kosiba 2010 Table C.4; Quave 2012: 262–275). Although much more archeological evidence is needed—particularly for elite Inca contexts—the emerging picture of variable food practices accords with chronicle accounts that describe diverse food production for the Inca heartland region. For example, when Betanzos (1999 [1550s Part I, Chap. 12]) relates how the legendary emperor Inca Yupanqui provisioned the city of Cuzco for the first time, he mentions storehouses for green and dried maize, chile peppers, beans, freeze-dried potatoes (*chuño*), quinoa, dried meat, and other foodstuffs.

Andean food sustained the early Spanish occupation of Cuzco, and the first European citizens of colonial Cuzco might have appeared to indigenous people as houseguests of the Inca nobility who overstayed their welcome, unknowingly eating their way into social obligations they had no means of discharging. In the earliest years, Inca nobles dominated the means of production and the surpluses produced on the

Fig. 12.2 A depiction of Inca nobles sowing a ceremonial maize plot in Cuzco. Noble Inca families controlled the best farmland in the Cuzco region at the time of the Conquest (From Guaman Poma de Ayala 1615, *folio* 250[252])

best lands in the imperial heartland, and they were the only individuals capable of underwriting public works, leading Andean armies, and reciprocating labor service. Spaniards accessed elite Inca power indirectly through *encomienda* grants, but over time, they purchased and reallocated attractive farmland and encouraged indigenous populations to join their estates as retainers.

Spanish Alliances and the First Tastes of Home in Colonial Cuzco

Spaniards attempted to introduce European domesticated animals into the highlands when they initially founded towns in these regions, although the breeding and adaptations of such species to these high elevations took quite a bit of time in some cases. Most Old World animals were probably brought to the Andes from the Caribbean (Earle 2010; Reitz 1992), but it is not understood how different species adapted to changes in elevation, seasonality, and forage. Differences in reproductive rates would have influenced the speed by which a species became economically viable in the Andes, and larger animals (e.g., horses, cows) seem to have remained relatively rare and valuable even as faster-breeding species (such as pigs and chickens) proliferated.[4] More archival work is needed to understand the dynamics of these processes, and early Colonial Period household excavations may provide direct evidence to clarify the movement and reproduction of such European domesticates, as well as their changing roles in indigenous subsistence and diet (e.g., deFrance 1996, 2003).

As non-Andean livestock populations grew, their social and economic significance changed, and they contributed in different ways to growing divides between "Spanish" and "Indian" foodways in colonial Cuzco. As Garcilaso de la Vega recalled (1960–1965 [1609, Book 9, Chaps. 16–23]), Spaniards in the city first focused on the growth of breeding stock, and Old World animals were not regularly slaughtered for meat or sold as commodities. The chronicler notes that individual Spaniards in Cuzco would acquire animals from other parts of Peru, and as they started to breed, the owners would share animals with their social allies as a way to cement alliances. This sort of faction-building speaks to the changing composition of Spaniards in early colonial Peru: the civil wars of the 1540s and 1550s led to the death of many of the original *conquistadors* who still remained in the Andes, and they disrupted subsistence and tributary economies of indigenous populations in significant ways. For example, royal accounts from Cuzco in 1555–1556 (AGI, Contaduría 1824 f. 10, 51) note that the rebel Francisco Hernández Girón seized local tributes and interrupted agrarian production in 1554. Rebellions also must have affected private Spaniards, and Diego Fernández (1571, Part 2, Book 2, Chap. 1) describes how Hernández Girón traveled to the Yucay Valley near Cuzco because of the abundance of food to be had

[4]It is also likely that the speed of introduction also depended on the extent to which an Old World animal complemented or competed with Andean domesticates as a source of labor (llama vs. donkey), meat (Muscovy duck vs. chicken), and wool (alpaca vs. sheep).

there (much of it on private lands). Royalist armies sent to suppress the rebellions brought new waves of Spanish migrants to Cuzco, men who had often come directly from Spain and did not share the hybrid food tastes of the original conquerors.

As leading Spaniards managed to introduce familiar crops and animals to their households, those with *encomiendas* also began to receive some of these foods from the Andean populations that served them, suggesting that indigenous populations quickly learned to produce them. Demand for Old World animals created extraordinary price bubbles in the 1540s and 1550s, and the rapid proliferation of some species led to dramatic decreases in the market values of these animals from the late 1550s. By the early 1600s, prices for some species had dropped by more than 90% (Table 12.1). Over time, Old World animals became commodities in the colonial marketplace, and Spaniards (and probably some Andean elites) were able to consume a kind of creole diet that increasingly included chicken (and eggs), pork, lamb, and rabbit, alongside indigenous birds, domesticated camelids, hunted game, and the turkey (which was introduced from Mesoamerica) (Garcilaso de la Vega 1960–1965 [1609]; cf. Acosta de Acosta 1894 [1590]).

As Old World animals became more common in the highlands, wheat (and to a lesser degree, barley) made similar inroads. Garcilaso de la Vega states that there was still no wheaten bread in Cuzco in 1547 (1960–1965[1609 Book 9, Chap. 24]), but by the time of Pedro de Cieza de León's arrival a few years later, bread was being produced and there were mills in several parts of the Cuzco region (1984 [1553 Chap. 93]). It is interesting to note Francisco López de Gómara's (1884: 278 [1552]) published claim from this time regarding the fertility of Peruvian lands when domesticated Old World plants and animals were raised: "one grain of barley shoots

Table 12.1 Anecdotal Prices for European Animals in Cuzco

Animal	Early price	Reduced price	References
Horse	10,000 pesos offered for a war horse in 1554 (on the eve of battle)	400 pesos for a good horse in early 1600s	Garcilaso de la Vega Book 9, Chap. 16
Donkey	400 pesos paid in Huamanga in 1557		Garcilaso de la Vega Book 9, Chap. 18
Cow	1,000 pesos paid in Lima for 10 cows in 1554	Same cows sold in 1559 for 17 pesos each; ~ 5 pesos per head in early 1600s	Garcilaso de la Vega Book 9, Chap. 17
Goat	100–104 ducats (~ 85 pesos) paid, perhaps in early 1550s		Garcilaso de la Vega Book 9, Chap. 18
Sheep	40 pesos per head in 1556	4 ducats or less (~ 3 pesos) in early 1600s	Garcilaso de la Vega Book 9, Chap. 20
Pig	10 pesos for a "good, fat" pig in 1560	6–7 pesos in early 1600s	Garcilaso de la Vega Book 9, Chap. 19
Chicken			Garcilaso de la Vega Book 9, Chap. 23

forth 300 spikes, and another of wheat 200 … the livestock multiply in like manner: a goat gives birth to five young … and if it had not been for the civil wars, there would already be infinite numbers of mares, sheep, cows, donkeys, and mules." Such hyperbolic expressions of Andean fecundity demonstrate the ideological connection between food and the "civilization" of the native landscape, but they are contradicted by documentary evidence for the slow introduction of many Old World species.

By the late 1550s and after, Spaniards began to plant vines and olives groves in the hopes of establishing domestic wine and oil industries.[5] Other European fruits and vegetables were also brought to Peru, and they circulated through social networks in the early days (see Lambaré et al., Chap. 4 for an analysis of the introduction of the peach to Argentina). Garcilaso de la Vega (*ibid,* Book 9, Chap. 25) recalls distributing bunches of grapes to the houses of the leading citizens of Cuzco in 1555. The grapes had been sent to his father by the *conquistador* Bartolomé de Terrazas, and around the same time, the royal treasurer García de Melo sent three asparagus stalks to the house, which the chronicler's father grilled and served to seven or eight of his closest associates (*ibid*, Book 9, Chap. 30). By the early 1570s, a wide range of European foods was available in the valleys surrounding Cuzco (e.g., López de Velasco 1894 [1574]: 482).[6] Decades later, Spanish writers such as Acosta and Garcilaso de la Vega claimed that many Iberian fruits and vegetables grew abundantly in the Andes, sometimes to incredible sizes. Indigenous populations produced almost all of these foods, and they incorporated some of them, such as garlic, into their local cuisines (Acosta 1894 [1590] Book 4, Chap. 18).

The Tribute Economy and Food Commodities

After numerous *conquistador* uprisings in the 1540s and early 1550s, royal officials began to issue more detailed tribute levies (e.g., Rostworowski 1993 [1983–1984]) (Fig. 12.3). By this time, European domesticates were well-established in many coastal areas and were beginning to be successfully transplanted to some highland valleys (cf. Cieza de León 1984 [1553, Chaps. 66, 80, 93; Gasca 1998 [1551–1553], Chap. 27–28). Highland tribute levies (*tasas*) issued in 1549 by Pedro de la Gasca following his defeat of Gonzalo Pizarro illustrate the integration of European plants and animals to indigenous agriculture, as well as the continuing consumption of Andean foods by European elites (Table 12.2).

These documents reflect administrative targets rather than subsistence realities, and in many cases, they were revised within a few years of being issued. The Cuzco *tasas* provided for substitutions of some items, acknowledging that the advent of

[5]Peruvian olive production came to focus on eating the fruit rather than oil production. In the highlands, pig lard was important for cooking until imports of Chilean olive oil began to increase decades later (Acosta 1894 [1590 Book 4, Chap. 32]; Garcilaso de la Vega 1960–1965 [1609 Book 9, Chaps. 19, 26–27]). See Rice 1997 on early Colonial Period viticulture.

[6]See Earle 2012 for broader discourses on American fertility and Spanish colonization.

Fig. 12.3 Felipe Guaman Poma de Ayala's depiction of Pedro de la Gasca's defeat of the *conquistador* Gonzalo Pizarro, which led to new tribute levies in the Andean highlands (From Guaman Poma de Ayala 1615, *folio* 424[426])

Table 12.2 Andean and Spanish plants and animals in selected highland *Tasas* from 1549

Repartimiento	Andean	European	Source
Qichuas mitimaes (Huánuco)	Maize (300 *fanegas*), potatoes (150 *fanegas*), beans (8 *fanegas*), birds (200), eggs (520 +), fish, fruit	Wheat (300 *fanegas*), pigs (12)	Murra 1972a
Yachas (Huánuco)	Maize (200 *fanegas*), potatoes (100 *cargas*), birds (200), eggs (520 +)	Sheep (30), pigs (15)	Murra 1972b
Soras (Huamanga)	Maize (600 *fanegas*), potatoes (60 *fanegas*), Andean camelids (156), birds (500), eggs (150/week), fish	Pigs (36)	Rostworowski 1983–1984
Lucanas (Huamanga)	Maize (600 *fanegas*), potatoes (100 *cargas*), llamas (163), birds (300)	Wheat (100 *fanegas*), sheep or alpacas (6), pigs (15)	Rostworowski 1983–1984
Amaybamba (Cuzco)	Coca (200 *cestos/mita*), fruit (24 *cestillos/mita*)		Rostworowski 1983–1984
Calca (Cuzco)	Maize (500 *fanegas*), potatoes (25 *fanegas*), "partridges" (200), eggs (20/week), fish	Wheat (50 *fanegas*), chickens (240), pigs (15)	Rostworowski 1983–1984
Yanahuaras (Cuzco)	Maize (50 *fanegas*), potatoes (150 *fanegas*), chuño (150 *fanegas*), ducks (100), "partridges" (200), eggs (25/week), fish	Wheat (30 *fanegas*), sheep (120), pigs (15), chickens (100)	AGI Justicia 472 [1559–1563] f. 7–9v.
Aymaraes (Cuzco)	Maize (600 *fanegas*), potatoes (50 *fanegas*), molle (20 *fanegas*), chiles (50 *cestillos*), llamas (83), ducks (150), "partridges" (200), eggs (25/week), fish	Wheat (60 *fanegas*), sheep or alpacas (3), pigs (20), chickens (150)	Rostworowski 1983–1984
Azángaro (Cuzco)	Llamas (133), llamas for caravans (50), dried llama meat (40), fish (10 *arrobas*), coca (60 *cestos/mita*)	Sheep or alpacas (3), pigs (24)	Rostworowski 1983–1984

(continued)

Table 12.2 (continued)

Repartimiento	Andean	European	Source
Chucuito (La Plata)	Maize (1000 *fanegas*), chuño (1200 *fanegas*), quinoa (50 *fanegas*), kañiwa (50 *fanegas*), dried fish (300 *arrobas*)	Wheat (12 *fanegas*)	Rostworowski 1983–1984

some European domesticates was not as advanced as the annual levies would seem to indicate.[7] Specifically, in the event that pigs could not be given, alternative animals would be accepted—for Calca, 12 chickens or ducks for each pig, and for Yanahuaras, Aymaraes, and Azángaro, a sheep.

The early *tasa* documents indicate some important perspectives on subsistence practices and identity building in the early colonial highlands. The repeated requirement that indigenous populations provide their *encomenderos* with fish and eggs for non-meat days indicates an administrative concern with using food to encourage more observant Christian life among colonial elites. The *tasas* indicate that the bulk of food consumption by Spaniards still consisted of Andean plants and animals, but that wheat and European animals had begun to be produced with varying degrees of success in different parts of the highlands. The availability of non-Andean crops and animals indicates that indigenous Andeans were learning how and where to cultivate and herd these new sources of food. The *tasas* differentiate between the indigenous presentation of food surpluses at the home of the *encomendero*, and the provision of labor to produce food in the *encomendero*'s fields and gardens (e.g., Vega 1896 [c. 1562]). In some instances, indigenous laborers were to be provided "appropriate" food by the *encomendero*, although it is unclear what kinds of food would have been offered, or in what quantities. It is plausible that the increase in the production of European domesticates affected local indigenous diets to varying degrees, although archeological evidence from household contexts is needed to evaluate such changes.

The *tasa* from Chucuito offers important information on food in the developing political economy of colonial Peru. Chucuito was held by the Crown, and royal officials managed its tribute. As Table 12.2 notes, the *tasa* was to collect almost exclusively Andean products, and these were to be transported to the burgeoning mining center of Potosí by royal llama or mule caravans. Although Chucuito had no *encomendero*, its *tasa* describes the material support for the person (presumably a Spaniard) overseeing the estate:

> every month two *fanegas* [bushels] of maize, and one of potatoes and two *fanegas* of *chuño*, and two sheep; and every three months a pig (or a sheep in its place) and an ounce of salt; and every week six ducks and partridges; and on every fish day, ten eggs and two *arreldes*

[7]Francisco Pizarro granted Diego Maldonado pasture lands for his pigs and mares in the Lucre Basin near Cuzco in 1539 (Espinoza Soriano 1974 [1539/1553]). However, Garcilaso de la Vega recalls that breeding stocks were still being acquired in the 1550s. Spanish civil unrest probably presented setbacks for the introduction of Old World animals—there are references to rival armies appropriating food while on the march, and destroying local food stores when pursued.

[about 3.68 kg] of fish; and wood to burn, and fodder for his horse; and every day a jug of *chicha* (Rostworowski 1993 [1983–1984] p. 312; author's translation).

Whether or not the royal overseer was a Spaniard, he ate a diet that still lacked wheat bread and included daily consumption of maize beer.

The intensification of royal administration of indigenous *repartimientos* resulted in the production of large amounts of tribute that had to be converted into other forms of value for the empire. Although some tribute in kind generated in royal *repartimientos* went to the mines, much of it was sold by Crown officials who produced records of the changing value of different subsistence commodities over time (e.g., AGI Contaduría 1824–1825). Before 1550, royal accounts from Cuzco tended to focus only on payments of the *quinto*, but the documents after that time mention royal sales of everything from cloth to staple goods to mangy alpacas. Prices for most commodities are not consistently recorded over time, but the available information offers evidence of the changing economic roles of Andean and European domesticates (Table 12.3). The fragmentary and variable nature of recorded prices discourages probing analysis of these data, but it is noteworthy that certain products seem to have held their value over time (e.g., potatoes, birds), whereas prices for others changed significantly. While the price of maize appears to have fluctuated over time, wheat became considerably cheaper after the mid-1550s, when it was first mentioned in the documents consulted. Sales of coca leaf were commonly recorded by royal officials in Cuzco, and the value of the leaf dropped by as much as 80% from the early 1550s to the 1570s, which will be discussed later in this chapter.

Maize continued to hold prominence in the colonial economy, assuming the rough equivalent of wheat in the administrative mind even as wheat came to be much more prominent in the *tasas* of the Toledan era. The two crops were seen as "bread-making" staples that belonged to distinct ecological zones, and the chronicler Acosta (1894 [1590] Book 3, Chap. 20, Book 4, Chap. 16) contrasts these two kinds of bread with

Table 12.3 Commodity prices recorded by royal officials in Cuzco

Product	1549–1551	1554–1557	1561–1562	Toledan (1570s)
Maize	1.6 pesos/fanega	1.8–1.9 pesos/fanega	2.3 pesos/fanega	1.0–1.5 ps/fanega
Wheat		2.1–3.2 pesos/fanega	2.0 pesos/fanega	1.0–1.5 ps/fanega
Potatoes		0.5–0.8 pesos/fanega		0.5 pesos/fanega
Chuño		1.9 pesos/fanega		2.0 pesos/fanega
Quinoa		0.7 pesos/fanega		
Ají		<0.7 pesos/cesto		0.8–2.0 ps/cesto
Coca	3.1–5.1 ps/cesto	4.0–5.3 pesos/cesto	2.0–3.3 ps/cesto	1.0–1.8 ps/cesto
Pigs		1–1.6 pesos		1.5 pesos
Birds		0.1 pesos		0.1 pesos
Alpaca	1 peso, 1 ducat			1.3–2.5 pesos
Sheep		4 pesos		

Sources AGI Contaduría 1824–1825; Cook 1975

supposedly inferior sustenance offered by root crops such as cassava and potatoes (*ibid.* Book 4, Chap. 17). Despite the enduring role of maize as a civilized food, wheat bread still held the highest food status for Spaniards, and chroniclers of the mid-sixteenth century note places where wheat was cultivated, ground to flour in mills, and made into bread (e.g., Cieza de León 1984 [1553]).

As maize, wheat, coca, and Andean camelids assumed central roles in the Colonial tributary economy in the highlands, other plants and animals were excluded from levies of tribute in kind. Andean crops such as *tarwi*, beans, *oca*, and quinoa seem to have been largely excluded from tribute, suggesting that these were not crops Spaniards typically consumed, or that indigenous populations sold in markets. European sheep and other domesticated animals became widespread in the highlands in the second half of the sixteenth century but not as tribute categories.

Indigenous Communities and Private Property

In the Cuzco region, the spread of crops and herd animals from the Old World intersected with existing patterns of agricultural production. Inca elites had developed new valley-bottom agricultural infrastructure in some parts, and rulers designated other areas as pasture lands for the nobility and the state religion (e.g., Niles 1999). Permanent retainers and labor colonists worked in the lands of the Inca nobles and the sun cult, increasing the population of the Cuzco region and making it more diverse. At the same time, local communities claimed access to some improved Inca lands, but they also possessed their own fields and pastures across the imperial heartland. As Spaniards attempted to transplant European agriculture to Cuzco, they faced challenges in directly accessing indigenous labor and the means of production. In principle, the *encomienda* system provided some Spaniards with the fruits of Andean labor, but it did not grant ownership of native lands. Access to land came through grants of "vacant" lands, or by purchasing plots from Andean elites willing to sell them, but the purchase of land did not guarantee the availability of labor to carry out agricultural labor when it was needed.[8] The *encomienda* system imposed demands on indigenous communities to produce European domesticates, but as the market for staple commodities grew—along with the population of Spaniards in Cuzco who did not have *encomiendas*—important changes in landholding and production practices occurred across rural Cuzco (Covey and Quave 2017).

From the 1530s, Spaniards sought out warm valley-bottom lands where their plants and animals would thrive, and the Yucay Valley to the northeast of Cuzco stands out in the ethnohistoric literature (e.g., Covey and Amado 2008; Cieza de León 1984 [1553] Chaps. 93–94; Vega 1960–1965 [1609] Book 5, Chap. 27; Huerta 1624: 242; López

[8] Some Spaniards purchased land near their *encomiendas* and had their tributary laborers work them, whereas others attracted retainers (*yanaconas*) or brought non-Andean slaves to their properties. Wage labor became more important over time as royal officials phased out uncompensated personal services for *encomenderos*.

Fig. 12.4 The warm, well-watered Inca terraces of the Yucay Valley attracted Spaniards, who established country houses and gardens in the sixteenth century. (Courtesy of Google Images)

de Velasco 1894 [1574]: 483; Villanueva Urteaga 1970) (Fig. 12.4). Within the valley itself, Spaniards acquired lands that had formerly been parts of royal Inca estates, where their retainers could raise livestock and grow Iberian fruits and vegetables. By 1550, the leading Spanish and Inca citizens of Cuzco had houses and garden plots in Yucay, and within a few years, attention turned to less desirable lands nearby. In the late 1550s, *encomenderos* on the nearby Maras plain purchased "vacant" lands that had formerly been pastures for royal Inca estates and the sun cult, converting them to productive wheat and barley fields that were worked by natives from their *encomiendas* (Covey and Quave 2017).

As Spaniards and Andean elites developed personal lands in and around tributary communities, they purchased and sought grants for herding stations (*estancias*) in higher elevation lands where they placed retainers to care for their animals. The growth of the herding economy created new demands for pasturelands, as well as frequent conflicts with indigenous communities who claimed the uplands for their own tuber cultivation and herding (e.g., Títulos de Eqquecco-Chakan [1569] f. 2). In the 1570s, the viceroy Francisco de Toledo designated large areas of the Xaquixaguana Valley as a grazing area for Cuzco, and he granted *estancias* to private individuals in that area. The reallocation of land occurred at the time that indigenous tributaries were being reduced into new colonial towns as part of a strategy to keep the "Indian" and "Spanish" republics separate. Visits to the valley in the late 1570s and early 1580s registered large numbers of European sheep and cattle, and there is evidence of pig herding as well (López de Velasco 1894 [1574]: 483).[9]

[9]Small Spanish herding concerns might have a few dozen cattle and several hundred sheep. By comparison, the 1582 will of the Inca prince Carlos Ynga listed several thousand head of sheep, more than 100 cows, 300 goats, 50 pigs, and 80–90 Andean camelids (BN Ms. 20193 f. 45 ff.).

Food, Health, and Caste-Making in Cuzco's Two Republics

The first *conquistadors* struggled to eat as Christians in the Andean highlands, but over time, colonial elites were in a better position to differentiate "Spanish" and "Indian" foodways. Despite administrative projects intended to set the colonized and colonizers apart, hybrid diets developed in Cuzco and its rural hinterland, as they did in other areas colonized by Spaniards (e.g., Reitz 1992; Rodríguez-Alegría 2005). Creole elites incorporated a range of Andean foods into their diets as staples, and by the end of the sixteenth century, it is clear that they had also found a place for these foods in their humoral approaches to individual health (Earle 2012; López Beltrán 2007). For example, Acosta (1894 [1590] Book 4, Chap. 16) notes that although maize is an important source of bread in the Americas, it is heavier and hotter, and influences the humors differently than wheat. He states that Spaniards and indigenous people use maize beer for its medicinal properties, to aid in kidney health. Acosta also (Book 4, Chap. 20) describes the healthful properties of chile peppers, which are a "cool" food despite their apparent heat, and should be taken in small amounts to avoid exciting one's sensuality.[10] Garcilaso de la Vega notes that the Incas had not recognized the humors (*op. cit.*, Book 2, Chap. 24), but he discusses how Spaniards incorporated Andean foods into their medical practices. The chronicler describes the healing properties of maize (Book 2, Chap. 25; Book 8, Chap. 9) and *molle* (Book 8, Chap. 12), and he notes that doctors encouraged invalids to eat the meat of young camelids as a more healthful flesh than chicken (Book 8, Chap. 16). It is important to note that Spaniards did not adopt Andean medicinal practices, but rather accepted certain foods and drinks as appropriate to maintaining a healthy humoral balance. Such practices challenge the easy racial distinctions between "Spanish" and "Indian" food practices and reflect the fact that early colonial Cuzco was much more diverse than colonial officials would have liked to admit.

As "Spanish" and "Indian" foodways diverged in the second half of the sixteenth century and commodity production stimulated new patterns of competition over the means of production, priests and colonial officials began to portray certain indigenous consumption practices in a negative light. Spanish writers associated the periodic mass consumption of maize beer and the daily use of coca leaf with the preservation of non-Christian practices, and they also blamed these food practices for keeping indigenous people from eating, leaving them vulnerable to sickness and death (e.g., Toledo 1867 [1572 Título 21]: 89–90). Although both practices were viewed as dangers to native bodies and souls, administrative responses were not the same for *chicha* and coca. Whereas the public consumption of maize beer was treated as a threat to public health and outlawed by royal administrators, the same officials helped to ensure that indigenous consumption of coca leaf continued, despite the incompatibility of the practice with Christian missionary work.

[10]Years earlier, Enríquez de Guzman (1886 [1543 Chap. 43]) reported using chiles to treat a toothache while in Lima. Garcilaso de la Vega (Book 8, Chap. 12) recalls that a Spaniard once told him that peppers improved eyesight, and he claims that all Spaniards who had lived in the Americas adopted chiles and preferred them to Oriental spices.

Maize Beer or *Chicha*

As wine became more readily available in Peru in the mid-1500s (Rice 1997), Spanish writers appear to have changed their stance on *chicha*, and on its mass consumption in public events mobilized by Andean elites (Cummins 2002). As noted above, the first conquistadors drank maize beer during years of exploration and conquest in the Caribbean and Central America, and they continued to do so in Peru. By 1550, Spanish writers understood that large-scale festive drinking was a core element in Inca political activities, religious ceremonies, and performances of dynastic histories. Some authors began to describe indigenous feasting and drunkenness as an impediment to missionary work.[11] Municipal authorities and priests might not have approved of such activities, but they did not take strong actions against them. Cieza de León (1988 [c. 1553 Chap. 11]) notes that a version of the Inca *purucaya* ritual was performed in Cuzco in 1550 to mark the first anniversary of the death of the Inca Paullu, and Solórzano Pereira (1648 Book 2, Chap. 25) recounted that the Second Council of Lima (1567) attempted to moderate public drinking organized by indigenous elites, but did not prohibit it. One reason for this may be that more discerning writers, such as Fernando de Santillán (1879 [1563 #77]) identified maize beer as an important source of sustenance in indigenous diets that were too often insufficient.

In the advent of viceroy Toledo's administrative reorganization, other writers portrayed indigenous drinking as a public health concern, and the viceroy issued orders to curb these practices (Toledo 1867 [1572 Título 21]: 89–90) (Fig. 12.5). It is noteworthy that Toledo also banned Spaniards, Africans, and people of mixed race from selling maize beer to indigenous people, and from drinking with them. Attempts to exclude Andean maize beer from colonial commerce coincided with the representation of traditional brewing practices as unhealthy. Whereas early authors (Zárate 1995 [1555] Book 1, Chap. 8) describe the practice of masticating maize to begin the fermentation of *muqu chicha*, this essential women's work came to be described as dirty and disgusting, and distinguished from the production of *wiñapu chicha* from malted maize (e.g., Krögel 2011: 90–92). As Mangan (2005) notes for colonial Potosí, changing values regarding maize beer reflect growing geographic and ethnic divides: Spanish women entered the *chicha* production market in urban areas, producing their ware with milled corn, whereas *muqu chicha* came to be associated with indigenous women of the surrounding countryside.

[11] Several writers mention heavy drinking activities in Inca times, and by Andean elites of that time (e.g., Tamara 1556 Book 3, Chap. 10). Francisco Hernández Girón (Hernández 1876 [1553]) claimed that leaving the Indian population unburdened by personal services to Spaniards would leave them free for drunkenness and idolatry, although the self-serving nature of his statement is obvious.

Fig. 12.5 Guaman Poma de Ayala's illustration of public drunkenness in colonial Peru (From Guaman Poma de Ayala 1615, *folio* 862[876])

Coca

Coca use was widespread in the Andes at the time of the European invasions, although it became considered as an elite Inca privilege as Crown officials sought to monopolize production in the 1550s. Indigenous people stated that chewing it reduced their hunger and thirst and gave them greater strength (Cieza de León 1553 Chap. 96). Spaniards do not seem to have adopted this practice, which Cieza de León describes as, "a corrupt custom," but nevertheless "convenient for people who are similar to how the Indians are."[12] This chronicler noted that by the early 1550s, demand for coca—including at the mines of Potosí—had made many Spanish *encomenderos* and merchants rich, and his contemporary López de Gómara (1884 [1552]: 276) claimed that coca leaf was more costly than gold or bread (Fig. 12.6). For decades, coca leaf also provided income and influence for the Inca nobles living in Cuzco. In 1565, the viceroy reported paying two descendants of the ruler Tupa Inca Yupanqui pensions from the coca of Tono, and he divided 350 *cestos* of coca among 47 descendants of Huayna Capac (García de Castro 1906 [1565]: 64–65).

Cieza de León mentions that from 1548–1551, the years immediately following the defeat of Gonzalo Pizarro's rebellion, coca fetched a particularly high price, and *encomiendas* that could produce the leaf were much sought-after. This is substantiated by Crown records of coca sales in the early 1550s (Table 12.4), although the intensification of royal authority saw a significant decline in coca prices, to the point that the leaf was valued at 1 *peso* per basket (*cesto*) or less in the 1570s Toledan *tasas*.

The decline in prices was not due to declining demand among the indigenous population, but rather to the successful increase in coca production in the Amazonian slope in Cuzco and other highland regions. Increasing coca production occurred at a time when Spanish writers expressed deep concern over the cultivation and use of coca by indigenous people (e.g., Santillán 1879 [1563 #115]; see Gagliano 1963). Highland populations suffered when they were sent to humid lowland areas to cultivate the leaf, and many died or fled their *repartimientos* as a result.[13] Native labor was also instrumental in transporting dried coca leaf to distant markets such as Potosí. Whereas the labor of producing and transporting coca threatened native bodies, the everyday consumption of coca leaf was seen as a vice that harmed body and soul.[14] Spanish writers claimed that regular coca consumption prevented Indians

[12]Other authors also note that Spaniards did not chew coca or use it for other purposes. For example, Bartolomé de Vega (Vega 1896: 127[c. 1562]) describes it as the most bought and sold product in Peru, but also "a thing that the Spaniards do not eat, nor do they make use of it for any other thing" (translation of author). In 1597, Baltasar Ramírez (1906: 325) stated that only Indians used coca, although he describes the smell and taste of the leaf.

[13]Bartolomé de Vega (1896: 128[c. 1562]) notes that indigenous highlanders returning from coca service were seen in the hospital in Cuzco with cancerous lesions on their noses (possibly *leishmaniasis*) and goiters the size of citrons.

[14]Baltasar Ramírez wrote in 1597 (1906: 324) that many treatises had already been prepared debating the benefits and ills of coca leaf, which had been presented to the Crown and the royal Consejo de Indias.

Fig. 12.6 Guaman Poma de Ayala's depiction of local Andean horticulture. A couple is sharing coca leaf in the dooryard of a house that has chickens and Andean and European fruit trees (c. (From Guaman Poma de Ayala 1615, *folio* 865[879])

Table 12.4 Records of crown coca sales in Cuzco, 1550–1562

Year	Price/*Cesto*	Reference
1549	5.1 pesos	AGI Cont. 1824 f. 2v.: 1130 pesos from the sale of 223 cestos from Abisca
1550	3.1	AGI Cont. 1825 f. 140: "dozientos y diez y syete p[es]os y tres tms y seys granos que valieron setenta y vn cestos de coca…"
1554	4.0	AGI Cont. 1825 f. 142: proceeds of 100 cestos of coca from the natives of Nuñoa, sold at 4 pesos/cesto on the road to the Collao
1554	4.5	AGI Cont. 1825 f. 143: 450 pesos for 100 cestos of coca from Nuñoa
1554	5.3	AGI Cont. 1825 f. 144v.: 1062 pesos and 4 tomines from 200 cestos of coca from the repartimiento that had belonged to the Licenciado de la Gama; sold for 5 pesos and 2 tomines per cesto
1561	2.0–3.0	AGI Cont. 1825 f. 76–76v.: 1185 pesos, representing the value of 395 cestos of coca from Toaima; 63 pesos, representing the value of 31 cestos of coca from Piccho; and 310 pesos and 4 tomines, representing the value of 144 cestos of coca from Amaibanba
1561	3.0–3.3	AGI Cont. 1825 f. 83: Coca of Tono: 609 pesos, representing 203 cestos of coca; 968 pesos, 3 tomines, 6 granos, representing 295 cestos. f. 83v: 1053 pesos, 5 tomines, representing 353 cestos
1562	2.6	AGI Cont. 1825 f. 180: Coca of Tono: 563 pesos and 6 tomines, value of 221 cestos of coca

Sources AGI, Contaduría 1824 [1535–1558] Caja de Cuzco. Cuentas de Real Hacienda desde 1535 a 1558. AGI, Contaduría (1825) [1532–1572] Caja de Cuzco. Cuentas de Real Hacienda desde 1532 a 1572

from eating properly, causing them to die (Vega 1896 [c. 1562]: 127–128). They also noted the use of coca as an offering to the Sun, to sacred places and objects, and to the demons that inhabited the Andes and led the native population astray (Vega 1896 [c. 1562]: 127–128). The viceroy Francisco de Toledo (1867) sent judges and medical men from Cuzco to areas of coca cultivation, and he used their reports to establish new ordinances for coca production in 1572. These laws made the Spanish Crown less culpable for the excesses of production by discontinuing the designation of coca as a form of tribute in kind and prohibiting coerced labor on lowland coca plantations.[15] The Toledan ordinances laid out labor practices for coca production and transport and stated that no new coca plots could be planted in the Cuzco region. Nevertheless, production continued to increase, and by the end of the sixteenth century, wage laborers from other regions worked in Cuzco's coca lands, producing hundreds of tons of the leaf for use in the Potosí mines (Lizárraga 1908 [c. 1595–1609, Chap. 44]; Ramírez 1906 [1597]: 325–326) (Fig. 12.7).

The consistent criticism of indigenous coca use—and concomitant administrative efforts to regulate, rather than outlaw, it—raises questions about why royal officials

[15]Despite the optimism of Spanish officials—in 1574, the royal chronicler Juan López de Velasco (1894: 480–481) stated that the greatest commerce of Cuzco was coca, "but it is already diminishing and will end with the ordinances of the viceroy don Francisco de Toledo"—coca production continued to increase, and to play a central role in Cuzco's economy in the decades that followed.

Fig. 12.7 A woodcut of the early city of Potosí. After the discovery of silver in 1545, the high elevation grew to a population of more than 100,000 in 1600, creating a new market for agropastoral commodities (From Cieza de León 1553: *folio* 122v.)

did not take action to address what they describe as a serious public health issue. The simple answer is that colonial administrators believed that for all its deleterious effects, coca use was necessary to maintain the supply of indigenous labor. In 1567, Juan de Matienzo (1967 [Part 1, Chap. 44]) noted that "if coca were taken from [the Indians], they would not go to Potosí, nor would they work, or excavate silver." Understanding that coca amplified the effects of unfair and unfree labor in the Andean highlands, royal officials nevertheless acted to ensure a steady supply of coca, and to keep the leaf cheap for native to purchase. When Philip II attempted to impose a 5% sales tax on coca in 1591 to help royal finances recover after the debacle of the Armada, his viceroys declined to impose it, as "they judged that, the price being raised, the Indians would not come to work in the mines" (Borja 1859: 129 [1621]). Keeping coca inexpensive and encouraging its regular consumption by the indigenous population helped to transform coca chewing from a rare ceremonial activity into a regular food practice that was strongly associated with "Indian" identity.[16]

Daily Bread at the Close of the Sixteenth Century

By 1600, the Inca heartland had been transformed from the seat of a royal dynasty to a center of commodity production that sent meat, grain, and coca leaf to Potosí, which had grown to be one of the largest cities in the world. The shift from royal Inca estates to Spanish *haciendas* was part of a broader set of processes that attempted to impose ambivalent colonial distinctions between "Spaniards" and "Indians" in

[16]See Duke et al. 1975 on the nutritional value of coca leaf. 100 g of coca leaf from Chapare, Bolivia, contained more than 300 calories, 19 g of protein, and substantial amounts of calcium, iron, phosphorus, vitamin A, riboflavin, and vitamin E. Allen (2002) documents the continuing significance of coca chewing with indigenous ethnic identity in rural Cuzco.

the former imperial capital region.[17] Changing foodways played an important role in this process, although the creation and performance of distinct dietary practices were about as consistent as the racial categories they were supposed to reinforce. Andean populations absorbed European plants and animals into their subsistence regimes, and Spanish *criollos* in Cuzco developed a taste for a wide range of South American foods that distinguished them from residents of Spain.

In the context of dietary overlap between colonial social groups, it is interesting that bread and wine, the foods that Spaniards considered essential to their Christian and European identities, served to stress the divide between them and indigenous populations they sought to convert to Christianity. In an everyday sense, tuber agriculture and traditional *chicha* offered a distinct rural "bread and wine" from what was set on the tables of urban Spaniards and Indians. In a ritual sense, indigenous populations elevated coca leaf to a sacrament, and they used maize beer as an offering in their own religious ceremonies, and in large-scale drinking events that subverted colonial authority throughout the colonial period. Centuries later, Catherine Allen (1981: 166) notes that her informants in rural Cuzco referred to coca as *t'anta* (bread) or *hostia* (Host), making explicit connections between their coca chewing practices and the body and blood of Christ with reference to Christian dogma surrounding Holy Communion. Such practices remain prominent in the ethnographic literature of rural Cuzco, where they are treated as essential Andean indigenous practices, along with the maintenance of foodways that urban populations consider to be "Indian."

Conclusions

Early chronicles and archival documents make it possible to describe the profound changes to subsistence practices, economic organization, and identity that evolving foodways brought to the former Inca heartland in the generations following the European invasion. Although archeological data from early households still need to be collected, and more could be gleaned from archival documents, it is possible to reconstruct aspects of the introduction of Old World plants and animals to the Andean highlands, and to see how Spaniards used these foods to assert clearer dietary distinctions between themselves and the indigenous population. Indigenous labor and agronomical expertise were essential for the production of these foreign species, but it remains unclear the extent to which rural agrarian changes led to significantly altered production practices. Just as Spaniards incorporated "Indian" foods into their foodways, indigenous populations probably consumed diets that reflected the new suite of plants and animals that they produced. Food practices across the colonial Andes were diverse and dynamic, cutting across colonial status and racial identities

[17]Much more attention could be given to the role of Inca and other Andean elites in the Cuzco region during this process, and their economic and dietary performances. For example, the 1582 will of Carlos Ynga mentions the large-scale herding of European animals, wheat and barley fields, a mill and a sugar plantation (BN, Ms. 20193).

in ways that we still do not understand perfectly, although recent work in the field and archives holds out the prospect of building stronger perspectives on this important question.

References

Archivo General de Indias (AGI), Contaduría (1824) [1535–1558] Caja de Cuzco. Cuentas de Real Hacienda desde 1535 a 1558

AGI, Contaduría (1825) [1532–1572] Caja de Cuzco. Cuentas de Real Hacienda desde 1532 a 1572

AGI, Justicia 472 [1559–1563] Visita hecha a la Audiencia de Lima siendo oidores el doctor Melchor Bravo de Sarabia, el licenciado Diego González de Altamirano, el licenciado Pedro Mercado de Peñalosa, el doctor Gregorio González de Cuenca, el licenciado Francisco Saavedra, el fiscal Juan Bautista Monzón y a los demás ministros de justicia, por el licenciado Diego Briviesca de Muñatones, consejero de Castilla y juez nombrado para este efecto

Biblioteca Nacional (BN) (Madrid) Ms. 20193 [c. 1626] Ascendencia de Melchior Carlos Inca. Títulos de Eqquecco. Copy of the Libro de Repartición de Tierras de la Comunidad Equecco-Chakan 1560–1730 maintained in the possession of the president of the community

de Acosta J (1894) [1590] Historia natural y moral de las Indias. Ramón Angles, Madrid

Allen CJ (1981) To be Quechua: the symbolism of coca chewing in highland Peru. Am Ethnol 8(1):157–171

Allen CJ (2002) The hold life has: coca and cultural identity in an andean community, 2nd edn. Smithsonian Institution Press, Washington, DC

Bauer BS (1996) Legitimization of the state in inca myth and ritual. Am Anthropol 98(2):327–337

Bauer BS (2004) Ancient cuzco: heartland of the Inca. University of Texas Press, Austin

Betanzos J (1999) [1550s] Suma y narración de los incas, Martín Rubio MC (ed). Universidad Nacional San Antonio Abad del Cusco, Cusco

Borja F (1859) [1621] Relación de d. Francisco de Borja. In: Memorias de los vireyes que han gobernado el Perú, durante el tiempo del coloniaje español, tomo I. Felipe Bailly, Lima, pp 71–146

Cañizares-Esguerra J (2006) Puritan Conquistadors: Iberianizing the Atlantic, 1550–1700. Stanford University Press, Stanford, CA

Cieza de León P (1984) [1553] La crónica del Perú, Ballesteros M (ed). Historia 16, Madrid

Cieza de León P (1988) [c. 1553] El señorío de los incas, Ballesteros M (ed). Historia 16 Madrid

Covey RA (2015) Kinship and the Inca imperial core: multiscalar archaeological patterns in the sacred valley (Cuzco, Peru). J Anthropol Archaeol 40:183–195

Covey RA, Amado DG (eds) (2008) Imperial transformations in sixteenth-century Yucay, Peru. Memoirs of the Museum of Anthropology, University of Michigan, 44. University of Michigan Museum of Anthropology, Ann Arbor

Covey RA, Quave KE (2017) The economic transformation of the inca heartland (Cuzco, Peru) in the late sixteenth century. Comp Stud Soc Hist 59(2):277–309

Covey RA, Quave KE, Covey CE (2016) Inca storage systems in the imperial heartland (Cusco, Peru): risk management, economic growth, and political economy. In: Manzanilla L, Rothman M (eds) Storage in ancient complex societies: administration, organization, and control. Left Coast Press, Walnut Creek, CA, pp 167–188

Cummins TBF (2002) Toasts with the Inca: andean abstractions and colonial images on Quero vessels. University of Michigan Press, Ann Arbor

deFrance SD (1996) Iberian foodways in the Moquegua and Torata valleys of Southern Peru. Hist Archaeol 30(3):20–48

deFrance SD (2003) Diet and provisioning in the high andes: a spanish colonial settlement on the outskirts of potosí, Bolivia. Int J Hist Archaeol 7(2):99–125

Duke J, Aulik D, Plowman T (1975) Nutritional value of Coca. Botanical Museum Leaflets, Harvard University 24(6):113–119
Earle R (2010) "If you eat their Food": Diet and bodies in early colonial Spanish America. The American Historical Review 115(3):688–713
Earle R (2012) The body of the conquistador: food, race and the colonial experience in Spanish America, 1492–1700. Cambridge University Press, New York
Enríquez de Guzman, A (1886) [1543] Libro de la vida y costumbres de don Alonso Enríquez, caballero noble desbaratado. In: Colección de documentos inéditos para la historia de España, t. 85. Miguel Ginesta, Madrid, pp 1–419
Espinoza Soriano W (1974) El habitat de la etnía Pinagua, siglos XV y XVI. Revista del Museo Nacional 40:157–220
Farrington IS (1983) Prehistoric intensive agriculture: preliminary notes on river canalization in the sacred valley of the incas. In: Darch JP (ed) Drained field agriculture in central and South America. BAR International Series 189. British Archaeological Reports, Oxford, pp 221–235
Fernández D, El Palentino (1571) Primera, y segunda parte, de la historia del Peru. Casa de Hernando Díaz, Seville
Gagliano JA (1963) The coca debate in colonial Peru. Americas 20(1):43–63
García de Castro L (1906) [1565] Carta del Licenciado Castro a S. M. dando cuenta, entre otras cosas, de haber dividido las poblaciones de indios por provincias, nombrándoles Corregidores, 23 de Septiembre de 1565. In: Maurtua VM (ed) Juicio de límites entre el Perú y Bolivia: Tomo Primero, Virreinato Peruano. Henrich y Comp, Barcelona, pp 57–74
Garcilaso de la Vega E (1960–1965) [1609] Obras completas del Inca Garcilaso de la Vega. In: de Santa María CS (ed) Biblioteca de Autores Españoles 132–135. Ediciones Atlas, Madrid
Gasca P (1998) [1551–1553] Descripción del Perú (1551–1553), Barnadas JM (ed). Centro de Estudios Regionales Andinos "Bartolomé de las Casas," Cusco
Hernández F (1876) [1553]) Carta de Francisco Hernández para la justizia y regimiento de la Villa de Plata. In: de Odriozola M (ed) Documentos Literarios del Perú, Colectados y Arreglados, tomo octavo. Lima, Imprenta del Estado, pp 147–152
Huerta G (1624) Historia natural de Cayo Plinio Segundo. Luis Sanchez, Madrid
Kosiba SB (2010) Becoming Inka: the transformation of political place and practice during Inka state formation (Cusco, Peru). Ph.D. dissertation, Department of Anthropology, University of Chicago, Chicago
Krögel A (2011) Food, power, and resistance in the andes: exploring quechua verbal and visual narratives. Lexington Books, Lanham, MD
Lizárraga R (1908) [c. 1595–1609] Descripción y población de las Indias, Romero C (ed). Imprenta Americana, Lima
López Beltrán C (2007) Hippocratic bodies: temperament and castas in Spanish America (1570–1820). J Span Cult Stud 8(2):253–289
López de Gómara F (1884) [1552] Primera parte de la historia general de las Indias. In Historiadores primitivos de Indias, tomo primero. Biblioteca de Autores Españoles 22, de Vedía E (ed). M. Rivadeneyra, Madrid, pp 157–294
López de Velasco J (1894) [1574] Geografía y descripción universal de las Indias, Zaragoza J (ed). Fortanet, Madrid
Mangan JE (2005) Trading roles: gender, ethnicity, and the urban economy in colonial potosí. Duke University Press, Durham, NC
Matienzo J (1967) [1567] Gobierno del Perú (1567), Villena GL (ed). Institut Français d'Etudes Andines, Lima
Murra JV (ed) (1972a) Tasa del presidente Gasca—Quichua mitimaes (1549). In: Murra JV (ed) Visita de la provincia de León de Huánuco en 1562, t. 2. Universidad Nacional Hermilio Valdizán, Huánuco, pp 280–284
Murra JV (1972b) Tasa del presidente Gasca—Yachas (1549). In: Murra JV (ed) Visita de la provincia de León de Huánuco en 1562, t. 2. Universidad Nacional Hermilio Valdizán, Huánuco, pp 285–290

Niles SA (1999) The shape of inca history: narrative and architecture in an andean empire. University of Iowa Press, Iowa City
Pizarro P (1986) [1571] Relación del descubrimiento y conquista de los reinos del Peru. Villena GL (ed). Pontificia Universidad Católica del Perú, Lima
Quave KE (2012) Labor and domestic economy on the royal estate in the inka imperial Heartland. Ph.D. dissertation, Department of Anthropology, Southern Methodist University, Dallas
Ramírez B (1906) [1597] Descripción del reino del Perú. In: Maurtua VM (ed) Juicio de límites entre el Perú y Bolivia: prueba peruana,t. 1. Henrich y Comp, Barcelona, pp 281–364
Reitz EJ (1992) The Spanish colonial experience and domestic animals. Hist Archaeol 26(1):84–91
Rice PM (1997) Wine and brandy production in colonial Peru: a historical and archaeological investigation. J Interdiscip Hist 27(3):455–479
Rodríguez-Alegría E (2005) Eating like an Indian: negotiating social relations in the spanish colonies. Curr Anthropol 46(4):551–573
Rostworowski M (1993) [1983–1984] La tasa ordenada por el Licenciado Pedro de la Gasca (1549). Ensayos de historia andina: Elites, etnías, recursos. Instituto de Estudios Peruanos, Lima, pp 293–348
Sancho de la Hoz, P (1968) [1534] Relación de la conquista del Perú. Ediciones J. Porrúa Turanzas, Madrid
Santillán H (1879) [1563] Relación del orígen, descendencia, política y gobierno de los incas. Biblioteca de Autores Españoles 209. Ediciones Atlas, Madrid, pp 97–149
Tamara F (1556) El libro de las cosumbres de todas las gentes del mundo, y de las Indias. Martín Nuncio, Anvers
Toledo F (1867) [1572] Ordenanzas que el señor Viso-Rey D. Francisco de Toledo hizo para el buen gobierno de estos reinos del Perú y repúblicas de el. In: Relaciones de los vireyes y audiencias que han gobernado el Perú, tomo I. Imprenta del Estado, Lima, pp 33–154
Valverde V (1959) [1539] Letter to the emperor, 20 March 1539. In: Barrenechea RP (eds) Cartas del Perú (1524–1543). Sociedad de Bibliófilos Peruanos, Lima, pp 311–335
Vega B (1896) [c. 1562] Memorial de fray Bartolomé de Vega al Real Consejo de Indias sobre los agravios que reciben los indios del Perú. In: Zabálburu F, Rayon JS (eds) Nueva Colección de Documentos Inéditos para la Historia de España y de sus Indias, tomo 6, Hijos de M. G. Hernández, Madrid, pp 106–131
Villanueva Urteaga H (1970) Documentos sobre Yucay en el siglo XVI. Revista del Archivo Histórico del Cuzco 13:1–148
Xérez F (1985) [1534] Verdadera relación de la conquista del Perú. In: Bravo C (ed) Verdadera relación de la conquista del Perú, Historia 16, Madrid, pp 57–166
Zárate A (1995) [1555] Historia del descubrimiento y conquista del Perú. Franklin Pease F, Martínez TH (eds) Pontificia Universidad Católica del Perú, Lima

Chapter 13
Fermented Intoxicants and Other Beverages Among Hispanic and Indigenous Cultures in the Audiencia De Quito, and Their Roles in Rituals and Rites

Juan Martínez Borrero

Abstract The colonial government or Real Audiencia de Quito (1563–1822) ruled over a culturally and ecologically diverse region. This research explores the role of fermented intoxicants and other beverages during the late Colonial Period in both daily and ritual uses, particularly the role played by wine among Hispanic populations and maize beer or *chicha* with regard to the indigenous population as well as sugarcane fermented products. The Colonial government and the Church were particularly concerned with regard to intoxication and the consumption of fermented intoxicants by indigenous communities. References to disorder and problems with indigenous societies often blamed the role played by alcohol. The Church tried unsuccessfully to suppress and even eradicate such consumption. Colonial accounts and testimonies emphasize how the consumption of alcoholic beverages in private parties and public celebrations resulted in social interaction among Hispanicized populations. Drinks such as chocolate, guayusa and yerba mate, are distinguished from the consumption of alcoholic beverages, particularly wine associated with Hispanic culture, and its relationship to status and prestige. These are contrasted to the role of *chicha* in indigenous rituals and rites, and the general perception that it was a low-status drink.

Introduction

The consumption of caffeine-rich stimulants and alcoholic beverages in eighteenth-century Real Audiencia de Quito (modern Ecuador) followed unique colonial cultural, social, and religious practices. In this analysis, I compare first-hand historical interpretations and experiences of writers with a deep understanding of Hispanic American culture with the ethical judgments by colonial administrators or by

J. Martínez Borrero (✉)
Cátedra Abierta de Historia de Cuenca, University of Cuenca, Cuenca, Ecuador
e-mail: juan.martinezb@ucuenca.edu.ec

J. E. Staller (ed.), *Andean Foodways*, The Latin American Studies Book Series,
https://doi.org/10.1007/978-3-030-51629-1_13

members of the Catholic Church. The cultural perceptions of the Spaniards and religious clergy were both the product of Enlightenment thought in Western European culture and Catholic orthodoxy (Gerbi 1993 [1960]). In this analysis, I emphasize the historic accounts by de Velasco and Pólit (1961b [1837]) and Serra (1994 [1799]) to provide a comprehensive understanding of late colonial culture in Ecuador (see also Cicala 1994 [1771], 2004 [1771]). Colonial beverage consumption was a highly important component of food habits in the Audiencia de Quito. Food consumption reconfigured the relationship between space and place and distinguished the spectacular and the seductive from the ordinary and mundane. Food provides a contrast between heterogeneity and homogeneity and inspires us to think about how our world functions (Goodman et al. 2010).

This research examines the social and cultural significance of several important beverages consumed in the Real Audiencia de Quito during the late Bourbon era (c. 1738–1810). These include yerba mate (*Ilex paraguariensis*), which is a caffeinated holly species presently associated with modern consumption practices in southern Latin America, especially in Argentina and Uruguay. During the late colonial period, yerba mate was also extensively consumed in Quito along with two other caffeine-rich stimulants, *guayusa* (*Ilex guayusa*) and chocolate (*Theobroma cacao*).

Additionally, in late eighteenth-century Ecuador maize beer called chicha and also known as azua, was consumed. Maize beer or *chicha* is closely associated with, but not limited to, traditional indigenous Andean cultural traditions. The origin, regions of production, and technologies associated with the distribution of maize are defined; however, the focus of this research is its consumption as *chicha* because of its many associations with distinct social practices (see also Staller and Covey this volume for discussions of *chicha* consumption in colonial contexts). In contrast to *chicha*, wine, a beverage introduced during the Spanish colonial period, was considered a decadent beverage. Decadent beverages at this time in Ecuadorian history were something that only high-status individuals could afford; although in the early colonial periods wine maintained greater cultural prestige. However, sugarcane-derived *aguardiente*, a high alcohol distilled beverage, was much more widespread. Sugarcane extract fermented in a jug was called guarapo while for mistelas sweeteners and flavourings were added.

I begin this analysis at the time that the colonial pact was broken by the Bourbon reforms in 1717. The Bourbon reforms included new economic rules and the installment of the Nueva Granada Viceroyalty, both of which profoundly affected the Real Audiencia de Quito. I conclude this research with the initial Hispanic Independence movements in Quito that coincide with the first Cry of Independence that took place on August 10, 1809. This period of colonial history is characterized by changing laws and political uprisings, as well as several natural events including volcanic eruptions and earthquakes that contributed to growing instability in the social environment.

The Quito jurisdiction extended from Popayán in the north, to the Túmbez coast in the south and included tropical forests of the upper Amazon, including Mainas, Jaén de Bracamoros, Quijos, Canelos, and Macas. The region also included the coastal regions of Tumaco, Esmeraldas, Guayaquil, Machala as well as the island of Puná in the Gulf of Guayaquil. The urban centers of Quito, Cuenca, Guayaquil, Riobamba,

Loja, Ambato, Latacunga, Ibarra, Zaruma, Túmbez, and Popayán distinctly classified during the colonial era as *ciudades*, *villas,* and *pueblos*, exerted economic and social control closely related to smaller centers and rural communities in the surrounding areas (Delgado 2011). Among these cities and towns, there were different systems of land tenure, ranging from small *mestizo* holdings to diminutive indigenous properties or *huasipungos* located inside vast *criollo* or Spanish-owned haciendas (Figs. 13.1 and 13.2).

Society was based on a quasi-caste system distinguished by Spanish-born *chapetones*, American-born *creoles*, *mestizos*, *mulatos,* and *zambos*, as well as native indigenous populations. African-born slaves were distinguished from American-born, and those from *Cimarrón*, and freemen who acquired liberty by various means. Concrete social experience defines the classificatory systems that social actors build (Jelin 2014). *Vecinos* and *forasteros*, as well as foreigners, had different rights and obligations that expanded to cultural practices (Herzog 2012). An analysis of food and culture in this colonial setting is essential for the comprehension of everyday practices whose multiplicity of traditions weaves a continuous tapestry of their cultural status, ethnicity, and gender relationships (de Certeau et al. 1999).

Ethnohistorical texts from this time period provide significant insights into colonial perceptions and culture. Fray Juan de Santa Gertrudis Serra (Serra 1994 [1799]) *Maravillas de la Naturaleza* in 1799 and Juan de Velasco finished *Historia del Reino de Quito* after 1767; however, it was not published until 1837. Cicala's (1994) *Descripción Histórico Topográfica* was completed in 1771. These historic chronicles provide a first-hand assessment of life in colonial Quito. In contrast, second-hand accounts are not always as accurate. Alcedo (1786–89), his father Dionisio de Alcedo y Herrera (1915 [1787]), de Ulloa and Juan (1748), and others provided contrasting viewpoints. Ecuadorian and foreign contemporary authors provide the context and background for this research. Some archival documents as well as graphic, artistic, and monumental sources provide a clearer understanding of aspects of beverage usage and consumption particularly during the late colonial period.

The geographic focus of my analysis is the Real Audiencia de Quito, or just the Audiencia, which refers to the vast and varied territory in the northern Andes ruled by the Spanish colonial government from the city of Quito. Quito was the capital of this part of the Andes. This research initially addresses beverages considered stimulants followed by a discussion of alcoholic beverages. I explore historical details regarding changes in consumption practices after the Spanish conquest as well as the consumption and cultural perceptions by mestizos and indigenous populations. In many instances, the popularity of particular beverages was accompanied by the production of distinct ceramic serving dishes. Additionally, I describe many of the plants that people used in the creation of various drinks because these botanicals became important commodities, subsequently introduced to regions across the globe. As shown in this analysis, the moral and status perceptions that different people living in Quito held regarding either native or introduced beverages contributed to creating cultural boundaries, however at the same time, the adoption and consumption of beverages by different classes and ethnicities in Quito had cultural importance because it provided a basis for breaking down status and social class barriers.

Fig. 13.1 Carta de la province de Quito (courtesy La Condamine and Pedro Vicente Maldonado. 1751, public domain)

Fig. 13.2 Corregimiento de Quito domestic landscape (courtesy of de Ulloa and Juan 1748 lámina XIII)

Stimulant Beverages: Mate

Snow-chilled beverages or herb-infused teas and tonic drinks together with refreshing fruit juice played a significant role in Spanish America societies (de Velasco and Pólit 1961a; Cicala 2004 [1771]). Coffee, tea, and chocolate were available in Europe during the sixteenth-century following colonial expansion to Asia and the New World. Both coffee and tea were consumed in Latin America as imported goods or from local plantations. Creoles, Mestizos and Spaniards amply consumed local chocolate and caffeine-rich drinks.

During the eighteenth century, private consumption of beverages (and food) was gradually replaced in cities by public consumption. Daily life gradually became associated with food establishments rather than home cooking and dining at home. Similar changes occurred in France and other countries (Flandrin and Montanari 2011; Freedman 2009). In the colonial Andes, taverns or *cantinas* and *chicherías* began to be perceived as rowdy places and gradually became less appealing while cafés were favoured.

In the Audiencia de Quito people consumed *yerba mate*, *guayusa* or *huayusa* (herbal tea), and cacao (*Theobroma cacao* L.) as well; however, the consumption of native herbal beverages declined through time as imported products became popular. On February 10, 1791, Rossi y Rubi (1791: 109) author of *Rasgo Histórico y Filosófico sobre los Cafeés de Lima* mentions that there was an enthusiastic response to coffee shops, so much so that subsequently six were opened in the city center

Fig. 13.3 Guayusa (*Ilex guayusa*). (Digitalization Project of drawings by the Real Expedición Botánica del Nuevo Reino de Granada (1783–1816), directed by José Celestino Mutis: www.rjb.csic.es/icones/mutis (courtesy of Real Jardín Botánico-CSIC)

of southern capital of the Spanish empire, Lima, between 1772 to 1788 (Rossi y Rubi 1791: 109). Increased coffee consumption led to a slow decline and eventual abandonment of beverages like *guayusa* and *yerba mate.*

The bombilla circulates among the groups, which displeased Ulloa and Juan (1748), who point out that the chapetones are little given to this custom that excites the criollos who always drink it at the end of the day and before dinner. The Italian Jesuit Cicala considered mate a healthy and tasty drink. He states its name probably comes from the fact that it was made of calabash or *mate*, frequently gilded in silver or gold. A popular recipe for *mate* states:

> …use one teaspoon of the famous herb from Paraguay, together with a burnt piece of sugar, and pour boiling water over it, then some sweet aromatic lozenges are added, one or a half ripe mountain papaya or chilguacan (Vasconcellea pubescens) is pressed over the preparation, the beverage is consumed with one or two drops of orange juice in the morning, sour lemon in the afternoon. (Cicala 2004 [1771] p. 431).

Prepared servings of mate were covered for a short period of time before drinking. Alcedo states that drinking mate was done, "…with a small cane that is called a bombilla or a silver or gold tube with the thickness of a writing pen that ends in the shape of a holley almond with a salivadera (strainer) that prevents passing of the herb" (1788 p. 119).

Yerba mate originates in Paraguay and was available as an imported commodity in the Audiencia de Quito. Once the leaves were charred over barbecues and reduced to small pieces, they were packed in bags or *zurrones* of seven or eight arrobas that were distributed throughout the Andes in yearly quantities larger than twelve thousand *arrobas*, equal to approximately three hundred thousand pounds. Several *mate* qualities were available ranging from sifted *caá-miri, camini* or *caá-mi* to coarse *de palos,* but there was little difference between them (Azara and Schuller 1904). Light *yerba* was preferred in Buenos Aires, while a stronger-flavored variety was more commonly consumed in Chile and Peru.

It was common in Andean culture to drink *mate* as a shared commensal gesture between relatives and friends (Jamieson 2002). Mate was appreciated as "a sort of western tea that was drank as a hot water infusion, which had enhanced color and aroma very similar to the best European tea" (Cicala 2004 [1771]: 409). Local beverages were preferred to the more costly introduced teas and beverages that weren't assumed to have a better taste.

People spent considerable money on gilded beverage containers or drinking vessels with decorations made of precious metals. In fact, they were often an object passed down in wills and legacies. In Cuenca, the southern territory of the Audiencia de Quito, dowry registers dated between 1700 and 1820 list thirty-three mate drinking vessels of which thirty-two were silver gilded and was copper gilded. The vessels had appraised values ranging from five to thirty pesos (Chacón, Soto, Mora 1993: 462). Seventy-four gilded drinking *bombillas* represent 5% of all metal goods listed in dowry registers between 1777 and 1820. More than three-quarters of testators own only one gilded metal drinking vessel because the *bombilla* was commonly shared with family members and visitors (Chacón, Soto, Mora 1993; de Ulloa and Juan 1748).

Mate was considered a healthy and mild stimulant because of the caffeine and theobromine, although it has only two-thirds of the caffeine of coffee. The common practice of continuous drinking could cause a deeper effect, particularly stimulating the central nervous system and cognition after physical exertion or emotional stress. Mate was also used to treat headaches or for regulating the bowel. These medicinal properties are confirmed by contemporary medical research and were well known during late colonial times. Some authors emphasized that Paraguayans "fortify themselves" by drinking *mate* to be ready for work, and this effect was the same for any native, African, or Spaniard (Valladares de Sotomayor 1788).

Guayusa, A Native Tea

Guayusa is another caffeine-rich stimulant beverage that was consumed widely during the late Colonial Period, although in lesser quantities than *mate.* Indians harvested the leaves of wild trees in the tropical forests near Quito and the upper Amazon forests and traded this product to the highlands (see Fig. 13.3). A Tiawanakoid shaman's tomb was discovered in the Bolivian Andes with a package of *guayusa* leaves, among other curative and psychotropic plants widely used by Pre-Columbian Andean cultures (Schultes 1972). *Guayusa* production is currently limited to eastern Ecuador, Peru, and Colombia. *Guayusa* in Colombian Putumayo and Ecuadorian Amazon is generally associated with shamanic ceremonies to access the spiritual realms. Lucero writes in a letter to the Peruvian Viceroy in 1682:

> …they gather this malign herb with guañusa and tobacco, also invented by the Devil, and cook them in a way that the scarce juice obtained becomes malice quintessence [,] and to the faithful drinking, the Devil corresponds with the fruit of right malediction and always with harm to many… (Patiño 1990 p. 22).

Traditional *guayusa* usage among the Shuar and Achuar in the eastern Amazonian highlands is associated with purifying rituals. There are widespread cultural beliefs that *guayusa* consumption contributes to success in hunting and that it induces altered states of consciousness when consumed together with *ayahuasca* (Patiño 1990; Arnalot 2007). When the beverage is consumed in high concentrations, it can cause vomiting for purification or stomach-cleansing ceremonies. *Guayusa* itself is important in early morning purifying rituals:

> This guayusa hour is unique. It is the time for the family; it is also the time to correct children, marriage agreements, war preparation, resolution of personal matters, travel planning and for visiting other zones. It is the time for critical decision making, and frequently it is a mindfulness moment (Arnalot 2007 p. 215).

The stimulating properties and medicinal uses of *guayusa* have early origins. It was widely consumed in the Audiencia de Quito during colonial times. The Jesuits sold *guayusa* in Quito (Serra 1994 [1799]: 32, and de Velasco and Pólit (1961b: 75) considered it "a delicate taste tea," while Cicala (2004 [1771]: 315–316) suggests "*it is used to comfort, strengthen and warm the stomach and when drunk after the meals, serves as a marvelous digestive and is considered an effective tranquilizer*." *Guayusa* infusions were used as a substitute for caffeinated tea, coffee, or chocolate beverages for breakfast where it was drunk together with one or two pieces of bread or taken after meals. Its stimulating qualities were good for older people, but not recommended for youngsters "…because it over excites them" (Cicala 2004 [1771]: 316). Various chroniclers mention the drinking *guayusa* facilitated conception and this was guaranteed when combined with the honey of the *apaté* bee (Serra 1994 [1799]). The expulsion of the Jesuits from Spanish territories decreed by Carlos III in 1767 affected the *guayusa* trade. However, small quantities were available in Quito markets, so its use did not disappear completely. In 1840 Fray Vicente Solano recorded the use of the *guayusa* in southern Ecuador and recommended that physicians study its properties since people had forgotten its medicinal qualities (Solano 1892: 428–429). Coffee and tea were frequently replaced by infusions or tisanes prepared with local medicinal plants (de Velasco and Pólit 1961a [1837]: 70–80).

Mediterranean herbs such as marigold, *melissa* or lemon balm, lemongrass (*Cymbopogon schoenanthus* L) or *yerba luisa,* mint, and peppermint were also consumed in the Real Audiencia de Quito.

Chocolate, The Princess of Plants

Chocolate was introduced to the Andes from Mesoamerica early in the colonial era. It became one of the most popular beverages in the Real Audiencia de Quito since the sixteenth century. The first *Libros de Cabildos* from Cuenca dated 1557 state that *vecinos* ask and receive lands for cacao harvesting to provide a stable provision of cacao beans for daily chocolate consumption (Martínez Borrero 2010). Recent archeological evidence in Mayo Chinchipe, Ecuador indicates that cacao was consumed by numerous Pre-Columbian indigenous cultures in this region of the Andes. However, its preparation originates among Mesoamerican civilizations. Chocolate was greatly appreciated and considered one of the most extraordinary New World beverages in Western Europe (Lavedán 1795; Alcedo 1786).

Europeans appreciated some New World products, particularly cacao, together with cochineal or grana, the fiber of vicuna and medicines as quina, sarsaparilla, calaguala, canchalagua and balsams as tolú (Alcedo 1786, 1787, 1788).

Alcedo (1786) mentions that "cacao de Guayaquil," a local variety, later called "*fino de aroma*" had a bitter and concentrated flavor and a penetrating aroma. It stimulated the senses and contained a larger proportion of cacao butter. However, in Spain they preferred "juicier and less bitter" cacao varieties, while Guayaquil and Caribbean cacao were generally preferred in Northern Europe, particularly Germany (Lavedán 1795: 210–211). Caracas, Soconusco, Upper, Tabasco, and Maracaibo cacao beans were popular in Spain, where they would mix Guayaquil cacao with the Caracas varieties (de Osasunasco 1789). Cacao plantations were present throughout the Real Audiencia de Quito extending down the cordillera to the north and central coast of Ecuador, and along the coast as far south as Túmbez. The Esmeraldas coast produced "*the best quality cacao because of its large content of butter, the bean size and delicate taste*" (de Ulloa and Juan 1748: 476. Other chroniclers praise southern Machala varieties (Alcedo 1787). Other areas distant from population centers, such as the jungles of Jaén, Mainas and Quijos in the Quito Amazon, produced wild cacao (*Theobroma bicolor*), which was almost impossible to exploit because of the permanent risk represented by free peoples such as the so-called Jibaros on the Amazon flanks of the Andes. (Fig. 13.4).

Where cacao was cultivated, a refreshing beverage was prepared with the mucilaginous seed. This liquid was treated as a cooling drink, and praised, but hard to find beyond the haciendas or plantations. Cicala (2004 [1771]) distinguishes cacao varieties by the bitterness, sweetness, and aroma. He developed a scale of five different kinds, from deep and bitter, to aromatic and sweet cacao. However, in Spain Lavedán (1795) states that the Caracas cacao is differentiated from the Cayenne or Guayaquil variety by its small beans and even smaller seeds. Alcedo identifies wild

Fig. 13.4 Cacao silvestre (*Theobroma bicolor*). (Digitalization Project of drawings by the Real Expedición Botánica del Nuevo Reino de Granada (1783–1816), directed by José Celestino Mutis: www.rjb.csic.es/icones/mutis (courtesy of Real Jardín Botánico-CSIC)

cacao as "…bitter and used by Indians…" while the cultivated variety was "…distinguished according to the terrain and the country…" (1789: 34). Large quantities of Guayaquil cacao beans were shipped to Panama, and ultimately to Seville (Boy 1840).

Indigenous cultures diluted cacao in water and mixed it with both piquant spices and corn flour (Lavedán 1795: 214). Norton (2006) considers that European incorporation of flavorings and foam was an attempt to recover the exotic and strange taste of cocoa that transformed their palate. Three or four pounds of cacao beans were placed in iron, copper, or bronze receptacles suspended over an intense coal fire; then almonds were mixed thoroughly in a continuous way until the skin begins to peel. Once cooked the beans were placed on a flat surface and covered with a clean cloth. They were then handled to detach the skin from the seeds, and at the same time grains were crushed into nibs and the skin was blown to lose it completely (Lavedán 1795).

Nibs were processed in a gently warmed—never over open fire. Then a grinding stone (*metate*) with a stone or iron rolling hand was used until the crushed almonds liquefied and could be set aside to cool and be formed. It was also recommended that when processing the cacao not to extract too much cacao butter because an adequate ratio between solids and butter improved the chocolate quality. A fair quantity of butter also resulted in the chocolate mixing better with sugar and favors of aromatic compounds (Lavedán 1795). When the mixture was still warm, aromatic substances were added in adequate proportions. The chocolate was then placed on paper sheets to cool (Blégny 1687), although it was more common to put it on *achira*, plantain, or *bijao* leaves.

Hot chocolate was prepared in the Audiencia de Quito with pure cacao mass in different shapes and thicknesses with full cacao butter contents. If the valued cacao butter was extracted previously, annatto was occasionally added for the preferred reddish color and extra grease. Flavorings including vanilla, cinnamon, *ishpingo* or cinnamon flower, and corn flour were ingredients added to the pot or *olla*. Occasionally egg yolks were used to add thickness as were beaten egg whites, whose foamy substance provided an elegant touch to exceptional social occasions. Vanilla was harvested in both Guayaquil and the Amazon missions, the second being of better taste and aroma, although the first was used widely for chocolate aromatizing (Cicala 2004 [1771]). Additionally, prime quality vanilla was cultivated along the north coast of Atacames (de Ulloa and Juan 1748).

Chocolate preparation required specialized equipment. *Ollas* or cooking pots, hand beaters, *jícaras* or cups were essential. The olleta had a spherical shape at the base and a long, narrow neck, with a handle or an ear, to make it easier to hold. While the wooden hand beater—a Spanish addition—was used in a circular and continuous movement to emulsify the mix and produce the highly appreciated chocolate foam. Once ready, the chocolate beverage was served immediately to avoid cooling and to impede the butter from separating from the water and forming a displeasing appearance. *Jícaras* described as "…the small gourds…to drink chocolate" (Alcedo 1789: 182) were the most frequently used for consumption, as well as the light and resistant gourds *totumos* and *pilches* commonly used by travelers. Sometimes those were gilded in silver or gold, like those used for drinking *yerba del Paraguay*. Ceramic vessels were common, but some wealthy families used either porcelain chocolate cups from China or silver and gold cups.

Travelers carried the necessary implements to prepare chocolate on the road (Cicala 2004 [1771]). It was common throughout the Andes to drink chocolate in the morning with breakfast. Accounts state that early in the morning, "…there is always a place to have breakfast with chocolate, mate or a light fried meal" (Concolocorvo and Gaibrois 1997: 41). Chocolate was also popular with lunch or supper (Serra 1994 [1799]), but it was generally consumed anytime.

Alcoholic Beverages

Pre-Columbian civilizations were familiar with *pulque*—the fermented beverages prepared with *penco* a juicy extract from maguey plant—. *Chicha* or maize beer was commonly consumed throughout the Caribbean and of central importance among various Andean cultures since pre-Columbian times. It was also referred to by Quechua speakers as *aqha* and by Aymara speakers as *kufa* while it was also called *azua*. However, it is commonly referred to as *chicha* throughout the Andes (Nicholson 1960). Other fermented beverages made of cassava, *molle* (*Schinus molle* L.), various roots, from seeds and fruits, tubers and certain plants from the tropical lowlands were also popular in this time period (Staller 2006).

A range of alcoholic beverages were also popular during the late colonial period. The introduction of wine early in colonial Latin America history altered the perception of beverages, especially because of its close relationship with Catholicism and the Catholic mass (Rice 2011, 2013; Weaver 2015). Despite the high production and cultivation of grapes, wine was costly and primarily associated with high-status individuals in the Spanish culture.

In addition to wine, grape *eau de vie* was distilled to produce a brandy known as *pisco*. Wine and pisco brandy were complemented by other alcoholic beverages. Early sugar cane plantations and the popularization of sugar cane alcohol made it the most common drink during late colonial times. Sugar cane extract, called *guarapo*, was also highly valued among African slaves, mestizos, *mulatos*, indigenous people, and impoverished whites primarily because of its low cost, availability, and taste. The various inhabitants of Quito associated these alcoholic beverages with distinct social groups, consumption practices, and behaviors.

Chicha or Azua, A Man's Privilege

In the Audiencia de Quito, there is considerable archeological evidence of ceramic containers associated with the processing and consumption of maize beer or *chicha*. Such evidence extends back to pre-Columbian time before the Inca expansion, extending as far back as the Formative periods (Staller 2008). *Chicha* was associated with traditional funerary rituals, rites, and commonly given as ritual offerings to the deceased buried in the surrounding landscape, sometimes for as long as three years following their passing (Staller 2006, 2008). Early chroniclers indicate a relationship between *chicha* consumption and agricultural planting and harvesting, and also with festivals or *raymis* and various political rituals associated with the annual agricultural cycle (de Ulloa and Juan 1748; Staller 2006, 2008).

Among the varieties of *chicha* consumed during the late colonial period, the creoles drank one made of toasted ground corn flour, cooked with honey and other species. These ingredients were then left to ferment for a few days and when they turned somewhat acidic, they were ready to drink. Only indigenous cultures chewed corn kernels and then cooked them in water, sieved them, and let them ferment until it acquired a slightly alcoholic strength (Staller 2006). Malted maize *chicha* or *jora*, also known as *yura*, was quite intoxicating. *Ticti* was a by-product of *chicha*, with the consistency of porridge, and used as a meal for children and elders (De la Peña 1771: 438–439). *Chicha* is consumed throughout the Andes, and in some countries it is still an important part of food and ritual practices among indigenous communities and groups, particularly in Ecuador (Einzmann and Martínez Borrero 1993; Moreno 2008). Ulloa and Juan discuss how *chicha* was made during the eighteenth century;

> … to make it put somewhat humid corn to sprout, until the stem springs, once born put it in the sun until it is thoroughly dry: after that they roast it lightly, and grind it, whose flour makes a decoction; then put in jars, and add a portion of water according to the need, it ferments the second or third day, and when concluded, which is in another two or three, is ready to drink. It is very fresh, according to common sense, but its strength intoxicates, when used with the Indians immoderation, they never cease while at hand, until they have finished a pot, and chicha with their sense; its taste is good, mimicking somehow cider, but has the defect of not being able to be kept for much time, because six or eight days after it has finished fermenting it becomes vinegar (1748 p. 398).

In the context of indigenous feasting, women, most commonly prepared *chicha*, and it was generally abundant. The Spaniards stated that an individual commonly drank a pot containing thirty or more liters. However, some descriptions infer that the indigenous communities were very frugal regarding meals. An eighteenth century account by de Ulloa and Juan (1748: 547) indicates that residents partook of daily meals that consisted of camcha (roasted corn) that was cooked along with some wild herbs in a small pot with water.

Chicha was a privilege of married adult men; indigenous women, either single or married, could not get drunk, nor did youngsters. Women offered the beverage in a calabash container or *totumo* and tambourine and flute music sounded, signaling the propitious moment for dancing, "… According to their custom… that is to moving from side to side haphazardly, in the meantime some of them sang songs in their language, with continuing joy and celebration" (Ulloa and Juan 1748: 547). In the eyes of Catholic clergy, *chicha* was associated with drunkenness, sexual promiscuity, as well as incestuous and adulterous relations. Therefore, parrish priests would often empty the jugs of *chicha* because it was asserted that drinking brought on behaviors that directly challenged Catholic morality. According to Catholic orthodoxy, it was necessary to eradicate drunkenness:

> …the root of infidelity, and innumerable evils… So the authority will seek to avoid it, and diligence and good tricks of priests will be applied, they must warn them three times, then they should correct those who they find drunk and, especially, pursue public drunkenness, as they are signs of infidelity and heresy (De la Peña 1771 p. 205).

An important practice, in which the chicha was involved, is the request to be coupled without the mediation of an ecclesiastical marriage. The groom sent the bride's parents firewood, flowers, like *chuquiragua* (*Chuquiraga jussieui* J. F. Gmel), the flower of the Andes, and maize beer or *chicha*. If these gifts were accepted, then marriage was assumed. *Chicha* was not limited to feasts or rituals because they "barely eat or dine without *chicha*" although they drink little water. From *azua* they can drink several pots, four to six for pleasure, without getting drunk (De la Peña 1771: 317).

Another writer noted: "When the party is over it still continues with *concho*, which means, "day when they drink the seats left over from the previous one…". Jars were filled with water and fermented, and a light *chicha* was obtained, it was drunk for three days, until "finishing the beer, the money, and the credit to buy it" (Ulloa and Juan 1748: 48).

In the celebrations by the *curacas* (local leaders), which sometimes were performed in the town square and in homes, there were offerings of maize beer. Rituals and rites were practiced, while outsiders and the poor were offered food and drink. Sometimes *curacas* take, "flocks of silly women, discarding their own wife" as concubines (De la Peña 1771: 205) as they come with fruit and *chicha* to celebrate leading men in front of their spouses (*ibid.*: 373). *Curacas* received the beer from the native Indians under their charge as tribute or payment in addition to eggs, chickens, certain vegetables, and even boys of service, which was an ancient Andean custom called *yanaconaje* (*ibid.*: 245).

Burials were also associated with *chicha* consumption; mourners and guests and anyone who passed by were asked to toast the deceased. C*hicha* was also consumed while traveling, sometimes mixing it with cornmeal or roasted barley to make a mush, or drinking it after swallowing a camcha spoonful. Because of poverty, Indians carefully stored "… few crock pots, jugs, jars and pitchers or *pilches*" (de Ulloa and Juan 1748: 550) some recipients were associated with the production and consumption of what was considered a precious beverage.

Chicha was offered to people on long journeys by foot or on mules, especially at *tambos* or storage facilities along the *Camino Real* or Inca highway in the Andean highlands. Near Quito, Santa Gertrudis finds that "… Every house is a store that sells bread, cheese, *chicha,* and *guarapo* liquor" (Serra 1994 [1799] p. 83). Also, "priorisas" and "mayordomas" sold *chicha* on Sundays to contribute to the cost of wax for candles and adornments of the church (De la Peña 1771 p. 205). *Chicha* was commonly consumed on Sunday by indigenous populations who came to towns where there were significant quantities of the beverage. In the cities, indigenous workers gathered at *chicherías*, places and spaces for native Andeans to socialize and drink to excess. Consequently, they were often economically and morally exploited (De la Peña 1771: 373).

Crimes attributed to excessive drinking were violence, beatings, abuse, and murders, thus drinking fermented intoxicants was denounced throughout the late colonial period (Martínez Donoso 2013). However, drinking *chicha* was also a joyful act that served in social gatherings, ties and associations to ancestors, and as a cause for celebration and dancing. Maize beer commonly accompanied meals and supported work among indigenous populations. However, clerics considered its ubiquity a pretext for brutalization, the origin of aggression and violence, and of exploitative practices.

Wine, A Decadent Beverage

During the colonial periods, European vineyards were widespread throughout the New World. The Spanish Crown made a concerted effort to cultivate various varieties of grapes (*Vitis vinifera* L.) and to safeguard the interests Spanish winemakers. Initially, there was a clear association between high status and wine consumption. A clear association was developed between social prestige and wine consumption, since the quality of being Spanish is inseparable from the latter, to the extent that many food products, such as cassabe, corn tortillas and arepas, are incorporated into the diet in the absence of wheat, but a clear preference for wine is maintained in the Caribbean and Tierra Firme areas, as well as in New Spain, despite the availability of drinks such as pulque, chicha and masato. (Mira Caballos 2015).

The Viceroy of Peru, Francisco Toledo fostered and encouraged the initial planting of vineyards and wine production to serve a high demand for wine in both Lima and in the silver mining city of Potosí (see Rice 2011, 2013). The Jesuits became important producers and traders who sold their wine and brandy in Lima, Cusco, Ayacucho, Huancavelica, and other places (Lacoste 2004; Soldi 2006; Weaver 2015). Even *señores indígenas* (indigenous landlords) planted grapes and produced wine and brandy (Soldi 2006). Creole people in the Audiencia de Quito prefer to drink sugar cane brandy, *mistelas*, and punch, but not wine, unlike *chapetones* or those of high status. Ulloa points out for Guayaquil:

> Sugar cane brandy and wine are two inexcusable things there; that for the Creoles and one another for foreigners, because ... the Creoles drink a little, and only foreigners use it as well as wheat bread (de Juan and Ulloa 2002 p. 14).

This distinction between cultural practices of the local population and visitors or residents was evident in many cases. For example, green bananas were eaten in abundance in Guayaquil, while wheat bread was not. Among those of Spanish culture, and high-status individuals, brandy, *mistelas*, punch, and wine generally accompanied meals. Only Europeans preferred wine, but they also drank grape brandy (Juan and de Ulloa 2002).

Small amounts of Spanish wine could be found in the Audiencia of Quito, and in Cartagena since it was smuggled from places like Novedad in Chocó (Cicala 1994 [1771]); however, in Lima wine was legally traded and maintained strong cultural associations with high status and prestige. Wine was very expensive in Quito; Santa Gertrudis (Seera 1994 [1799]) mentions that a bottle of wine from Chile was worth 3 or 4 pesos, while grape brandy was 7 or 8. Consequently, because of the high cost of wine, sugar cane brandy became popular. The Andean and colonial perception was that "decent people" consumed brandy moderately—preferably as *mistelas*—and drank this instead of wine. Spaniards also consumed rum, because wine from Lima was so expensive, and scarcely available, "but they are not inclined to cane *aguardiente*, but of grapes, which also come from Lima" (de Ulloa and Juan 1748: 373).

Despite early efforts to plant vineyards in the Audiencia, wine production never achieved the scale it did in other Andean regions, and the practice was discouraged. Perhaps this was due to the low consumption and popularity of wines from Peru and Chile as well as Spain. For example, in *Libro de Alcabalas de Cuenca*, tax payment for only ten jars of wine was registered in 1791, suggesting there was little wine trade. Nevertheless, the El Carmen de la Asunción murals depict feasting with glasses and bottles of wine (Martinez Borrero 1983) (Fig. 13.5). Moreover, indigenous societies never adopted wine, but rather, consumed *chicha* and sugar cane brandy instead. However, Spanish governing officials or *corregidores* forced them to buy *botijas* (ceramic vessels) of wine—as well as olives and olive oil—at very high prices. Indians used to resell those products to mestizos and tradesman at a substantial monetary loss (Juan and de Ulloa 2002 [1826]).

Wine was consumed for medicinal infusions with *quina* (*Cinchona officinalis*) for headaches, tobacco for stomach ulcers, or *ipecacauna* for emetic usage. However, more often *espíritu de vino* or wine alcohol was perceived as an excellent astringent (Alcedo 1789). Some preparations, especially sweet treats, were prepared with wine. Pineapples cooked with sugar and red wine "taste like strawberries" (Alcedo 1789: 149). Sweet *camotillo* (mashed sweet potato) was also prepared with wine (Cicala 2004 [1771]); wine and sugar were mixed to "make the most gifted drink … something like the horchata drink in France" (Serra 1994 [1799] p. 45).

Fig. 13.5 Banquet. Mural painting at the Monasterio del Carmen, Cuenca, Ecuador, 1801 (courtesy of Juan Martínez Borrero)

Trapiches: Sugar Cane Mills

Sugar cane mills or *trapiches* were essential for the production of cane brandy, sugar cane juice or *guarapo*, as well as *panela*, brown and white sugar. The presses of the mills in the Real Audiencia de Quito were primarily made of wood from the guayacán tree (*Tebebuia guayacan*). They used guayacán wood because they knew the mill would last for many years (Cicala 2004; Alcedo 1789; Serra 1994 [1799]; de Velasco and Pólit 1961a [1837]). In larger scale industries, *trapiche* (mill) presses were made of bronze, whose hardness Cicala (2004 [1771]) considered of lesser resistance than *guayacan*, as bronze tended to wear out and break easily. Sugar mills were designed, "in the manner of a printing press" (*ibid*.), i.e., with vertical presses, usually rotated by two people, one on each side moving arms in opposite directions to squeeze the cane. Hydraulic power was common to wheat mills and felt presses, and when applied to sugar cane mills produced very high yields, lowering the cost of sugar cane brandy, and significantly increasing the profits and revenue for owners.

Cane juice or *guarapo*, syrup, *panela*, *alfandoque*, *alfeñique*, and *melcocha* were the products ground up at the mill (Serra 1994 [1799]). *Guarapo* was a drink made of sugar cane by indigenous populations that they drunk instead of wine. It was consumed as either a refreshing drink to quench thirst or as a fermented alcoholic beverage (Terreros and Pando 1786). Before becoming syrup, the juice was poured into large containers made of cow leather (*odres*) or crocks and fermented for two to four days until it lost its sweetness and increased the acidity. The Italian Jesuit Cicala considered this drink as "tasty and refreshing, as well as healthy" when not drunk excessively because it intoxicates (Cicala 1994 [1771]: 272). Ulloa shares this perspective and considers *guarapo*, "tasty, between sweet and sour, and healthy" (de Ulloa and Juan 1748:418). Fermentation for longer than four days, changed *guarapo* into vinegar, and it was then used in the kitchen.

Chicha and *guarapo* often appear together in many colonial descriptions. They were fermented intoxicants commonly offered at indigenous and mestizo parties and facilitated dancing. In warmer regions of the Audiencia de Quito, *guarapo* was the most commonly drunk beverage, and in the absence of wine, it was regarded as appetizing when fermented. It often accompanied stews, ham, eggs, and roasts (Serra 1994 [1799]). Its consumption was most frequent "among ordinary people" (de Ulloa and Juan 1748:418). In the sugar cane mills, travelers were often offered *guarapo mates* for their journey as a gesture of hospitality.

Trade of *guarapo* was common along the coast, in the highlands, and in regions surrounding cities where light mills were installed for the sole purpose of producing *guarapo* rather than for sugar or alcohol production. In some cities, *guarapo* replaced *chicha* among some indigenous populations and farmers. Spaniards and creoles also consumed it because of its abundance and ubiquity, and it subsequently became a standard drink for clergy.

Aguardiente, The Worst Poison

Developed initially in the West Indies, the frequent references to sugar cane alcohol during the eighteenth century indicate its importance and show the spread of these beverages beyond the Caribbean. Because of its low price and low taxation, aguardiente was more widely consumed than grape brandy. Grape brandy importers paid tribute or *alcabala* taxes to storehouses at Babahoyo, Naranjal, Caracol, and Yaguachi. Each flask or *botija* of Peruvian *pisco,* contained about 20 L. Although taxed with one real could reach a price as high as 60–70 pesos in Quito. A different variety called *aguardiente de la Nazca* sold in Quito costing 20–25 pesos was taxed with half a real per *botija* (de Ulloa and Juan 1748: 329). In Quito, the average price of a bottle of *aguardiente* was half a *real*, while Peruvian or Chilean grape brandy reached eight *reales*. It is not surprising that sugar cane *aguardiente* was the most popular beverage among the mestizos and indigenous communities. It was popular not only for its low price, but also because it was believed to better suit the humores of the indians (Spanish colonial perception of body composition).

Sugar cane, a crop particularly productive in warm valleys, could be cultivated in small mestizo or indigenous farms as well as large, private *haciendas* and by different religious orders. Brandy *de caña* or *criollo* was produced in warmer river valleys around Quito along with raspadura syrup, and cane juice (de Ulloa and Juan 1748: 417). Brandy production to the north of Quito included networks of sugar cane farms that generated permanent conflicts with the authorities regarding its distribution and consumption. Despite issues regarding brandy production, its consumption was culturally associated with leisure activities, such as poetry and traditional music, (de Ulloa and Juan 1748).

Around 1780 in the Corregimiento de Cuenca, thirty-five sugar cane farms were registered between Alausí, Cañar, Cuenca, and Girón. The highest concentration was in Gualaceo and Paute about 40 km. from Cuenca (Chacón 1990). It was also cultivated along the coast, but on a smaller scale. Different varieties of brandy were obtained from distinct types of cane. The most common was known as *white*, and it was aromatic and very pleasing to the palate. Ulloa states, "…when it is not *resacado*, nor so vigorous and strong as the one from grapes, it is not as harmful to health, according to M. de Jussieu, who sent it to the king of France, because, besides its lower strength, it is not so dry and more balsamic" (Juan and de Ulloa 2002 [1826]: 327). Sugar cane brandy was popular among the entire population, and commonly sold at *pulperías* or grocery stores along with "substantial portions of fresh bread… jars of sweets and pickles; wine, brandy and dried fruit …" (de Ulloa and Juan 1748: 104–105). It was bought directly from producers and sold in food markets—*cato* or *gatu*—by vendors, usually indigenous women. Brandy was consumed in various social contexts, straight or as punch. The French scientist Jussieu believed cane brandy should be drunk after burning it with a lump of sugar, a practice also recommended for the preparation of yerba mate (Juan and de Ulloa 2002 [1826]).

The Audiencia de Quito asserted that brandy encourages disorder and growing dishonesty. Moral standards became relaxed until it was impossible to describe, "…the disbanding and the nonchalance…" reached (Juan and de Ulloa 2002 [1826]: 497). It was common to find that visitors, alongside some priests, observed with extremely critical eyes liquor consumption in the regions around Quito. It was asserted to be a primary cause of evil and a defining feature of mestizos and indigenous communities. For example, Cicala (1994 [1771]), speaking of the vices of the inhabitants of Quito, stated that, drunkenness became common around 1745 and then increased in more recent times until such behavior was present in cloisters and among "individuals most conspicuous for their dignity… Over the past 15 or 16 years in every house… and those of the noblest families no other drink of refreshment than pink or sweet wine or stiff brandy with some sweet *bizcochito* (cookie) is offered" (Cicala 1994 [1771]: 237).

Primarily responsible for drunkenness was the spread of punch, a drink made with aguardiente, water, and acid lemon juice, called *limón sutil* today. The drink was introduced to Portobelo, Cartagena, and Panama by English and French pirates and popularized in Guayaquil, Quito, and Lima. Ulloa referring to Guayaquil notes that the punches were very popular to the extent that it was the drink of people of distinction who took it at 11 in the morning and evening (de Ulloa and Juan 1748: 373).

The beverage *punci* (punch) was prepared "In a large beaker, usually of silver, (or in a large glass or silver jar), almost filled with water, three or four *limoncitos* called *zentil* whose acidity was very strong and efficient were squeezed, then one or half a nutmeg is scraped, others used it powdered, together with a teaspoon of sugar. Finally, tiny little glasses of brandy called *ron*, a very high alcohol, distilled three times in the alembic with crushed ferment were added" (Cicala 1994 [1771] p. 238). Changes in its preparation and a gradual increase in alcohol content were associated with decreased water and lemon (*ibid*). According to accounts from the late eighteenth century, the effects were horrific: men became beasts and lacked reason. The desire for consumption escalated everywhere "… even among women of rank, widows, and married! How many squandered large amounts of their capital, for drunkenness and to encourage an increasingly abominable vice… which *gentilísimamente* or being a caring person (a local term) had been held to a new standard, some women even let their children perish of famine!" (Cicala 1994 [1771] p. 238–239).

Another common preparation were *mistelas*, a term that first appeared in the Spanish Academy dictionary in 1817, and defined as, "a drink made from a mixture of brandy, water and sugar and some cinnamon." It was associated with brandy production since the eighteenth century in Jaén, Guadalajara, and Cantabria. It was cane brandy infused with various fruits and sweetened with sugar or simple syrup flavored with diverse spices. In Loja, Ecuador Fray Juan de Santa Gertrudis prepared *mistela* from "…brandy and an infusion of clove, cinnamon, nutmeg, ginger, juniper berries and mint, all mixed with syrup" (Serra 1994:232 [1799]). In Lima, *mistelas* were sold in cartons or *cajones* (a sort of street vendor's box) along with sweets and ice cream by street vendors.

A popular revolt demonstrating the great importance that brandy had in Quito society began in May 1765 and ultimately prevented authorities from levying brandy and tobacco taxes, particularly because such taxes were viewed as an abuse against the poorest. This *Rebelión de los Barrios* became a distant precursor to the Independence of Quito and part of the great popular revolts in the second half of the eighteenth century along with the revolt of the Socorro communards of New Granada and the huge uprising led by Tupac Amaru (MacFarlane 1994).

Discussion and Conclusions

A variety of both alcoholic and non-alcoholic beverages were important to social interaction in colonial Ecuador. A range of historical sources indicate that the spread of fermented intoxicants both indigenous to the Andes and introduced through Indigenous-European contact, played a central role to Late Colonial Period cultures in the Real Audiencia de Quito. The transformation of indigenous Andean cultures through the imposition of the Catholic Church and colonial governments was to varying degrees fostered by fermented intoxicants. Open spaces, opportunities for social negotiation, escape from the pressures of economic disparities reified by almost servile work, by moments of cultural resistance and recovery were some topics writers highlighted when addressing the wide variety and diversity of beverages consumed in the Real Audiencia de Quito during the late colonial period. In many instances shared social practices involving beverage consumption crisscrossed both Andean social classes and Andean geography to become common cultural practice. When this expansion of food practice occurred, the colonial inhabitants often altered their social perceptions, particularly of each other.

Europeans claimed that they did not consume alcohol in such large quantities as did the indigenous cultures. Native Andeans were perceived as rather strange and wild, with unusual and sometimes incomprehensible customs. Perceptions of excess drinking by native Andeans in Ecuador were used by Europeans to reaffirm their cultural superiority, foster their criticism of indigenous cultures, and provide a rational for the subjugation and colonization of indigenous Andean populations. But local travelers, mestizos, and Spaniards whose dealings with indigenous guides, porters or *guandos,* and pilots took place in a cultural environment of confidence and shared cultural spaces were perceived more favorably by the state and upper classes.

Some non-indigenous peoples incorporated traditional practices and forms of beverage preparation into their consumption practices. In the cities and scattered villages, a set of imagined codes based upon particular forms of commensality including drinks, such as brandy or *guarapo,* began to play a key role in social negotiation and connections. However, they also played major roles in forms of close contact, seduction, and open eroticism—surprising and disgusting Catholic clergy that heard about it or witnessed it.

These fermented intoxicants also generated a sense of belonging, fellowship, and generosity within their cultures. However, at the same time, historical accounts suggest that the excessive consumption of intoxicants led to conflict, violence, loss of control, misery, adultery, and abuse of power. Social negotiation, through sharing fermented intoxicants did not always end in agreement, but rather could end in shock and estrangement. Moreover, the consumption of alcoholic drink was denounced in sermons and legal proceedings as well as definitive judgments. Among some parties, alcoholism was considered a unique character trait of the seventeenth and eighteenth century inhabitants of Quito. The consumption of large amounts of brandy was combated, and the doors of taverns or cantinas were closed, even at the risk of violence being manifested in the streets.

Although colonial residents did not sanctify maize beer consumption or ignore its risks, the historical record indicates that *azua* or maize beer consumption was one of the few practices that generated a sense of carefreeness or freedom, although it sometimes involved drunken rolling on the ground, or in some cases, actual drunken death. *Chicha* drinking created spaces of resistance, from the planting of maize to its cultivation, indigenous highland women tried to make the best beverage with a few herbs mixed with roots, tubers, and grains or ground *camcha*. The cultural obligations to work, offer housing, to guide others through the mountains, feed people and horses, etc., were all associated with *chicha*.

In contrast, *guarapo*, with its slight acidity and its sparkling alcohol, momentarily granted black slaves psychological escape from captivity and allowed them to unleash the performative power of their music, song, and dance. In an environment where opportunities were earned freehand and where it was not common to succeed without the support of skin color and origin, alcoholic beverages became an escape door for a life organized around inhumane work.

Yerba mate provided a social reason to gather that permitted communication among indigenous populations on how to deal with and adjust to class differences and the imposition of colonial restrictions and the Church. *Guayusa* from the Amazon rainforest may have reminded consumers of its temporal depth related to native traditions associated with the eastern cordillera. Plantations, exploited by the indigenous population of the jungle, allowed leaves to reach the cities where they were transformed into drinks representing new symbolic and cultural contexts that were unknown in earlier pre-Columbian times.

Cacao (*Theobroma cacao* L.) beans and cocoa butter were used to make drinking chocolate which was a complex drink that inspired commensalism. Loads of cacao beans were and continue to be harvested in the tropics, and they have spread throughout the Andes on mules or by *serrano* (native highland) carriers, and on the rivers in rafts to villages or deposited in *tambos* or storehouses waiting for a boat that leads the precious cargo to the port. Cacao drinks were made with maize, chili, vanilla (*Vanilla planifolia*), and honey.

Two of the most important fermented intoxicants during the Colonial Periods were wine and brandy made from introduced grapes. Wine had both religious and cultural associations with the Spanish conquest. Wine was of central importance to the veneration of Christ in the sacrament of the mass and had strong associations with the colonial elite and upper classes because of its relatively high cost. Despite of the strong association with Spaniards and Catholicism, neither wine nor brandy was adopted as beverages of choice by the seventeenth and eighteenth century indigenous inhabitants of Quito.

The diverse inhabitants of the Royal Audience of Quito during the late colonial period used a range of alcoholic and non-alcoholic beverages to create social and class distinctions, but at the same time the consumption of some beverages (e.g., yerba mate and chocolate) were also used to loosen social barriers and create greater tolerance. Indigenous peoples, mestizos, and Spaniards associated particular beverages with their real or imagined homelands. Although many residents of Quito linked specific beverage consumption practices with behavioral traits, many of which were unfavorable, introduced species particularly grapes and cane sugar used for the production of fermented intoxicants played important roles in colonial life and through time. Colonial customs resulted in the emergence of beverages and consumption habits that reflected the confluence of Andean, Amazonia, and Indigenous-European practices.

References

Alcedo AD (1786) Diccionario geográfico-histórico de las Indias Occidentales ó América : es á saber : de los reynos del Perú, Nueva España, Tierra-Firme, Chile y nuevo reyno de Granada Tomo 1. En la impr. de Manuel González, Madrid

Alcedo AD (1787) Diccionario Geográfico-Histórico de las Indias Occidentales ó Américanes á saber:de los reynos del Perú, Nueva España, Tierra Firme, Chile, y Nuevo reyno de Granada Tomo 2. En la impr. de Manuel González, Madrid

Alcedo AD (1788) Diccionario Geográfico-Histórico de las Indias Occidentales ó Américanes á saber:de los reynos del Perú, Nueva España, Tierra Firme, Chile, y Nuevo reyno de Granada Tomo 3. En la impr. de Manuel González, Madrid

Alcedo y Herrera D (1915 [1787]) Descripción geográfica de la Real Audiencia de Quito. Hispanic Society of America, Madrid

Arnalot J (2007) Lo que los Achuar me han enseñado. Mundo Shuar, Centro de Documentación, Investigación y Publicaciones. Abya Yala, Quito, Sucua, Morona Santiago, Ecuador

Delgado J (2011) Iglesia y Educación Pública en Quito y en Cuenca a Fines del Periodo Colonial (1750–1809). Universidad de Cuenca, Cuenca Ecuador, Licenciatura

Azara FD, Schuller RR (1904) Geografía, física y esférica de las provincias del Paraguay y Misiones Guaraníes. Museo Nacional, Montevideo

Blégny N (1687) Le Bon Usage du thé, du Caffé, et du Chocolat pour la Preservation & pour la Guerison des Maladies. Chez Estienne Michallet, Paris

Boy J (1840) Diccionario Teórico, Práctico. Histórico y Geográfico de Comercio, Imprenta de Valentín Torras, Barcelona

Chacón J (1990) Historia del Corregimiento de Cuenca, 1557–1777. Banco Central del Ecuador, Quito

Cicala M (1994 [1771]) Descripción Histórico-Topográfica de la Provincia de Quito de la Compañia de Jesus, traducido por Julián Bravo. Biblioteca Ecuatoriana Aurelio Espinosa Pólit, Quito

Cicala M (2004 [1771]) Descripcion Historico-Física de la Provincia de Quito de la Compañía de Jesús. Julián Bravo trad. Bibl. Ecuatoriana Aurelio Espinosa Póli, Quito

Chacón J, Soto P, Mora D (1993). Historia de la gobernación de Cuenca, 1777-1820: estudio económico-social. Cuenca, Ecuador: Universidad de Cuenca, Instituto de Investigaciones Sociales.

Concolorcorvo, Gaibrois MB (1997) El Lazarillo de Ciegos Caminantes. Agencia Española de Cooperación Internacional, Madrid, Spain de Terreros y Pando E, Meseguer y Arrufat F, Manuel y Rodríguez MD (1786) Diccionario Castellano, con las Voces de Ciencias y Artes y sus Correspondientes en las Tres Lenguas, Francesa, Latina é Italiana. Impresa de la Viuda de Ibarra, Hijos y Compañía, Madrid

de Certeau M, Giard L, Mayol P (1994) Historia de la Gobernación de Cuenca, 1777–1820 Estudio Económico-Social. Cuenca, Ecuador. Instituto de Investigaciones Sociales. Universidad de Cuenca, Ecuador

de Certeau M, Giard L, Mayol P (1999) La invención de lo cotidiano, habitar, cocinar 2. Pescador A. trad. Universidad Iberoamericana, Departamento de historia México

de Osasunasco D (1789) Observaciones Sobre la Preparacion y Usos del Chocolate. Felipe de Zúñiga y Ontiveros, Mexico

De la Peña, A (1771). Itinerario para párrocos de indios, en que se tratan las materias mas particulares tocantes a ellos para su buena administración. Madrid, Marín.

de Ulloa A, Juan J (1748) Relación Histórica del Viage a la America Meridional Hecho de Orden de S. Mag. para Medir Algunos Grados de Meridiano Terrestre, y Venir por ellos en Conocimiento de la Verdadera Figura, y Magnitud de la Tierra, con Otras Varias Observaciones Astronómicas, y Phisicas. A. Marín, Madrid, Spain

de Velasco J, Pólit AE (1961a [1837]) Padre Juan de Velasco, S. I.: Primera Parte. Puebla, editorial J. M. Cajica, Jr., México, D.F.

de Velasco J, Pólit AE (1961b [1837]) Padre Juan de Velasco, S. I.: Segunda Parte. Puebla, editorial J. M. Cajica, Jr., México

Einzmann H, Martínez Borrero J (1993) La Cultura Popular en el Ecuador Azuay. Centro Interamericano de Artesanías y Artes Populare, Cuenca, Ecuador

Flandrin J-L, Montanari M (2011) Historia de la Alimentación, Segunda edición (trans: Carvajal D, Carvajal N, Le Lanchon C, Martínez A, Pérez L). Trea. Gijón, Spain

Freedman PH (2009) Gastronomía: La Historia del Paladar. Universitat de València. València, Spain

Gerbi A (1993 [1960]) La Disputa del Nuevo Mundo: Historia de una Polémica, 1750–1900. Fondo de Cultura Económica, México D.F.

Goodman MK, Goodman D, Redclift MR (2010) Consuming space placing consumption in perspective. Farnham, Ashgate, Surrey, England

Herzog T (2012) Beyond race: exclusion in early modern Spain and Spanish America. In: Hering M, Martinez ME, Nirenberg D (eds) Race and blood in the Iberian world. Springer, Berlin

Jamieson RW (2002) Domestic architecture and power the historical archaeology of colonial Ecuador. Kluwer Academic Publishers, New York

Jelin E (2014) Desigualdades de Clase, Género y Etnicidad/raza: Realidades Históricas, aproximaciones analíticas. Freie University Press, Berlin

Juan J, de Ulloa A (2002 [1826]) Noticias secretas de América. Dastin, Madrid

Lacoste P (2004) La vid y el vino en América del Sur:el desplazamiento de los polos vitivinícolas (siglos XVI al XX). Universum (Talca) 19(2):62–93

Lavedán, A. (2010). Tratado de los usos, abusos, propiedades y virtudes del tabaco, café, té y chocolate extractado de los mejores autores que han tratado de esta materia. Mairena del Aljarafe, Sevilla, Extramuros.

MacFarlane A (1994) La Rebelión de los Barrios, una insurrección urbana en el Quito borbónico. Banco Central, Quito

Martínez Borrero J (2010) Alimentación y cocina en Cuenca de 1557 a 1607, los orígenes remotos de una cocina regional. In: Pazos Barrera J (ed) Cocinas Regionales Andinas: Memorias del IV Congreso. Corporación Editora Nacional, Quito, Ecuador

Martínez Borrero J (1983) La Pintura Popular del Carmen: Identidad y Cultura en el Siglo XVIII. CIDAP, Cuenca, Ecuador

Martínez Donoso M (2013) Violencia imaginada, violencia real en Cuenca en el siglo XVIII. Boletín de la Academia Nacional de Historia (Quito) XCI n° 188 :13-40

Mira Caballos E (2015) Vinos y élites en la América de la conquista. Iberoamericana (57):7–23

Moreno S (2008) La Cultura Popular en el Ecuador: Bolívar. CIDAP, Cuenca, Ecuador

Nicholson GE (1960) *Chicha* Maize Types and *Chicha* Manufacturing in Peru. Econ Bot 14:290–299

Norton M (2006) Tasting empire: chocolate and the European internalization of Mesoamerican aesthetics. The Am Hist Rev 111(3):660–691
Patiño VM (1990) Historia de la Cultura Material en la América Equinoccial. Instituto Caro y Cuervo, Bogotá, Colombia
Rice PM (2011) Vintage Moquegua: history, wine, and archaeology on a colonial Peruvian periphery. University of Texas Press, Austin
Rice PM (2013) Time-space perspectives on early colonial Moquegua. University Press of Colorado, Boulder
Rossi y Rubí J (Hesperiófilo) (1791) Rasgo Histórico y Filosófico Sobre los Cafés de Lima. Mercurio Peruano I (12):108–111
Schultes RE (1972) Ilex Guayusa. From 500 A. D. to the Present. In: A medicine-man's implements and plants in a Tiahuanacoid Tomb in Highland Bolivia, 1st edn. Göteborgs Etnografiska Museum, Göteborg, Sweden, pp 115–138
Serra J.D.S.G. (1994 [1799]) Maravillas de la Naturaleza. Biblioteca V Centenario Colcultura, Bogotá, Colombia
Solano V (1892) Obras. La Hormiga de Oro, Barcelona
Soldi AM (2006) La vid y el vino en la costa central del Perú, siglos XVI y XVII . Universum (Talca) 21(2):42–61
Staller JE (2006) The social, symbolic, and economic significance of *Zea mays* L. in the Late Horizon Period. In: Staller JE, Tykot RH, Benz BF (eds) Histories of Maize: multidisciplinary approaches to the prehistory, linguistics, biogeography, domestication, and evolution of Maize. Routledge Taylor & Francis, NY, pp 449–467
Staller JE (2008) Dimensions of place: the significance of centers to the development of andean civilization: an examination of the *Ushnu* concept. In: Staller JE (ed) Pre-Columbian landscapes of creation and origin. Springer, New York, pp 269–313
Valladares de Sotomayor A (1788) Semanario Erudito, que Comprehende Varias Oobras Ineditas de Nuestros Mejores Autores Antiguos, y Modernos. don Blas Román. Madrid, Spain
Weaver BJM (2015) "Fruit of the Wine, Work of Human Hands": an archaeology and ethnohistory of slavery on the jesuit wine Haciendas of Nasca, Peru. Unpublished Ph.D. dissertation, Vanderbilt University, Nashville, TN
Zarrillo S et al (2018) The use and domestication of *Theobroma cacao* during the mid-Holocene in the upper Amazon. Nat Ecol Evol 2:1879–1888

Chapter 14
Introduced Fruit Species as Food Heritage in the Quebrada de Humahuaca, Jujuy Province, Argentina

D. Alejandra Lambaré, Nilda D. Vignale, and María Lelia Pochettino

Abstract European conquerors introduced peaches (*Prunus persica*) into northwestern Argentina early in the colonial period. Among communities in the Quebrada de Humahuaca peaches represent an important crop and food. This contribution explores whether peaches may be considered food heritage in this part of the Andes. The perception that this exotic species is perceived as heritage is founded in the traditional strategies of management in selecting for certain characteristics that have cultural associations and certain characteristics that dramatically reduced peach diversity at a regional scale. Thus, culture, crop, and environment are interlaced to form—at present—a set of 9 ethno-varieties with typical features and defined uses. People selected peaches with regard to preferences of consumption, and in recent times their importance to commercialization. Peaches, as in the case of different Andean crops, such as potatoes and corn, are important to local consumption and ethnic identity resulting in networks of exchange—fairs and markets—enabling these communities to share in the fruits of their ethnic communities. Heritage is characterized as the capacity to symbolically represent the identity of a community, through phenotypic distinctions of the local variety of peaches or "*duraznos de la* Quebrada" (peaches from the Quebrada).

D. Alejandra Lambaré · N. D. Vignale
Laboratorio de Botánica Sistemática y Etnobotánica, Facultad de Ciencias Agrarias, Universidad Nacional de Jujuy, Alberdi 47, 3600 San Salvador de Jujuy, Argentina

M. L. Pochettino (✉)
Laboratorio de Etnobotánica y Botánica Aplicada, Facultad de Ciencias Naturales y Museo, Universidad Nacional de La Plata, Calle 64 nº 3, 1900 La Plata, Argentina
e-mail: pochett@fcnym.unlp.edu.ar

D. Alejandra Lambaré · M. L. Pochettino
CONICET, Buenos Aires, Argentina

J. E. Staller (ed.), *Andean Foodways*, The Latin American Studies Book Series,
https://doi.org/10.1007/978-3-030-51629-1_14

Introduction

The conquest of the New World in 1492 and subsequent colonization of Argentina was accomplished through Catholic orthodoxy imposed by various religious orders who settled the Argentine territory. Pre-Columbian natural environments once occupied by indigenous communities were transformed dramatically following the introduction of Indo-European plants and animals. Chickens, pigs, sheep, and cattle as well as domesticated food staples transformed the lifestyles and traditional indigenous foods and cuisines of this region (see Covey, Chap. 1 deFrance, Chap. 3, and Sayre and Rosenfeld, Chap. 7 for additional discussions of colonial changes in food animals). The introduction of Old World food crops and animals, different types cultivation, rearing, or production, also dramatically altered the ecology and environment of this region of the Andes (Capparelli et al. 2011; Capparelli and Raffino 1997; Giovannetti 2005; Giovannetti and Lema 2005; Lema 2009b).

Contact between both the Old World and Argentina was established within a framework of long processes of change in social structure, caused in part by the exploitation and the expropriation of land along with the annihilation of native populations by colonists and colonial governments (Capparelli et al. 2005). Consequently, one of the objectives of colonists and colonial governments was to discover new lands that could be developed by an indigenous workforce to recreate Spanish agricultural spaces in the new colony. After Spanish colonization and alternation of landscapes in both Peru and Bolivia to fulfill Spanish demands, further expansion of agricultural and pastoral space was no longer possible due to the large populations in these regions, especially in lowland coastal settings most suited to Old World crops and animals. Thus, in order to expand systems of production, Spanish colonists established a connection between the Central Andes and northwestern Argentina, in the region of Tucumán province (Sica and Ulloa 2010).

European colonists not only reorganized the agricultural and pastoral economy of northwestern Argentina according to their own preconceptions, favoring crops in low and flat areas, instead of the foothills as Andeans commonly did, but they also worked to tenaciously introduce and impose many European foods and drink on local indigenous communities. Documentary evidence of these transformations of food and culture was primarily focused on several varieties of wheat (*Triticum* spp. L.) as the primary staple, rather than other crops, such as barley, or fruits, such as peaches (Capparelli et al. 2005; Giovannetti 2005; Sica 2005, 2006). This greater historical documentation of wheat over other crops reflects the commercial and/or political importance of grains such as wheat to the colonial political economy, as well its role in Catholic religious practice. Thus, ethnohistoric documents related to other plants are scarce. However, Jesuits left extremely detailed documents regarding citrus fruits introduced from the Old World to the northeast region of Argentina (Stampella et al. 2013).

This research focuses primarily on the history of the peach species (*Prunus persica* Batsch) in Argentina as an example of the local appropriation of an introduced species. Our analysis uses several methodological strategies, including archaeological analysis of peach remains, firsthand accounts in colonial documents and ethnohistoric literature, and modern ethnobotanical research. We document the colonial establishment of the fruit and the development of multiple local varieties and show how the peach was incorporated into local diets as a traditional food.

Eurasian Crops: The Peach in the Land of Maize and the Potato

Peach (*Prunus persica* Batsch) is a species belonging to the botanical family *Rosaceae*. The genus, *Prunus*, is subdivided into several subgenera according to fruit morphology. The origin of the peach is in the mountainous regions of Tibet and northwestern China (Rivera-Núñez et al. 1997). The fruit was brought to the Mediterranean by early European explorers and was established by the Roman period. However, details regarding the origins, biogeography, and early domestication of peaches are still not well-documented with regard to dispersal centers in Western Europe and the Spanish colonial empire (Burger et al. 2011; Depypere et al. 2007).

Spanish colonists introduced peaches to the Americas during the sixteenth century (see Ruhl 1990). Analysis of ethnohistorical accounts suggests the first arrival of peaches in Argentina during the colonial period was via various routes from Peru to dry valleys of the Jujuy province (Fig. 14.1). There is documentary evidence that peaches may have been introduced from Chile through contacts between Argentine Cuyo and both the northwest regions and the trans-Andean region (Báez 1947, 1949; Capparelli et al. 2005; Giovannetti 2005). However, Sica (2010) suggests the northwest extreme of Argentina (Jujuy province) had a stronger relationship with Peru than with Chile. Taking into account that the province was part of the colonial government, the economic relationship can be seen mainly as a product of a geographical location that facilitated exchange and commercialization (Farberman and Boixadós 2006).

The first conquistador that explored the northwestern Argentina was Diego de Almagro. In 1535 he entered in the old path of Inca Collasuyo—that is the southern portion of Inca Empire, from south of Cuzco to Maule River in Chile—looking for precious metals and lands suitable for cultivation. Almagro went as far as the valley of Copiapó in Chile and arrived in Tupiza, which is 80 km from the Bolivia-Argentina border. Subsequently, he chose to follow a coastal route through the Atacama Desert on his way inland back to Cuzco (Salas 1945 in Ruiz 2004). As chronicled in the "Carta A.S.M. del Oidor de los Charcas; Licenciado Juan de Matienzo" dated 2 January 1566, these routes brought Almagro to what is today Jujuy in 1536 (Vitry 2007). During his expeditions in these regions, Almagro sent military governors or *adelantados* information on the surrounding landscape as well as the encounters with indigenous populations so Spanish forces could identify the safest roads to travel that would also provide pasture and water for their animals, and towns and villages

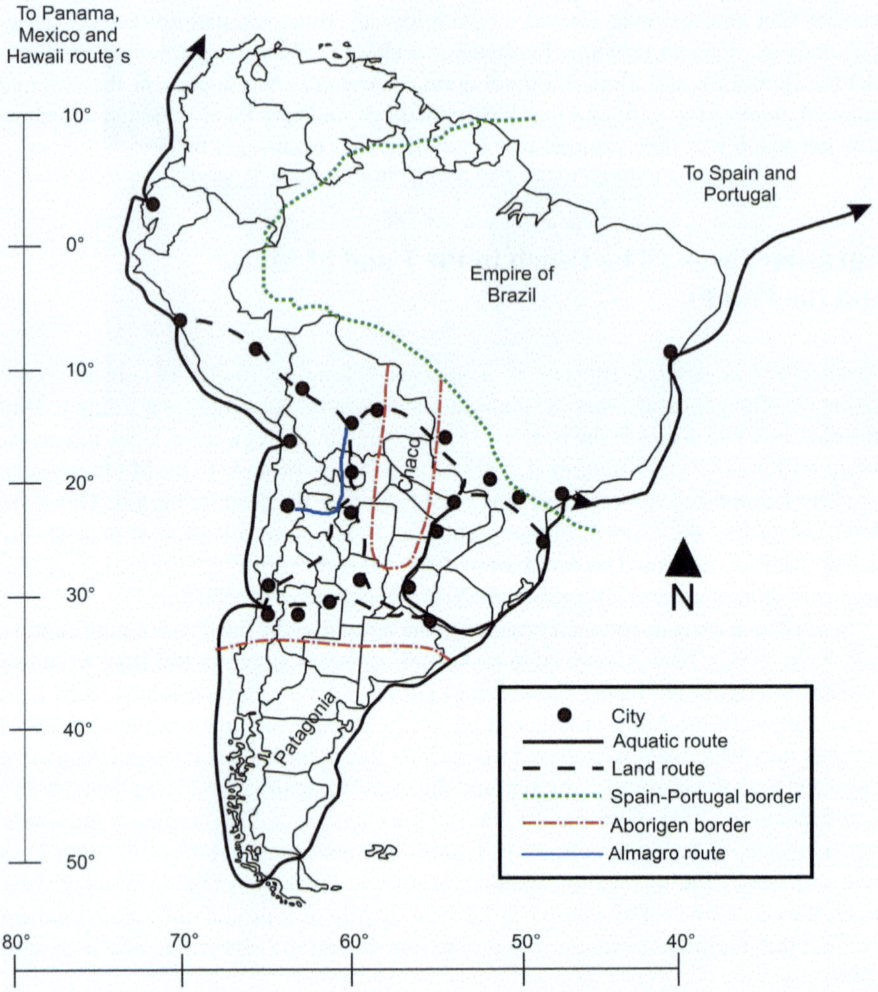

Fig. 14.1 The Iberian conquest and colonization of American continent: routes to terra firma of the sixteenth Century. (Courtesy of Stampella et al. 2013 and modified by D. Alejandra Lambaré)

where they could obtain food (Vitry 2007). Accounts and information related to this expedition described the environment and arid desert landscape. Consequently, towns and various Tambos (Inca way stations) located along the Inca highway or *Camino Real* were plundered for water and food. Almagro entered Argentina from Bolivia through the Tambo of Kalahoyo, the plains of Pozuelos and the Tambos of Queta. Moreno as well as other Tambos were also encountered on the journey to the Faldas del Acay (Salta) (Matienzo 1566). Subsequently, Almagro traveled to the valley of Calchaqui through mountain passes until he reached Copiapó in Chile (Ruiz 2004). Few European crops were consumed during this period (Capparelli et al. 2005). Later, colonial authorities in Peru ordered Juan Núñez de Prado to make the same trip in 1549 with a mandate to find a town in Tucumán (Levillier 1939). The founding of Tucumán took place the following year in 1550 at Barco de Ybatín in northwestern Argentina.

According to Giovannetti (2005) it would be reasonable to expect that the colonial founders of Tucumán had the basic necessities for their survival, i.e., cattle and plant species, especially wheat and barley. The founding of the first Spanish settlement consisted of three villages: Barco (1550 A.D.) in Tucumán province, Santiago del Estero (1553), and Londres de Quimivil (1558) in Catamarca province. The Peruvian Spaniard Núñez del Prado established the first one, while the other two were founded by Spaniards from Chile: Francisco de Aguirre and Juan Pérez de Zurita; respectively (Capparelli et al. 2005). Written accounts regarding the founding of the colonial government of Tucumán indicate the establishment of these towns in 1550 and 1553 coincided with the introduction of Indo-European cereal grains such as wheat and barley along with exotic species of fruit (e.g., peaches, cherries, apples, etc.).

Giovannetti (2005) suggests Spanish colonists obligated indigenous communities to cultivate Indo-European crops, resulting in large-scale monocrop cultivation in some regions. Several ethnohistoric accounts state that this is reflected in the agricultural cultivation in the present-day landscapes (Capparelli et al. 2011). Spanish colonists may have also introduced peaches in the same way. However, local history suggests that the introduction of the peach and its establishment in this region was the result of the conjunction of the inherent characteristics of the crop along with local, indigenous cultural decisions. Peaches acclimatized rapidly to the local environment, and the cultivation and human selection for certain desirable phenotypic characteristics contributed to the success of the fruit. Sica (2006) proposed that after the conquest during the colonial period there was a mix of local and foreign knowledge regarding cultivation and domestication of peaches in a changing socio-economic and cultural era.

Pre-Hispanic Agricultural Strategies

The agricultural landscape resulting from pre-Hispanic culture at the time of Spanish settlement is reflected by the existing cultivation in Northwestern Argentina (Capparelli et al. 2011). The local history of the societies that inhabited the dry valleys of Jujuy indicate a large majority of the pre-Hispanic indigenous communities resided in the Quebrada de Humahuaca and on the *puna*. Here, an agricultural-pastoral economy developed in which camelid livestock rearing of llamas (*Lama glama* L.) was mixed with agriculture at different intensity levels depending on where access to irrigation water was greatest. Moreover, traditional Andean economies based upon ecological complementarity were carried out to obtain nonlocally available food resources. In general, houses were very close to fields that contained a variety of plant and animal species adapted to the heterogeneous environmental conditions of the mountainous northwestern Argentina. In some cases, the soil demanded a modification in the landscape requiring the construction of several agricultural features such as large terraces, retaining lines, square foot gardening, platforms, stone mounds, and silos. These constructions prevented crop loss due to draft, excessive rainfall, frost, and other climate factors (Lorandi 1984).

The use of fertilizers was a common practice to improve agricultural output. One of the potential organic fertilizers used by native groups living in the Valle de Tafí was the exocarp of the pumpkin (*Cucurbita* sp.) to improve the yields of tarwi (*Lupinus mutabilis* Sweet) and maize (*Zea mays* L.), all of which have been identified through the study of microbotanical remains (Franco Salvi et al. 2014). Additionally, the use of llama manure was used as fertilizer. Camelid dung is present in Peruvian archaic sites and identified through the analysis of coprolites (Lema 2009b). Also, evidence for the presence of portable corrals or *majeo* suggests that this ancient technique was used to concentrate the camelid herd in one area of the farm to facilitate the accumulation of a thick layer of manure that could be collected more easily (González 2009).

With regard to pre-Hispanic hydraulic construction, Albeck (1984) mentions six different types of irrigation canals in the Casabindo region, and some earthen canals are still present in the region. The different canals are probably a reflection of the increased population density that would have demanded greater crop production, and thus, greater irrigation with canal-fed water systems in this dry valley (Albeck 2011; Lanzelotti 2011). The development of agricultural sectors during the Formative period (1000 BC-900AD) followed a progression from the valley bottom areas followed by the foothills and the alluvial fans and finally on the slopes of the hills (Albeck 2003–2005).

After several centuries of local agricultural production, Inca domination (1471–1536 AD) resulted in widespread changes in the hydraulic agriculture practices in Andes. The Inca Tambo or storage facility (*collcas* or *pirhuas*) held maize (*Zea mays* L.) and varieties of potatoes (*Solanum tuberosum* L.). The Inca also introduced metallurgic techniques in the form of smelting ovens or *wayras* (Raffino 1999). Amplification of the flow of goods leaving the region by Inca administrators necessitated the establishment of main and secondary routes of transport. There existed two means of communication in northwestern Argentina at this time: one that spread through the region transversally and another one through the Quebrada de Humahuaca longitudinally (Raffino 1996). The use of part of this road system by Europeans during their first expeditions into northwestern Argentina helped to form the foundation of the new colonial cities. Following Spanish colonization, some routes were reoriented for traffic of people, goods, and animals within the new economy and the increasing activity of the mining markets (Sica 2010).

During colonial times, the indigenous communities in the Quebrada de Humahuaca had strong direct contact with the Spaniards. Consequently, the native inhabitants were aware of and adopted European plants and animals (Sica et al. 2010). By the late fifteenth century among the routes of exchange for goods coming from the Government of Tucumán, the current territory of Jujuy was the ascension point to the highlands with travel passing through the ravine area and the Puna. Beginning in the seventeenth century indigenous populations increased food production and began to expand production of foodstuffs into the valley bottoms. It is also at this time that the colonial regime consolidated its control in the northwestern region of Argentina as well (Proyecto Cultivos Andinos 2007; Sica 2010; Sica and Ulloa 2010).

The agricultural practices of this period reflect colonial-era changes that resulted in a new biocultural landscape (Lema 2009b). Peaches, together with other above-mentioned Eurasian crops (e.g., wheat and barley) entered northwestern Argentina rapidly after the Spanish conquest and were defining elements of the Hispano-Indigenous or early Colonial period (A.D. 1535–1700). Introduced food crops functioned similarly to other products of European origin, such as iron tools and glass beads, which even though they were foreign elements they were rapidly incorporated into native cultures (Capparelli et al. 2007). Likewise, within the agricultural cycle, these new elements became associated with preexisting native crops such as maize and potato both of which are considered seasonal guide crops in the Andes, and thus define the agricultural cycle (Schulte 1996) (see Valenzuela and Moyano, Chap. 10 for an analysis of continuity of ancient ritual practices related to the annual agricultural cycle).

The Agricultural Calendar and Current Management of Peaches in Juella

Modern ethnobotanical research largely complements the archaeological and historical data, regarding the local knowledge and practices surrounding peaches (*Prunus persica* Batsch). Our field research in Juella (Jujuy province) and the surrounding areas established this locality as the main production center of peaches in the Quebrada de Humahuaca, particularly Juella, a small valley in one of the ravines intersecting the Grande river (23°31.315′ S and 65°24.091′ W) (Fig. 14.2). This region is located in Prepuna province and is characterized by scrubby steppe vegetation (Cabrera 1976). The population of 200 people (INDEC 2010) consists of small-scale farmers dedicated to commercial production of peaches as well as other food crops including maize and potatoes along with the rearing of livestock for personal consumption, as well as goats for cheese and meat.

In Juella, small peach trees grow in family fields along with traditional species native to the Andes, thus creating an agricultural space dedicated to the provisioning of food and also providing an economic alternative for a household. The joint presence of peaches and Andean crops following specific planting location patterns reflects long-standing agricultural traditions in these spaces. Commercial peaches are adapted to a mild climate with cold rainy winters, dry springs, hot dry summer, and mild cool autumn. Winter cold is essential for them to grow and sprout satisfactorily. In addition, this plant is sensitive to strong winds (Scorza and Sherman 1996). In the case of the Quebrada de Humahuaca peaches thrived because temperature conditions were favorable despite the differences in the rainy season and annual summer rains. In addition, human manipulation through time linked the peach to the growth cycle of

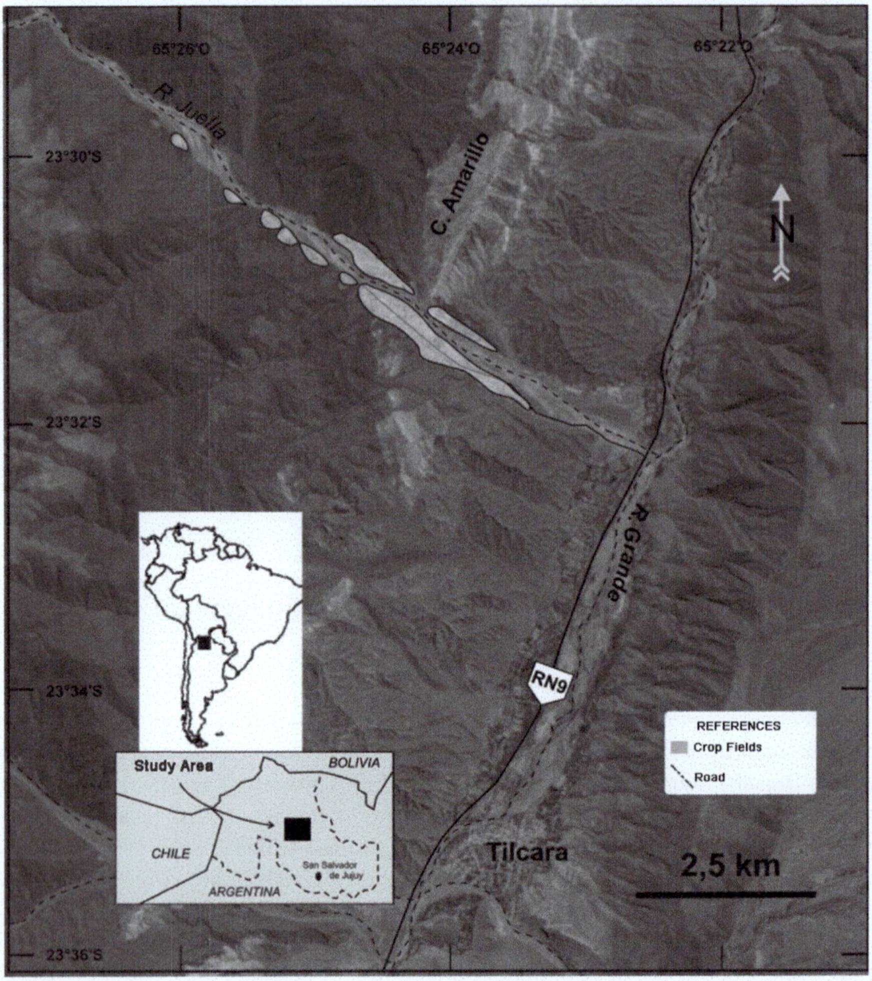

Fig. 14.2 Study area. Map available from https://www.google.com.ar/maps/@-23.5122261,-65.4265783,16516m/data=!3m1!1e3 and modified by D. Alejandra Lambaré

traditional Andean crops such as maize. Figure 14.3 illustrates the agricultural cycle of the peach crop, its phenology, and its relationship to cultural events. As previously commented, the cycle of maize—the most important crop of the region—and that of peaches can be seen to be closely related because both are important parts of traditional production by the inhabitants of dry valleys.

After the months of strongest annual frost (i.e., the dry cold austral winter months of June and July) activity returns to those cultivated fields that were unused. In August most Andean peoples celebrate and perform the ritual of the *Pachamama* which is associated with several ceremonial expressions connected to agricultural/livestock cycles that occur throughout the year. It is a thanksgiving ritual that also promotes

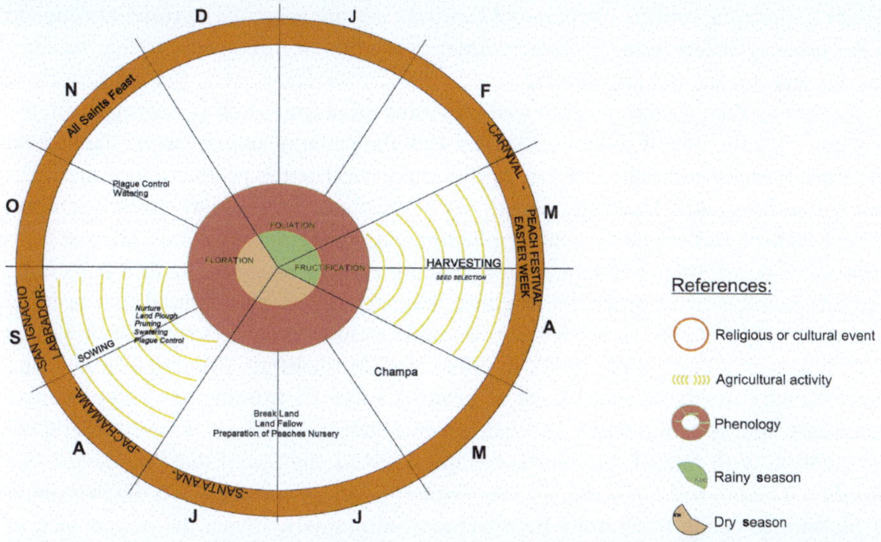

Fig. 14.3 Agricultural cycle of the "peach" crop; its phenology and relation to cultural events. Graphic: (Courtesy of Lambaré, Vignale, and Pochettino)

fertility and multiplication of food crops; it is usually called *corpachada* or *challada* (Podjajcer and Mennelli 2009). Concurrently with this celebration that involves "feeding" the land and smoking the crop fields, daily crop maintenance work begins to ensure a good harvest.

In the case of peaches, the local management of trees includes activities such as pruning, harvesting, and preparation of seedbeds. These tasks are done mainly by women, whereas land preparation (e.g., plowing and digging furrows) is carried out by males. Sometimes families share these tasks with other members of the families; in other cases, another person from the community is temporarily hired for chores such as cleaning canals or plowing, (i.e., furrows and turning over or aerating the soil). Soil preparation is carried out before adding the fertilizer and sowing the seeds with iron-point plows pulled by draft animals. In contrast, other activities related to land preparation involve the use of other tools such as shovels, picks (to break soil clods), and walking a plow pulled by mares or donkeys to break up the soil.

Soils are fertilized using sheep or goat manure, occasionally fermented. The most valued fertilizers include the *churcal* (i.e., manure from goats which feed on the leaves and fruits [*choloncas*] of the *churqui* [mesquite] trees (*Prosopis ferox* Griseb). The organic matter collected from under these trees is also used to nourish the soil. This type of manure is obtained from the farmyards owned by families or recollection or bought from a third party. Due to its high cost, some families do not have access to fertilizer. *Urea* (from both human and horse urine) is another source of nutrients for the soil. Additionally, we observed that after the harvest season, some people use a mix of peach epicarps from peaches skinned to make *pelones* (sun-dried peaches) and leaf sheaths or *chalas* from maize to improve the soil. Given that urea usage occurs in the area, reports of chemical fertilizer use are scarce. The residents engage in the process of fertilizing cultivated areas every two or three years so that peaches develop good taste and appearance.

Regarding the phenology, each tree blossoms three times before bearing fruit: the first bloom at the end of winter coincides with the pruning season, while the second and third blooms coincide with leaf sprouting as well as the production of the fruits that will be harvested. Due to the sensitivity of the plant, growers pay special attention to pest control and negative weather phenomena during these periods, such as hail. Pruning of peach trees begins in August. Some settlers take care to do it before the tree has blossomed in order not to cause damage. Residents reported that pruning is done: "*before the flowers come up, we need to prune the peaches and thin them out*"; however, some people prefer to prune after blossoming. Pruning is done using commercially available tools, such as pruning scissors, pruning saws to eliminate secondary shoots (*chupones*), and ladders to reach the treetops. Our ethnographic information with one of the people we interviewed suggested that "*the small tree should not be too tall… the shape of the crown should be rounded.*" At the beginning of blooming, peach trees must be watered continuously, hence the importance of irrigation canals derived from pre-Hispanic agricultural practices.

The first peaches begin to ripen after the last austral summer rains. The harvest season starts in March or April. The producers are aware of the different harvest seasons of each ethno-variety and agroecological zone, and they respect these differences. The criteria used to decide the time of harvest varied among the interviewees. Some prefer to gather the fruit when it is very ripe and the peaches are soft to the touch or some harvest the fruit based on color changes "*when the face becomes colored.*" Others prefer to harvest peaches before they ripen or get damaged by the sun which causes dehydration and loss of taste, or as they put it "*the fruit is sun-stricken.*" Harvest is done by hand, using wicker baskets, market bags, plastic buckets, or wooden boxes after which the dry fruit is stored for a short period of time inside these same containers. The storage containers may be deposited indoors (e.g., in the home) or outdoors (e.g., in yards or cultivated fields). The harvested fruit is destined for both consumptions and for local trade. Any ripe peaches that are not harvested and remain in the trees will be used to make *pelones* (sun-dried peaches).

At the end of the maize and peach harvest season (May–June), the cultivated fields are typically covered with accumulated cut maize plants with maize ears to peel (*chalcar*). These plant remains are left to dry in the fields and then used to make flour or as fodder for cattle. This practice provides both protection and organic fertilization to the soil.

The selection of seeds, called *pepitas*, is done after the harvest season with a focus on the morphological and phenotypic characteristics of the fruit. The first consideration is the taste, that is, if the mesocarp has a pleasant flavor, those seeds are selected. Pleasant color, size, and scent, as well as the peach being disease-free are also factors considered in seed collection for cultivation of the fruit. The selected seeds are kept and then sown when a peach tree is lost to old age, disease caused by pests, or simply when the producer finds it necessary. Other characteristics for seed selection include appearance—if the peach is aesthetically pleasing then this is a major consideration as criteria for storage. Flavor is deemed rather subjective. Flavor essentially is based on preference by other family members or friends in the community for a given flavor or taste. Seeds of the cling peach ethno-varieties (*priscos*) are cultivated with greater interest because of their pleasant taste and their high demand for informal exchange among communities such as Juella. In some cases, the selected pits receive special treatment (i.e., the mesocarp and endocarp of the fruit are peeled off using a sharp tool to prevent contamination of the seed).

The growers propagate peaches from their seeds and in recent years some producers have begun experimenting with grafting. According to local knowledge, the endocarp and seeds must be sun-dried for a long time before planting, and thus the tree will produce more flowers. Some people place the peach seed inside a piece of open weave fabric and then immerse it in running water to clean it. Seeds are sown directly in the stubble or indirectly in seedbeds from which they are later transplanted to the stubble. During visits to the fields in the harvest season, we observed that many settlers throw the seeds of eaten fruits anywhere in the stubble if they find those peaches to be tasty.

Generally, the producers renew their crops using their own seeds or seeds they exchange with neighbors or relatives. The locals say that during the first month after sowing, when seedlings are growing, they must be watered continuously. The plant's age influences its flavor as well as its quality and sensitivity to pests; contaminated trees are replaced. However, there is no general rule for the frequency of tree replacement. Some old trees (up to 30 years old) that no longer bear fruit are kept in the fields, possibly because they have symbolic value to the community or cultivator. It takes new plants between 3 and 5 years to bear fruit. No hybridization issues among varieties were noted, which suggests the occurrence of self-pollination, a common trait of cultivated plants resulting from cultural selection.

Among the factors that may cause diminished harvest and lower peach quality are crop-damaging pests. These include the fruit fly (*Anastrepha fraterculus* Wiedemann). In the words of one of the villagers: "*the worm appeared because of the intense heat; there is a mosquito that leaves the poison and the worm grows*". Local residents also mentioned damage caused by birds. Leaves and other aerial parts are affected by a local vine known as *barba de chivo* or goat's beard (*Clematis* sp.): They also mentioned that: "*It is a vine that causes the plant to dry up.*" Another is the *araña roja* or red spider mite, *Tetranychus urticae* Koch. The roots of peach trees may be infested by round worms or nematodes that cause galls. The locals state that: "*When the trunk of the peach tree dies, it is because the roots are sick; one must dig the roots up and remove the swellings that form in the roots; these must be burned so the fungus does not contaminate other plants. Then, the damaged tissue must be coated with mud.*" Some villagers mentioned that "*The peach plant gets sick from this fungus when it is old about 20 years old, and that young plants rarely ever parasites.*" Locals believe that climatic changes such as prolonged windy seasons, increases or decreases of temperature, as well as a delay of the onset of cold months, as well as an increase of hail and rainy periods, are factors that bring on contamination by pests. They have found that older trees are more susceptible to pests. It is important to highlight that peach producers highly value their trees not only because of the fruit they provide, but also because they tolerate drought and adverse climate conditions characteristic of the region.

Both agrochemicals and homemade products or "remedies" are simultaneously used for pest control in this region. The latter include ash, cattle, and fowl manure from chickens, goats, and sheep, as well as manure mixed with milk wood shavings and sugar, mud and garlic juice, and powdered soap. One local person commented that for the remedy (ash, in his case) to be effective it is preferable to spread it very early in the morning before the dew lifts. He used this remedy against insects such as ants. Another popular home remedy used in the past included a mixture of garlic, sugar, and lemon that was allowed to ferment for about six months before use.

Scarcity of water is one of the major problems faced by producers; therefore, each year starting in the month of August a high-status individual called the Water Judge (*Juez de Agua*) is chosen by locals to control the irrigation networks for three years. This encompasses their maintenance, organization, and detailed accounts regarding water usage and distribution. Water is distributed from the main irrigation canal located approximately 6 km from the network, and secondary irrigation canals which collect this water and distribute it to third-order canals that extend into the crop fields or irrigated areas. These water-transporting structures are built above the ground as they were in pre-Hispanic times, and therefore they demand continuous maintenance from the locals for their correct functioning. The Water Judge sets the irrigation schedule for each cultivated field on the basis of the zones where each field is located. In contrast, on Sundays the water running in the canals is used to water lawns and vegetable gardens. Irrigation is fundamental to ensure that blossoms do not fall off the trees and that peaches reach good size, appearance and are flavorful when they are harvested. The producers can tell if the fruit has not received the necessary amount of water because the harvested fruits are small, rugose, and lack flavor. This environmental restriction has similar effects on all the peach ethno-varieties.

Peach Diversity in the Quebrada de Humahuaca

This research takes an *emic* or indigenous perspective to differentiate and characterize the diversity of peaches existing in cultivated areas (Table 14.1). There are two groups of ethno-varieties and 9 distinct varieties perceived by indigenous communities in the Quebrada de Humahuaca (Figs. 14.4 and 14.5). They are all cultivated and correspond to the botanical species *Prunus persica* Batsch. Characterizations were primarily based on their morphology and phenotypic characteristics, as well as their consumption by local populations. These ethno-varieties refer to peach varieties that share the same common name. A group has diverse characteristics that result from the sum of attributes corresponding to the different criteria that are recognized, such as adherence of the pit. This is a trait that allows initial separation into two groups: common peaches and cling peaches (known also as *prisco*, *frisco*, or "easy-split" by the locals). An additional subcategory defines the ethno-varieties according to individual morphological characteristics such as (e.g., color of the epicarp and mesocarp)

Table 14.1 Groups of ethno-varieties and ethno-varieties (from Lambaré and Pochettino 2012)

	Ethno-varieties group	Ethno-variety	Epicarp color	Mesocarp color
Prunus persica L	Common peach	Whole yellow	Yellow	Strong yellow
		Yellow; red center	Yellow with little spots	Yellow with red center
		White	White with little spots	White
		Pink	Light yellow with little pink spots	Light yellow
		Jorge (=cholo cholito)	White; white with little spots	White
		Alancate (=olancate)		
	Cling peach (=frisco; to split)	Yellow	Yellow with spots	Light yellow
		White	Light yellow with little pink spots	White with pink center
		Pink	Light yellow and pink	Yellow with pink center

and sensory characteristics (e.g., taste, consistency, and flavor). A secondary criterion for classification are attributes related to agricultural traits, such as when they are ready to harvest (Lambaré and Pochettino 2012).

The diversity of peaches reveals a corpus of knowledge linked to their conservation and phenotypic characteristics. This is related to local knowledge regarding their local history, both previous and subsequent to their entry in the region. Knowledge is also tied to the diverse related activities associated with agriculture and continuously incorporates new information and experimentation by locals and/or external cultivators. This information provides analytical evidence regarding the dynamics of local botanical knowledge associated with both long-standing precolonial, traditional activity and to current cultural perceptions regarding an introduced crop incorporated into the local landscape. Additionally, local residents informed us about the entrance of germplasm of modern commercial peaches from other areas of Argentina. Such varieties were incorporated or brought from other regions of the Quebrada, such as Uquía (Humahuaca Department) and Iruya (Salta province), as well as Río Negro and/or Mendoza; the latter referred to southern or lowland peaches. These varieties were not observed in cultivated fields. In contrast, the Juella peach is present in all of the areas we visited. Furthermore, the traditional varieties differ from commercial peaches by having characteristics that are undesirable for the latter (small size, marked fuzziness, higher sugar content that favors quick fermentation) and presumably are similar to those that were introduced early during the Spanish conquest and colonization. This has resulted in an assemblage of local varieties that locals recognize as their own. They are well-differentiated from commercialized peaches in other

Fig. 14.4 Different ethno-varieties of common peaches: **a** branch with entirely yellow fruit; **b** yellow fruits with red heart; **c** branch with pink fruits; **d** branch with white fruits (Jorge ethno-variety); **e** branch with white flowers; **f** branch with white fruit (Jorge ethno-variety). (Courtesy of Alejandra Lambaré)

Fig. 14.5 Different ethnovarities of cling peaches: **a** yellow fruit (bottom right) and pink (top left); **b** white fruit. Pictures: D. (Courtesy of Alejandra Lambaré)

areas, so these may be considered as exclusive to these communities, particularly in the greater Juella region.

In terms of germplasm management, we observed a permanent balance interplay between selection for diversification on one hand and strategies that promote stabilization of the newly obtained diversity on the other. Unlike the conditions recorded elsewhere in the world, for instance in the Mediterranean where archaeological sites have shown endocarps with greatly varied morphologies (Sadori et al. 2009), the varieties in the Quebrada de Humahuaca are morphologically homogeneous (Capparelli and Raffino 1997). In addition, ethnohistorical references mention only "peaches" not varieties (de la Puente y Olea 1900; López de Gómara 1922). From these accounts, we infer that the introduced varieties were phenotypically homogeneous and probably cultivated and selected by traditional strategies of human selection, which also favored conspicuous local diversification with few varieties.

Local Consumption, Differentiation, and Food Heritage

Different uses for the fruit and other parts of the peach tree are known around the world. In addition to its importance as a fruit, peaches are mentioned as natural laxatives. Indeed, the flowers and leaves are used in infusions and/or syrups and recommended as laxatives for children. In addition, the flowers have also been described as having diuretic, sedative, and vermifuge effects, and may be toxic due to amygdalin, which is more abundant in some varieties. In some communities, water distilled from peach flowers is applied for removal of skin spots and freckles (Font Quer 1988; Rivera-Núñez et al. 1997). Peach leaves are applied on the stomach to combat parasites, and on ulcers and herpes lesions because they are believed to be anti-inflammatory and to facilitate healing. Infusion of the leaves in milk is also mentioned in some communities as an anti-parasitic. The dry and pulverized leaves are used to heal zits and boils, while a mixture of seeds ground with egg whites is used to stop hemorrhages (Manfred 1947). The leaves and bark are used as antipyretic, astringent, expectorant, or as a diuretic. Oil is produced from the seeds, and the flowers are commonly consumed raw in salads and used to make a tea-like infusions in this region. The fruits are consumed fresh, dried, canned, and processed into jams, juices, liquors, and moonshines. From a nutritional perspective, 100 g of peach (wet weight) contains 89% water, 10% carbohydrates, fiber, vitamins A, B, and C, potassium, phosphorus, iron, iodine, and calcium, with an energy value of 35 kcal (Hurrell et al. 2010).

In northwest Argentina, archaeological evidence points to the ceremonial use of peaches in ritual contexts by the native peoples in Catamarca during the Colonial period, around A.D. 1600 (Capparelli et al. 2005). Records of regional-scale use and ethnobotanical studies in the area of the eastern Salta valleys, at intermediate levels of the cloud-forest (*yungas*), also highlight the medicinal uses of peaches (Hilgert 2001, 2004). The literature indicates that a decoction prepared with dried flowers and grounded seeds is used to treat intestinal diseases (diarrhea, meteorism, intestinal inflammation). This beverage may be taken up to three times a day, and its use repeated for two consecutive days if the discomfort persists. Likewise, according to hot–cold syndrome, the buds are macerated in vinegar and applied by rubbing on the body to eliminate fever (Hilgert 2004). A peach infusion ingested three times is also used for *chuscho*, a condition characterized by very high fever and sweating (Hilgert 2001). The fleshy part of peaches and apples are ground and mixed for use as agglutinant in the preparation of *yista*. This is a solid mixture called *coquear* made from plants used for coca chewing (Hilgert 2000; Hilgert et al. 2001). Additionally, in *Mapuche* pharmacopeia peaches that grow in Patagonian gardens are used to treat gastrointestinal conditions (Molares 2010).

In the Quebrada de Humahuaca peaches are first and foremost food. They are consumed fresh during the harvest season and as processed products and preserves throughout the year. These include *pelones*, a sun-dried fruit made using common and cling varieties. Peaches are also in sweet preserves, peach jams or syrups, and in liquors. Other preparations that include peaches, such as fruit salads or pastries, are constantly tested and prepared. In addition to being consumed in the home, they represent an additional source of income for families, as the peaches and derived products are sold or traded locally in fairs and/or festivals where there are often a large number of attendees. Once again, this economic use in addition to food use highlights the importance of this crop for the producers in the Quebrada de Humahuaca. Likewise, during the pruning season the secondary shoots (*chupones*) and non-floriferous lateral branches are removed from the trees and then used as firewood or for grafting. The epicarp that is removed from the fruits is used as fertilizer in cultivated fields.

In terms of phenotype, many of these fruits do not comply with all the phenotypic characteristics required by the fruit and vegetable market for commercialization. Among other issues, these local peaches vary widely in size and their harvest is more intensive. Nevertheless, such peaches continue to be selected by family members, friends, or people from other communities as well as by foreigners that participate in fairs and festivals where they are able to try the peaches from the Quebrada. In this sense, terms such as "peaches from the Quebrada" or "peaches from Juella" identify these peach varieties both within and outside the community. Those labels summarize the characteristics selected and/or maintained by the locals through centuries of interaction, and resistance to the introduction of other varieties. The naming process of various varieties is extremely important because it represents possession, both symbolic and real to the community or region (Sica 2005). Heritage-making—that is the social construction of heritage—entails knowledge with regard to human and natural selection, cultivation, and a series of practices. These local peach varieties are the result of human selection for particular characteristics and constitute an active heritage that is associated with a cultural landscape and territory. The process of using peaches for heritage-making is in a continuous state of innovation in relation to the phenotypic characteristics of various varieties. This process is ongoing because peaches can be modified by appropriation, exchange, adaptation, new food preferences, consumption practices, and long-term associations with the identity of communities and societies in the region (Álvarez and Sammartino 2009; Lema and Pochettino 2012; Remedi 2002). Through their use and preferences for particular types of peaches, individuals identify themselves with a given variety and at the same time differentiated them from others. Local culinary knowledge is understood as part of the food heritage of a community and is both communally and individually managed in unique ways. The peaches from the Quebrada de Humahuaca are the product of these long-term social and historical factors that have guided decisions regarding conservation, hybridization, or extinction of biological diversity.

Conclusions

The results presented here reflect the dynamics of traditional botanical knowledge regarding agricultural management and cultural selection by inhabitants of Juella on local peach crops. We were able to identify a local peach type (peach from Juella or peach from the Quebrada) consisting of nine varieties that locals differentiate based on morphological, phenotypic, and organoleptic characteristics. Local inhabitants also use these criteria in establishing their consumption practices and uses of peaches as food, which includes processing peaches for self-consumption, exchange, and trade. These characteristics allow the local residents to differentiate the "peach from Juella" from commercial varieties. The continued propagation and contemporary social value placed on early colonially introduced crops and the establishment and permanence of the peach as an element of the local agricultural space resulted from the Spanish conquest. Their introduction into this region impacted multiple aspects of the local circumstances of indigenous societies. The use of the peach as a form of food heritage demonstrates how the people of northwestern Argentina responded to the dramatic social and biological changes that accompanied Spanish colonization. The cultivation of peaches is evidence of the ability of these peoples to adapt to changing situations. Local peoples adapted to a changing colonial reality by incorporating exotic botanical and cultural practices into existing social and individual behavior. The adoption of the peach and behaviors to ensure the success of the crop occurred in conjunction with existing practices of large-scale agriculture and knowledge of the surrounding ecology. Additionally, the peach was integrated and materialized into the practices and knowledge of the locals, some of whom have strong associations with Andean ethnic identity. And third, there is the presence of a designed cultural landscape where in situ local management promotes the conservation/permanence of exclusive varieties recognized as native to the area, and which are incorporated as components of the food heritage of the community.

This analysis of the history of the peach in Juella contributes to our understanding of the role of this fruit in the creation of local heritage. If the value of a heritage is associated with the consolidation of the cultural identity of a local community, then local heritage gains value according to its role as an element of identification and appropriation of the immediate environment by the people who live in that region. When local heritage is linked to a food crop the designation of a geographic space with the name of the food item as a toponym denotes belonging. In agreement with Guglielmino (2007), we consider that theories and methods for both determination and management of cultural goods lead to practices of rescue and conservation that are creative, i.e., dynamic, rather than stationary. In order to recognize an active and sustained program of agricultural conservation, researchers need to examine how people understand perceptions regarding heritage or its objects (material and immaterial), in other words, researchers need to examine the emic perspective of the individuals that recognize the environment and landscape of their sociocultural group.

As shown in this analysis, human management of the peach in the Quebrada de Humahuaca since the introduction of the fruit tree and over 500 years of history has included the incorporation, local diversification, and stabilization of the variability of this crop by the local inhabitants. The decision to incorporate this foreign food, the peach, must be understood within the context of the worldview of these native Andean groups, many who are still undergoing strong cultural transformations, while maintaining their protagonist role in decisions that contribute to cultural reproduction (Hilgert et al. 2014). Indeed, the inhabitants of Juella actively participate either rejecting, incorporating, reinforcing, or abandoning cultural customs, objects, and products within the constant flow of social and historical dynamics. Within this cultural context, the appropriation of peaches transforms these originally exotic fruits into elements of local identity that contribute to define the communities that cultivate them and the scenarios of their daily lives in the Quebrada de Humahuaca, Jujuy Province.

Acknowledgements We are indebted to all the people who shared with us their knowledge and time. Universidad Nacional de La Plata, Universidad Nacional de Jujuy, and CONICET provided financial support for this research. We thank Cecilia Morgan for her help with the English version and the editors of this volume for their suggestions to improve the manuscript.

References

Albeck ME (1984) Riego prehispánico en Casabindo (Provincia de Jujuy). Rev Mus Plata Antropol 8(60):265–278

Albeck ME (1992/93) Áreas agrícolas y densidad de ocupación prehispánica en la Quebrada de Humahuaca. Av Arqueol 2:34–51

Albeck ME (2003/05) Sitios agrícolas prehispánicos: la búsqueda de indicadores cronológicos y culturales. Cuad Inst Nac Antropol Pensam Latinoam 20:13–26

Albeck ME (2011) Estudios de agricultura prehispánica en Casabindo (1980–1993). In: Korstanje MA, Quesada M (eds) Arqueología de la Agricultura. Casos de estudio en la región andina argentina. Magna, Santiago del Estero, pp 12–47

Alejandra Lambaré D, Pochettino ML (2012) Diversidad local y prácticas agrícolas asociadas al cultivo tradicional de duraznos, *Prunus persica* (Rosaceae), en el Noroeste de Argentina. Darwiniana 50:174–186

Álvarez M, Sammartino G (2009) Empanadas, tamales y carpaccio de llama. Patrimonio alimentario y turismo en la Quebrada de Humahuaca-Argentina. Estudios Perspect Tur 18:161–175

Báez JR (1947) La primera colonia agrohispana en el Tucumán (Siglo XVI). Rev Argent Agron 14:85–93

Báez JR (1949) Breves apuntes sobre la migración de las plantas agrícolas Euro-Indianas en el momento del descubrimiento. Lilloa 18:311–360

Burger P, Terral J-F, Ruas M-P, Ivorra S, Picq S (2011) Assessing past agrobiodiversity of *Prunus avium* L. (Rosaceae): a morphometric approach focused on the stones from the archaeological site Hôtel-Dieu (16th Century, Tours, France). Veg Hist Archaeobot 20:447–458

Cabrera AL (1976) Regiones fitogeográficas argentinas. In: Kugler WF (ed) Enciclopedia argentina de agricultura y jardinería, 2nd ed, vol 2(1). Acme, Buenos Aires, pp 1–85

Capparelli A, Giovannetti M, Lema V (2007) Primera evidencia arqueológica de cultivos europeos (trigo, cebada y durazno) y de semillas de algodón en el NOA: su significación a través del registro arqueológico de El Shincal de Quimivil. V. In: Marconetto B, Oliszewsky N, Babot P (eds) Paleoetnobotánica del Cono Sur: Estudios de casos y propuestas metodológicas. Ferreyra, Córdoba, Argentina, pp 25–48

Capparelli A, Hilgert N, Ladio A, Lema V, Llano C, Molares S, Pochettino ML, Stampella P (2011) Paisajes culturales de Argentina: pasado y presente desde las perspectivas etnobotánicas y paleobotánicas. Rev Asoc Argent Ecol Paisaje 2(2):67–79

Capparelli A, Lema V, Giovannetti M, Raffino R (2005) The introduction of old world crops (wheat, barley and peach) in Andean Argentina during the 16th century A.D.: archaeobotanical and ethnohistorical evidence. Veg Hist Archaeobot 14:472–484

Capparelli A, Raffino RA (1997) Arqueoetnobotánica de El Shincal I: tallos finos, frutos y semillas. Tawantinsuyu 3:40–57

de la Puente y Olea M (1900) Los trabajos geográficos de la Casa de la Contratación. Escuela Tipográfica y Librería Salesianas, Sevilla

Depypere L, Chaerle P, Mijnsbrugge KV, Goetghebeur P (2007) Stony endocarp dimension and shape variation in *Prunus* section *Prunus*. Ann Bot 100:1585–1597

Farberman J, Boixadós R (2006) Sociedades indígenas y encomienda en el Tucumán Colonial. Un análisis comparado de la visita de Luján de Vargas. Rev Indias 66:601–628

Font Quer P (1988) Plantas medicinales, el Dioscórides renovado. Labor, Barcelona

Franco Salvi V, López ML, Molar RM (2014) Microrrestos vegetales en campos de cultivo del primer milenio de la era en el Valle de Tafí (Prov. de Tucumán, República Argentina). Arqueol Iberoam 21:5–22

Giovannetti M (2005) La conquista del noroeste argentino y los cultivos europeos. Front Hist 10:253–283

Giovannetti M, Lema V (2005) Cultivos introducidos por los europeos en el Shinkal de Quimivil: La presencia de los hispanos en la supervivencia ritual. In: Cetti A, Re A, Rindel D, Valeri P (eds) Entre pasado y presente. Buenos Aires, Instituto Nacional de Antropología y Pensamiento Latinoamericano, pp 410–429

González NM (2009) Producción agrícola Prehispánica en el Alfarcito, Departamento de Tilcara. Undergraduate Dissertation, Facultad de Humanidades y Ciencias Sociales, Universidad Nacional de Jujuy, San Salvador de Jujuy, Argentina

Guglielmino MM (2007) La difusión del patrimonio. Actualización y debate. Rev Electrón Patrim Hist 1. https://www.revistadepatrimonio.es/revistas/numero1/difusion/estudios/articulo.php. Accessed 5 Aug 2016

Hilgert N (2000) Especies vegetales empleadas en la insalivación de hojas de "coca" (*Erythroxylum coca* var. *coca*, Erythroxylaceae). Darwiniana 38:241–252

Hilgert N (2001) Plants used in home medicine in the Zenta River basin, Northwest Argentina. J Ethnopharmacol 76:11–34

Hilgert N (2004) Las plantas en las festividades religiosas de la selva andina argentina. Bull Soc Suisse Am 68:37–49

Hilgert N, Reyes S, Schmeda-Hirschmann G (2001) Alkaline substances used with coca (*Erythroxylum coca* var. *coca*, *Erythroxylaceae*) leaf insalivation in northwestern Argentina. Econ Bot 55:325–329

Hilgert N, Alejandra Lambaré D, Vignale ND, Stampella P, Pochettino ML (2014) Especies naturalizadas o antropizadas? Apropiación local y la construcción de saberes sobre los frutales introducidos en época histórica en el norte de Argentina. Rev Biodivers Neotrop 4:69–87

Hurrell JA, Ulibarri E, Delucchi G, Pochettino ML (2010) Frutas frescas, secas y preservadas. LOLA, Buenos Aires

Instituto Nacional de Estadísticas y Censos (INDEC) (2010) Censo Nacional de Población, Vivienda y Hogares de 2010. Buenos Aires

Lanzelotti S (2011) ndicadores para el reconocimiento de represas arqueológicas. Relac Soc Argent Antropol 36:177–197

Lema V (2009a) Criterios de selección en los procesos de manipulación vegetal: el aporte de la etnobotánica a la interpretación de restos arqueológicos de *Curcubita* sp. Darwiniana 47:35–55

Lema V (2009b) Domesticación vegetal y grado de dependencia ser humano- planta en el desarrollo cultural Prehispánico del Noroeste argentino. Facultad de Ciencias Naturales y Museo, Universidad Nacional de La Plata, Tesis doctoral

Lema V, Pochettino ML (2012) Cambio y continuidad al plato: los saberes culinarios y su rol en la dinámica de la diversidad biocultural. In: del Pilar Babot M, Marschoff M, Pazzarelli F (eds) Las manos en la masa. Arqueologías, Antropologías e Historias de la Alimentación en Suramérica. Universidad Nacional de Córdoba, Córdoba, Argentina, pp 25–45

Levillier R (1939) Historia de la Nación Argentina desde los orígenes hasta la organización definitiva en 1862, vol 2. El Ateneo, Buenos Aires

López de Gómara F (1922) [1552] Historia general de las Indias, vol 2. Calpe, Madrid

Lorandi AM (1984) Pleito de Juan Ochoa de Zárate por la posesión de los indios ocloyas. Un caso de verticalidad étnica o un relicto de archipiélago estatal? Runa 14:132–142

Manfred L (1947) Siete mil recetas botánicas a base de mil trescientas plantas medicinales. Kier, Buenos Aires

Molares S (2010) Flora medicinal aromática de la Patagonia: características anatómicas y propiedades organolépticas utilizadas en el reconocimiento por parte de la terapéutica popular. PhD Dissertation, Universidad Nacional del Comahue, San Carlos de Bariloche, Argentina

Podjajcer A, Mennelli Y (2009) La mamita y pachamama en las performances de carnaval y la fiesta de nuestra señora de la candelaria en Puno y en Humahuaca. Cuadernos de la Facultad de Humanidades y Ciencias Sociales Universidad Nacional de Jujuy 36:67–90

Proyecto Cultivos Andinos (2007) La quebrada en cifras. Estudio cuantitativo marco del perfil económico- productivo de la Quebrada de Humahuaca. Fundandes (Colección Documentos EcoComerciales), San Salvador de Jujuy, Argentina

Raffino RA (1996) Integración sur andina en tiempos del Tawantinsuyu. In: Albó J, Arratia MI, Hidalgo J, Núñez L, Llagostera A, Remy MI, Revesz B (eds) La integración surandina. Cinco siglos después, Cusco, Perú. Centro Bartolomé de Las Casas, Universidad Católica del Norte, Taller de Estudios Andinos, Cuzco, Perú, pp 63–80

Raffino RA (1999) Las tierras altas del Noroeste. In: Nueva Historia de la Nación Argentina, vol 1, Academia Nacional de la Historia. Planeta, Buenos Aires, pp 83–108

Remedi FJ (2002) El consumo alimentario como realidad plural. Similitudes, divergencias, cambios y permanencias dietarias en la Córdoba de entre siglos. In: Mariani V (ed) La cocina como patrimonio (in)tangible. Temas de Patrimonio 6. Comisión para la Preservación del Patrimonio Histórico Cultural de la Ciudad de Buenos Aires, Buenos Aires, pp 75–86

Rivera-Núñez D, de Castro CO, Ruiz SR, Fernández CS, Colmenero FM, López AV, Trigueros FC (1997) Las variedades tradicionales de frutales de la Cuenca del Río Segura. Catálogo Etnobotánico (1): Frutos secos, oleaginosos, frutales de hueso, almendros y frutales de pepita. Universidad de Murcia, Murcia, España

Ruhl DL (1990) Spanish mission paleoethnobotany: an overview and some speculations for the 16th and 17th century La Florida. In: Thomas DH (ed) Columbian consequences, archaeological and historical perspectives on the Spanish Borderlands East, vol 2. Smithsonian Institute, Washington, DC, pp 560–580

Ruiz M (2004) Articulación del territorio en el altiplano jujeño. Argentina. Período de Desarrollos Regionales (1000–1430 d.C). Magister Dissertation, Facultad Latinoamericana de Ciencias Sociales sede Ecuador, Colegio Andino, Centro de Estudios Regionales Andinos Bartolomé de las Casas, Cuzco, Perú

Sadori L, Allevato E, Bosi G, Caneva G, Castiglioni E, Celant A, Di Pasquale G, Giardini M, Mazzanti M, Rinaldi R, Rottoli M, Susanna F, Francesca (2009) The introduction and diffusion of peach in ancient Italy. In: Morel J-P, Mercuri AM (eds) Plants and culture: seeds of the cultural heritage of Europe. Edipuglia, Ravello, Italy, pp 45–61

Schulte M (1996) Tecnología agrícola altoandina. El manejo de la diversidad ecológica en el Valle de Charazani, Plural Editores- CID, La Paz

Scorza R, Sherman WB (1996) Peaches. In: Janick J, Moore JN (eds) Methods in fruit breeding. Purdue University Press, West Lafayette, pp 66–73

Sica G (2005) Maíz y trigo; molinos y conanas; mulas y llamas. Tierras, cambio agrario, participación mercantil indígena en los inicios del sistema colonial. Jujuy. Siglo XVII. In: Santamaría D (ed) Jujuy, arqueología, historia, economía y sociedad. CEIC-Ediciones El Duende, San Salvador de Jujuy, Argentina, pp 106–124

Sica G (2006) Del pucará al pueblo de indios. El proceso de construcción de la sociedad indígena colonial en la jurisdicción de San Salvador de Jujuy (Noroeste de Argentina), siglo XVII. Ph Dissertation, Universidad de Sevilla, Sevilla

Sica G (2010) Del tráfico caravanero a la arriería colonial indígena en Jujuy. Siglos XVII y XVIII. Rev Trans Territ 3:23–39

Sica G, Bovi MT, Mallagray L (2010) La quebrada de Humahuaca: de la colonia a la actualidad. In: Teruel A, Lagos M (eds) Jujuy en la historia de la colonia al siglo XX. Universidad Nacional de Jujuy, San Salvador de Jujuy, Argentina, pp 351–372

Sica G, Ulloa M (2010) Jujuy en la colonia. De la fundación de la ciudad a la crisis del orden colonial. In: Teruel A, Lagos M (eds) Jujuy en la historia de la colonia al siglo XX. Universidad Nacional de Jujuy, San Salvador de Jujuy, Argentina, pp 41–85

Stampella P, Alejandra Lambaré D, Hilgert N, Pochettino ML (2013) What the Iberic conquest bequeathed to us: the fruit trees introduced in argentine subtropic, their history and importance in present traditional medicine. Evid Based Complement Altern Med 2013. Article ID 868394. https://dx.doi.org/10.1155/2013/868394. Accessed 5 Aug 2016

Vitry C (2007) La ruta de Diego de Almagro en el territorio argentino: un aporte desde la perspectiva de los caminos prehispánicos. Rev Esc Hist 6(1). https://www.unsa.edu.ar/histocat/revista/revista0615.htm. Accessed 5 Aug 2016

Chapter 15
Commercializing the "Lost Crop of the Inca": Quinoa and the Politics of Agrobiodiversity in "Traditional" Crop Commercialization

Emma McDonell

Abstract In the past decade, quinoa transformed from a traditional "peasant food," produced and consumed almost exclusively in the Andes, into a global consumer "superfood." Renowned for its exceptional nutritional profile, the "lost crop of the Inca" is now a staple in the diets of health-conscious consumers in the US and Europe. In Peru, now the world's leading producer, quinoa exports increased 12-fold between 2009 and 2014, and thousands of smallholder agriculturalists converted subsistence farms into export-oriented quinoa operations. Quinoa's remarkable success in global markets has made quinoa a poster child for multicultural development's project to integrate cultural difference into paths to "progress" and "development." Yet quinoa's commercialization requires a radical reconfiguration of socio-natural landscapes of production, as well as patterns of consumption. Drawing on ethnographic fieldwork with quinoa producers and development practitioners in the Puno region of Peru, this chapter examines how quinoa producers and development practitioners negotiate local production systems and corresponding regimes of value and knowledge that emphasize intra- and inter-species diversity and logic of capital that emphasizes homogeneity and standardization. The quinoa boom is a case study to point to broader paradoxes of neoliberal multiculturalism and the politics of commercialization traditional Andean foods.

Introduction

In the past decades, quinoa has transformed from a local "Indian food," produced and consumed almost exclusively in the Andean highlands, into a global "superfood." High-end supermarket shelves across North America, Europe, and Asia are now lined with a rapidly expanding array of new products featuring the "lost crop of the Inca," (National Research Council 1989) while quinoa (*Chenopodium quinoa*) increasingly appears on menus at some of the most exclusive restaurants in the world.

E. McDonell (✉)
Anthropology, University of Tennessee, Chattanooga, TN, USA
e-mail: emma-mcdonell@utc.edu

J. E. Staller (ed.), *Andean Foodways*, The Latin American Studies Book Series,
https://doi.org/10.1007/978-3-030-51629-1_15

In Peru, the world's leading quinoa producer,[1] exports increased 12-fold and prices quadrupled between 2009 and 2013 (MINAGRI 2014) (Fig. 15.1). As thousands of highland subsistence agriculturalists took up market-oriented quinoa production, quinoa commercialization became a model of multicultural development to refashion

Fig. 15.1 "Easy Quinoa Mushroom and Vegetable Medley"—one of many new products in the US supermarkets featuring quinoa. Photo credit: Emma McDonell 2015

indigenous identity for "modernity." The United Nations declared 2013 the International Year of Quinoa, as the Peruvian state, along with local and international NGOs and development organizations, framed the quinoa boom as a long-awaited opportunity to transform the agricultural practices of highland subsistence *farmers*, long lamented as "culturally backward" and "economically unproductive," into "modern" capital-accruing *producers*, i.e., large-scale monocrop cultivation, which would contribute to the export-focused economic developmental model. Since the boom, "improved" quinoa varieties are able to thrive outside the Andean highlands. Quinoa production produces far greater yields along the Peruvian coast as well as in Asia, Europe, and North America. As global production mushroomed, prices plummeted. The farm gate price by 2015 was less than half what it was in the previous year, and prices are predicted to continue to fall in the foreseeable future. Smallholder farming communities that benefited at the boom's outset, now struggle to maintain a stake in the growing global quinoa market.

This research is focused upon the quinoa boom–bust as a case study to examine the politics of agrobiodiversity in "traditional"[2] crop commercialization, and more broadly, to identify some of the prospects and challenges of commercializing traditional foods as rural "multicultural" development. Biodiversity is defined as the variety of genetically distinct populations and species of plants, animals, and microorganisms. *Agro*biodiversity is the subset of biodiversity that feeds and nurtures humans, including genetic resources for food and agriculture such as harvested crops, livestock breeds, fish species, which are thus managed by humans more explicitly and intensively than other living organisms (Ehrlich and Ehrlich 1992; Kameri-Mbote and Cullet 1999; Shand 1997). Agrobiodiversity is generally divided into inter-specific diversity, meaning diversity among species, and intra-specific diversity, which is diversity within a single species, which is the primary focus of this analysis.

While scholarship on the politics of agrobiodiversity has examined conservation politics (Brush 2013; Zimmerer 1996), issues of "bioprospecting" and "biopiracy" (Shiva 1997; Aoki 1998; Kloppenburg 1988) and intellectual property rights conflicts in relation to genetically modified organisms (GMOs) and the pharmaceutical industry (Hayden 2003; Mugabe 1996; Schiebinger 2004; Soto Lavaega 2009; Sunder Rajan 2006), the politics of agrobiodiversity in traditional food commercialization projects has received comparatively little attention in social science literature. As consumer demands for traditional crops, prominent examples of which include quinoa, goji berries, acai berries (Brondizio 2008), river algae (Van Esterik 2006), and chia seeds, surge and the commercialization of traditional crops is increasingly espoused as a model for a "multicultural" rural development that respects cultural difference.[3] It is critical that we understand the tradeoffs of this strategy for the agriculturalists involved in nascent traditional food markets, particularly the small farmers such projects purport to help. This study addresses the institutional origins of some of the drawbacks of this rural development strategy.

While numerous thematic entry points exist from which to begin a critical analysis of traditional crop commercialization, including nutritional, labor, and economic dimensions, understanding how commercialization affects the traditional crop cultivation of farm-level agrobiodiversity is critically important. Moreover, commercialization conditioned by the dynamics of movement and control over plant genetic resources (PGRs) at the global level has dramatic impacts upon biodiversity among varieties. Exploring how global-level institutions governing genetic resources, local-level customs regarding the exchange of seeds, and a booming international market interact provides insight into the prospects and challenges of traditional crop commercialization for smallholder farmers. The institutional arrangements governing PGRs at different scales are key factors determining the ability or inability of local communities to earn sustained benefits from traditional crop commercialization and maintain control over the crop post-commercialization. The quinoa boom reveals that existing PGR governance institutions, and more broadly, relations of capital, enable short-lived benefits for smallholder communities producing a traditional crop, a result of farming communities' loss of control of the traditional PGRs of quinoa and other cultigens such as potatoes.

Focusing on the governance of quinoa seed exchange and seed provisioning in producer communities in the Peruvian highlands and the governance of quinoa germplasm at the global scale, I analyze two primary patterns or trends: (a) transformation of long-standing local-level seed management institutions that actively cultivate intra-specific biodiversity into a state-backed commercialized seed industry selling single varieties of intensively bred and cross-pollinated quinoa seeds and (b) expansion of quinoa production outside its native Andes as enabled by international exchanges of quinoa germplasm adapted to different environments and ecologies in new production zones.

These trends illustrate major challenges with traditional food commercialization as a rural development model. They demonstrate the difficulty of ensuring that the crop's original cultivators[4] maintain control over their traditional crop's PGRs and that they continue to benefit after a traditional crop "booms." Moreover, these trends indicate that farmers' vulnerability to climate and market shocks may increase with reductions in farm-level and species-level agrobiodiversity related to increasing cash crop production. The shift to purchased seeds alongside quinoa's introduction to new production zones results in highland farmers' loss of control over quinoa commercialization, which ultimately undermines their capacity to benefit from quinoa's market growth while increasing their dependence on purchased inputs and cash incomes. By providing insight into the politics of agrobiodiversity in quinoa commercialization, this paper seeks to inform the design of institutional formations that offer the traditional crop's original cultivators long-term benefits from its commercialization.

This analysis is based primarily on ethnographic fieldwork conducted in 2014 and 2015 during quinoa harvest season in the Puno region around the Titicaca Basin of Peru, the production hub of quinoa.[5] My ethnographic research included both quinoa farmers and buyers, primarily during quinoa buys. I also carried out semi-structured interviews with farmers, agronomists, quinoa buyers, quinoa processers, and policymakers at the United Nations Food and Agriculture Organization (FAO) office, the Peruvian Ministry of Agriculture, the National Agricultural Innovation Institute (INIA), and the Regional Agrarian Agency (DRA) in Puno, as well as a handful of NGOs. It is important to point out that, while this study could reasonably focus in a number of production areas, including parts of Bolivia that have also been highly affected by the quinoa boom, the general trends examined in this research occurred across highland quinoa production regions.

Prior to the recent geographical expansion of quinoa production, approximately 80% of the nation's quinoa was produced in the altiplano of Puno. As such, a study centering in Puno offers insight into the ways farmers who cultivated quinoa since pre-Columbian times are affected by its commercialization. Puno is located in the Southeastern corner of Peru, adjacent to Lake Titicaca. This region is considered the location where quinoa was originally cultivated and domesticated c. 3,500–5,000 years ago. Such assertions in the literature are supported by the high concentration of genetic diversity among quinoa varieties in this region as well as the presence of ancestral and wild varieties (Tapia 1979; Christensen et al. 2007; Bruno 2006; Eisentraut 1998; Bruno and Whitehead 2003). One of the most altitudinous regions in Peru, most of Puno's land area is located between 3000 and 5000 masl. The climate is exceptionally arid, and the extreme altitude creates extreme diurnal variation in temperature. As a result, the commodity crops that now dominate Peru's coastal agriculture such as rice and asparagus cannot be cultivated in such conditions, while hardy native crops adapted to these highlands such as quinoa, kañiwa, and potatoes thrive and remain central to the highland agricultural economy.

Puno is dominated by small-scale agriculture, and as of 2014, approximately 60% of the region was engaged in subsistence agriculture (for auto-consumption) (del Carpio and Yupa Vereau 2014). Likewise, this is one of the poorest regions in Peru. According to the Ministry of Agriculture, in 2014, 32.4% of its inhabitants lived in poverty (INEI 2014a, b). Within Peru's racialized spatial hierarchy, this part of the Andean highlands is generally seen as a culturally backward bastion of indigenous "tradition," a region seen as obstructing national "progress" (Zimmerer 2011; Rénique 2004).

The indigenous populations and communities are generally concentrated in the highlands. However, the cultural variability of racial and institutional histories of indigenous peoples in Peru has resulted in a lack of national-level indigenous movements. Most communities and individuals which might be classified as indigenous, do not identify as such due to its connotative link to the derogatory term, "Indian." Outsiders have come to define this region in terms of deficiency, be it environmental (e.g., "nutrient-poor" soils, lack of oxygen) or social (e.g., "under-developed"). Colonial accounts often invoke images of wild peoples and empty landscapes to justify their conquest and "civilizing" particularly with regard to the religious conversion of native Andean populations (see also Covey, Chapter 1, Martinez Borrero, Chapter 2, deFrance, Chapter 3 for similar analyses of Andean populations). Thus, frontiers have not only been characterized as lacking but are also coveted sites of potential (Li 2014: 13).

The quinoa boom was believed to have been the long-awaited opportunity to transform such "economically unproductive" regions into a capital-accruing "productive" landscape. It was believed to transform the unruly subsistence farmers into obedient cash crop producers. It is one of many agricultural modernization efforts made in the region in recent decades, but while past efforts focused on introducing profitable non-native crops, quinoa commercialization sought to make a native crop profitable. As the highland's harsh climate has, in a very tangible way, prevented highland agriculturalists from adopting commodity crops that dominate Peru's coastal agriculture, the opportunity to generate substantial income from a highland crop was an important and rare opportunity for highlanders.

Quinoa: Nutritional and Agronomic Aspects

Quinoa is among a small handful of crops that survive in the Andean Altiplano's extreme altitudes (quinoa is commonly cultivated above 3,900 masl) and is adapted to the highlands' saline soils, extreme cold, and periodic drought (Tapia 1983). With over 100 recognized varieties and 3000 "ecotypes," quinoa's intra-specific diversity is nothing short of extraordinary (MINAGRI ca. 2016). While quinoa diversity is highest near the Lake Titicaca where it was domesticated, quinoa varieties survive in diverse climates including coastal regions and lowland pampa landscapes. Quinoa's adaptability and resistance to severe climates and environmental conditions make it particularly important in the Andean highlands where harsh and fickle climates leave agriculturalists with few crop options.

While development experts and agronomists have long been enamored with quinoa's agronomic aspects (McDonell 2015), quinoa's nutritional dimensions have in large part driven quinoa's sudden popularity outside the Andes. High in fiber (compared to other grains), an excellent source of protein, and rich in many vitamins and minerals, particularly calcium, phosphorus, and iron, quinoa's nutrient profile is exceptional (Repo-Carrasco et al. 2003) (Fig. 15.2). Much of quinoa's nutritional acclaim in the United States (US), however, relates to its protein content. While the amount of protein in quinoa is only slightly higher than wheat and other grains', quinoa's protein quality distinguishes it from other grains. The proteins in quinoa mainly belong to albumin and globulin, which have a balanced composition of essential amino acids much like milk's protein, casein (Repo-Carrasco 1991). Likewise, quinoa is one of only a handful of plant-based foods considered a "complete protein," meaning it contains roughly equal proportions of all nine essential amino acids that the human body does not make. Some researchers argue that the consumption of quinoa has substituted for relatively low availability of animal protein in the highlands, and in many rural areas, quinoa is still the principal protein source (Tapia 1997; McDonell 2016).

A Brief History of Quinoa Consumption

Quinoa served as a culinary staple and cultural cornerstone for numerous pre-Columbian Andean ethnic populations, including the Inca. Quinoa played a central

Fig. 15.2 Multiple varieties of quinoa prior to harvest; Cusco, Peru 2015. Photo credit: Emma McDonell

role in dietary practices and ritual activities and was consumed throughout the Inca Empire. In fact, quinoa cultivation, prior to the recent expansion, maps onto the contours of the sixteenth-century Inca Empire (National Research Council 1989). During the periods of Spanish colonization, quinoa production and consumption was suppressed in part because of its importance to rituals and rites as well as a close association to ethnic identity among different populations.[6] While never successfully eradicated, quinoa came to be harshly stigmatized as a marker of indigenous identity (Tapia 2012).

In the Spanish strongholds of Lima and Arequipa came the culinary traditions promoted by the state in large part mimicked European tastes. Highlands Andean cuisines continued to revolve around native crops such as quinoa and potatoes. Likewise, quinoa remained a staple throughout much of the Andes even as "national cuisines" developed that mostly rejected indigenous cultigens and cuisines.

In the 1960s and 1970s, quinoa was exported to laboratories and seed banks across Europe, and by the following decade, quinoa was cultivated in experimental plots in the US. Collaborations between Andean universities and some in Western Europe facilitated its introduction throughout European countries and Scandinavia (Galwey and Risi 1984; Carmen 1984; Risi and Galwey 1991). However, the transformation of quinoa into a global commodity really began in the early 1980s. Two American entrepreneurs formed the Quinoa Corporation, an export business. They shipped Bolivian *Quinoa Real* (a variety unique for its large, white seeds grown exclusively in the Uyuni Desert) to health stores across the US (Thier 2010). Quinoa demand outside the Andes increased incrementally through the 1980s, and by 1988, 750 tons were sold in the US (Gonsalves 2010). As Bolivia's export market grew, and Peru's quinoa market remained domestic. In fact, Peru's National Food Program, PRONAA, was the largest buyer of Peruvian quinoa throughout the 1990s (Hellin and Higman 2003). Interestingly, Peru was the primary importer of Bolivian quinoa exports in the 1990s, much of which was illegally trafficked, making it unclear how much quinoa was actually produced in Peru in the 1990s.

US consumer's interest in the grain continued to increase and by the late 1990s, US annual imports reached 900 tons (MINCETUR 2013). Peru began exporting quinoa, though Bolivia still commanded 83% of worldwide exports in 2002. Peru's export capacity increased rapidly, increasing at an annual average of 47.8% between 2002 and 2013 (from 250 total tons in 2002 to 18,341 tons in 2013) (FAO 2013). In 2011, Peru's quinoa exports exceeded Bolivia's and between 2012 and 2013, Peru reported a 71% increase in quinoa exports, totaling $78.3 million, compared to 2012.[7] Peru now exports quinoa to 35 different markets (43% to the US; 6.4% to Canada, 5.9% to Australia) (FAOSTAT 2014; INEI 2014a, b; Laguna 2003).

From Cultivated Diversity to Enforced Homogeneity

Quinoa's conversion from a local highland staple into a globally traded commodity has required radical reconfigurations of socio-natural landscapes of quinoa production in the Andes and of quinoa's materiality itself. While traditional highland agricultural systems and their institutional formations have long encouraged, and indeed created, inter- and intra-species agrobiodiversity through human selection and cultivation (Mayer 2002; Zimmerer 1996), commodity production and its institutional apparatuses require specialization and homogenization. Likewise, quinoa's commercialization is in large part an effort to make the manifold and heterogeneous quinoas found throughout the Altiplano into a single *uniform* commodity quinoa. This institutional approach seeks to provide understanding to recent changes in quinoa's intraspecific diversity in the Peruvian highlands. The development of a commercial quinoa seed industry and the emergence of stringent quality standards for quinoa, both of which are critical aspects of quinoa's transformation around commodity logic, are explored. These shifts brought about a displacement of the locus of decision-making power (and thus control over) quinoa PGRs, from farming communities and individual agriculturalists to the Peruvian state, plant breeders, field techs working for NGOs or processors, and quinoa buyers. As indigenous farming communities lose control over quinoa's genetic resources they become consumers of seeds created in state and private laboratories and no longer are they the key decision-makers regarding quinoa PGR management.

As Karl Marx noted, "Animals and plants we are accustomed to consider as products of nature, may be, in their present form, not only products of, last year's labour, but the result of a gradual transformation continued through generations under human control, and through the agency of human labour" (Marx 1867:287–288). Agrobiodiversity then is not a "natural" resource that is "conserved" when humans stay away or "depleted" when humans use it; it is actively created and maintained by collective decisions of humans and institutional formations and thus dependent upon human actions. Quinoa's exceptional intra-specific diversity has been created and maintained by past and present agriculturalists and corresponding institutional regimes and agricultural practices that encourage species differentiation.

Prior to the quinoa boom, typical highland agriculturalists cultivated four or five varieties of quinoa each year, each of which had a unique purpose (i.e., use-value) to them. Some quinoa varieties are resilient to droughts, others to frost or pests and the per acre yields of some quinoas are triple that of others. It was the cultivation of multiple varieties of quinoa that provided a sure harvest in such extreme highland environments and the often unpredictable climate (Hernández Bermejo and León 1994; Mayer 2002; Zimmerer 1996, 2003). Taste characteristics and cooking uses also vary considerably, and farmers' selection of quinoas is influenced by culturally situated ideas about what is "tasteful" in the context of their regional cuisines. Moreover, some quinoas are deeply embedded in family histories or linked to distinct regions and ethnic groups, and their cultivation and consumption form a material and symbolic link between people and places, evoking sensory memories critical to collective identities (Sutton 2001; Allen 2002).[8] Quinoa's genetic diversity then is produced through practices deeply embedded in a regime of value (cf. Appadurai 1984) that prizes individual quinoas for diverse uses, and rewards diversity in general. These practices were embedded in institutional apparatuses intended to actively create and manage this diversity.

Seed-exchange and seed-saving practices foster new agrobiodiversity (Collins and Qualset 1998). The central institutional apparatus responsible for managing quinoa's exceptional intra-specific diversity in recent centuries is the *feria* system and individual farmer's active work to improve and experiment with their quinoa. *Ferias* are coordinated, but generally informal (i.e., not state-sanctioned) marketplaces where farmers bring farm products to exchange in public spaces (e.g., town plazas) (Larson 1995). *Ferias* take place across the Andean highlands and while most individual communities and towns have them weekly, some farmers travel longer distances to participate in distant *ferias* to acquire nonlocal goods and resources, including seeds not available in the region. *Ferias* are seed-exchange hubs where farmers exchange seeds they have been saving and improving in order to acquire new varieties (Salomon 1986). By saving, replanting, and exchanging seeds farmers create new genetic diversity (Collins and Qualset 1998). Since prior to the boom quinoa seeds had a very low market value, exchanges were primarily direct exchanges of seeds for seeds or "*treque*" in Quechua. Many scholars consider the *feria* system a remnant of the "altitudinal zonation" strategy practiced during Inca and pre-Inca times. As climactic and environmental conditions vary radically along with different locations on the Andes' steep altitude gradient, many communities controlled landholdings at different elevations along the "vertical archipelago" (Murra 1956, 1972) and engaged in sustained exchange relations with communities at different altitudes. Both the altitudinal zonation and the contemporary *feria* system allowed and encouraged the exchange of germplasm from different agroecological zones, and the constant creation and adaptation of new varieties.

In addition to seed exchanges, on-farm seed selection and seed-saving practices among quinoa farmers are critical to managing quinoa PGRs. This practice generally involves saving the best seeds of a given harvest for the subsequent year ("mass selection"), a practice that, over time, will change the variety itself (Zimmerer 1996). As previously mentioned, different varieties are created and maintained with diverse use values, such that farmers do not select for single ideal quinoa, but rather aim to maintain multiple quinoas with diverse characteristics. My ethnographic research with farmers in Puno indicates that farmers strategically cultivate quinoas with complementary characteristics. For instance, a farmer might cultivate a variety resistant to drought, another resistant to mildew (which occurs in wet years), and yet another that susceptible to adverse climate and pests, but nevertheless exhibiting high productivity and large grain size. While most farmers practice some seed-saving and selection, many communities in the Peruvian highlands have an established *conservacionista*, a special seed-saver farmer who cultivates up to 100 kinds of quinoa each year and conserves varieties not currently in use. The seed savers rarely sell quinoa for consumption and instead sell quinoa as seeds.[9] In conserving and maintaining a repository of genetic diversity, these highly esteemed figures provide an integral public good for the local community.

Institutional Changes with Quinoa Commercialization: Developing "Improved" Quinoas

When demand for quinoa began to surge in 2007, numerous institutional changes were enacted regarding quinoa PGR management and the procurement of seeds. In general, these changes sought to transform the hundreds of regionally differentiated quinoas into a single homogenous commodity. The state, non-government organizations (NGOs), private businesses, and development organizations such as FAO began to play important roles developing new systems of seed procurement, germplasm management, and metrics to judge its quality. These changes wrested control of genetic diversity and decision-making about what will be sold from farmers and put it increasingly in the hands of the state, plant breeders, and quinoa buyers. The quinoa boom resulted in the loss of control over the PGRs and undermined highland agriculturalists ability to benefit from its commercialization.

While a handful of plant breeders had been interested in quinoa for decades, the work of "improving" quinoa took on new import as quinoa's market value skyrocketed (McDonell 2015; Rojas et al. 2014; Tapia 2014). Breeders at Peruvian universities and the nation's agricultural innovation agency, *Institución Nacional de Innovación Agraria* (INIA) intensified efforts to create, procure, and disseminate "improved" quinoas, while international researchers began working towards developing new quinoa varieties adapted to different climates (Fig. 15.3). Breeders sought to create quinoas that exhibited higher productivity and to increase seed size, replacing the heteroglot quinoas of varying sizes, resistances, and productivity levels. This logic of uniformity was in direct opposition to highland agriculturalists who sought to foster diverse varieties. While quinoa PGRs were previously managed and procured through the described *feria* system and seed-saving practices by "users" of the resource (approximately 90% of agriculturalists in Puno used "native" seeds in 2009) when quinoa's market value surged, efforts to define and propagate "*semillas de calidad*" ("quality seeds")[10] intensified (Marca et al. 2009) and "native" seeds came to be seen by many as subpar remnants of the past.

As thousands of farmers who had been producing relatively stable amounts of quinoa attempted to increase quinoa production, a seed shortage ensued that made conditions ripe for the development of a commercialized quinoa seed market (Marca et al. 2009). The "seed shortage" helped catalyze a shift from saving and exchanging seeds to purchasing market-oriented seeds. INIA promoted their line of improved quinoas to farmers, which they made available for purchase at "experimental stations" in each region. To further disseminate these seeds, INIA sent pickup trucks around quinoa-producing regions selling quinoa seeds directly to farmers while NGOs and buyers also purchased INIA seeds to them sell to the farmers they worked with. At

Fig. 15.3 Hundreds of quinoa accessions in the Puno Germplasm Bank. Plant breeders use these accessions to develop commercial varieties of quinoa. Photo credit: McDonell (2014)

present, INIA sells 12 quinoas adapted to different regions of Peru. This selection includes all-white quinoas except two, and all but one was selected for large seed size. Most of these seeds are bred through crossing quinoas or mass selection. At present, there are no genetically modified (GM) quinoas available for sale, though a recent successful effort to map the quinoa genome indicates the GM quinoas are not far off (Murphy and Matanguihan 2015; Jarvis et al. 2017). INIA's "certified" seeds were considerably more expensive than seeds bought from other farmers in ferias, but as they had been specifically designed in accordance with exporter's preferences and were supposedly more uniform, they were seen to guarantee a higher likelihood of complying with buyers' standards. Likewise, many farmers, even those vocally opposed to the idea of a uniform, large-seed quinoa as "good quality" quinoa, replaced their diverse quinoa plots composed of native seeds with plots entirely composed of certified seeds, or at least reduced the variety of seeds they planted in favor of commercial varieties. They justified the expense of these seeds with the expected high monetary returns promised by the skyrocketing quinoa prices. Quality standards favoring large-seed, uniform quinoa thus served to enforce the uptake of the improved quinoas.

Unsurprisingly, the *feria* system as the social institution underpinning quinoa diversity is in decline. Very few farmers exchange seeds at ferias now and even the *conservacionista* farmers prefer to exchange seeds directly with each other instead of using *ferias* to locate new kinds of quinoa.

The promotion and dissemination of improved quinoas are parts of a broader effort to shift the Peruvian seed landscape away from the Altiplano and its informal seed markets based on exchanges between native farmers, toward a commercial seed industry in which farmers purchase inputs from private companies or, in this transitional moment, state entities. This effort was set in motion in 2000 by the passage of the General Seed Law (*Ley de Semillas* No. 27,262), which established standards for production, certification, and commercialization of *semillas de calidad*. While establishing INIA as the authority to regulate seeds and explicitly promoting the private sector as the ideal seed purveyor, creating a privatized seed industry. INIA now had the ability to sell seeds in the interim between informal and commercial seed procurement landscapes to ease the transition to a commercial seed regime. While the commercial seed industry had taken hold along coastal Peru, which is dominated by industrial plantation-style agriculture, the highlands have remained dominated by seed exchange and saving. This push toward certified seeds and private sector dissemination sought to change that, and the quinoa boom came to be seen as an inroad for the industrial commercial seed expansion into the highlands. In 2012, the Supreme Decree *S No. 26-2008-AG* and *Resolución Jefatural No. 0014-2012-INIA* set specific protocol for the production, certification, and commercialization of quinoa seeds, and formalized a partnership with the FAO's *Semillas Andinas* project. Their goal was to improve the dissemination of Andean cultivars through the privatization of the seed procurement. While INIA (i.e., a state agency) is the main provider of certified seeds at the moment (57% of the total), these institutional shifts are intended to be transitional and ultimately stimulate the development of a private sector commercial seed industry. By the 1990s, the move towards a commercialized seed industry was supported by changing rules governing PGRs at the national level.

While GMOs are often designed so farmers cannot reproduce them, hybrid seeds exhibit a severe reduction in productivity in subsequent generations. In general, the main barrier to the development of commercial seeds is the fact that there is no guarantee that farmers will not reproduce the seeds on the farm (and thus farmers do not need to keep buying seeds in subsequent years) (Kloppenburg 1988; Muller 2006). As obligated by the Free Trade Agreement with the US, Peru signed onto the 1991 version of the Union for the Protection of New Varieties of Plants (UPOV) that protects "plant breeder's rights" in an effort to incentivize breeders to create new seeds by ensuring they profit from their efforts.[11] Within this system, plant breeders effectively receive weak patents for the varieties they create, prohibiting others from reproducing the seeds (Bosselmann 1996). The 1991 version of the UPOV limited the "farmer's exemption" that had previously made small-scale farmers exempt from these rules, and while the farmer's exemption was reinstated in the language of "farmer's rights" during the establishment of a new plant breeder's rights regime in 2003, the trend in national-level legislation supported establishing a commercial seed industry (Muller 2006).

From a Marxian perspective, seeds are both a product and a means of (re)production for farmers (Kloppenburg 1988; Marx 1867), and likewise, a shift from farmers creating and reproducing seeds, i.e., owning them, to purchasing seeds from a company (or the state). The state took over the means of production from farmers, a move critical to establish capitalist relations of production. While quinoa farmers were not forced to use certified seeds, quinoa buyers increasingly favored the homogenous grains certified seeds yielded. Thus, over time farmers who wanted to participate in quinoa production for commercial markets have increasingly replaced native seeds with certified seeds to meet quality standards.

Quality Standards and Quinoa's "Purification"

While the rise of commercial quinoa seeds has been critical to quinoa's homogenization, enabling farmers in distant regions to have access to identical seeds, the creation and implementation of quality control standards have perhaps been an equally impactful institutional change in affecting farm-level agrobiodiversity. In 2009, the Peruvian government instituted national-level regulations that defined classes of quinoas based on their "purity." Each class is specified by the percentage of "impurities"[12] allowed such that "quality" is synonymous with "purity." This understanding of quality is directly at odds with highland agriculturalists' valorization of difference and strategic cultivation of variety in quinoa seeds. Environmental historian William Cronon's (1991) investigation of the nineteenth-century commodification of wheat in the American Midwest provides a template for the study of quality standards in grain commodification. Cronon draws attention to the centrality of classification systems, standards, and techniques of inspection in the commodification of grain crops specifically, and the ways "grades" allow for the mixing of different grains into a single "homogenous" commodity. Critically, Cronon argues new forms of measurement and classification complete a "process of the alienation" of the wheat from farmers. They erase any connection between a sack of grain and its conditions of production. Previously, a farmer brought grain directly from farm to market, and their product reflected the unique labor and ecological conditions of the farm (Fig. 15.4). The creation of grain classes allowed for the mixing of all farms' grain together, thereby erasing each farm and farmer's distinctive characteristics. As standards became more stringent and certain categories received better prices than others did, farmers tailored their production to the standards and homogenized their grain. This dynamic of rendering the uniqueness of the labor and cultivation irrelevant underscores how grades and attendant standards and inspection practices, legal shifts in ownership, institutional innovations (e.g., the Chicago Grain Board) affects crop diversity. In a similar fashion, the quinoa quality standards used to ensure "quality" is largely about assisting buyers to guarantee that quinoa from disparate farms is commensurable and mixable—in other words, ensuring the uniformity the commodity form entails.

These quality standards are primarily enforced at the point of sale at the farm and in negotiations between local quinoa processors and international quinoa buyers. Ethnographic fieldwork with buyers and farmers during harvest season in Puno revealed that at times, an entire harvest was rejected, because the quinoa exhibited too many impurities, such as small grains or grains of contrasting colors. Buyers recommended

Fig. 15.4 Quinoa harvested, on the *panoja* awaiting processing; Puno, Peru. Photo credit: McDonell 2015

changes to production practices and buying certified seeds as a solution for farmers whose quinoa was rejected. Consequently, many farmers radically changed their production practices in order to meet these standards. While no highland farmers converted their farm into a monoculture, which is all but impossible given the Altiplano's fragile soils and a culture of diversity as risk management, many farmers replaced space previously dedicated to other crops with quinoa destined for markets (Soto et al. 2012).

Reduction in species diversity brings to the fore questions of farmers' vulnerability, both to climate and market shocks. While buyers favored the improved quinoas, many farmers found them more susceptible to pests and frosts even while they promise larger harvests and more uniform grains in years with amenable weather. Most farmers believed a complete conversion to improved quinoas would increase the likelihood of losing an entire harvest to pests or adverse climate, and all farmers had tales of losing large parts of their harvests to inclement weather even while practicing a diversified planting strategy. Previously, farmers were protected from such losing entire harvests through strategic selection of quinoas with varying resistance profiles and from market behavior by diverse crop portfolios, yet the intensification of quinoa production leaves farmers at the whim of fickle boom–bust market dynamics and an already mercurial climate exacerbated by climate change.

The push toward commercial seed procurement and the ascendancy of quality standards based on measures of uniformity, encourage reductions in farm-level intraspecific diversity, are not unique to the quinoa commercialization process. Instead, institutional shifts that enable homogenization and a broad remaking of a traditional crop through logic of capital are likely to be general trends in commercializing traditional crops, as is the attendant reduction in farm-level intra-specific diversity of commercialized traditional crops. The rise of commercial seed procurement moves the locus of decision-making power regarding and management of agrobiodiversity from farming communities to seed corporations and the technocrats at national agricultural decision-making bodies. The dominance of commercial quality standards enforces the use of commercial seeds by mandating homogeneity for commercial grade quinoa while disciplining farm practices.

Quinoa Germplasm Governance and the Quinoa Bust

As highland quinoa farmers increasingly purchase seeds to remain competitive in increasingly crowded markets, quinoa production was expanding into new regions. The movement of its germplasm out of the Andes and into US and European scientific laboratories and new production zones has intensified since an increase in the market value of quinoa. National and international institutional agreements and informal institutional affiliations enabled this movement, and international agreements regarding PGRs further inhibit rural farming communities from maintaining control over PGRs. By revealing a direct linkage between the quinoa bust (or price drop to production glut) and institutional arrangements regarding agrobiodiversity, this research suggests that agrobiodiversity governance institutions may play a central role in ensuring (or preventing) that a traditional crop's original cultivators (i.e., cultivators prior to commercialization) accrue sustained benefits once a traditional crop is commercialized.

Prior to the mid-twentieth century, quinoa was mostly unknown outside the Andes. Yet, when European scientists "rediscovered" it in the mid-twentieth century, it quickly made its way into a handful of scientific laboratories and germplasm banks across Europe and the US. Researchers were intrigued by its unique agronomic characteristics and nutritional content. They sought to create varieties that could be cultivated outside the Andes. Initially, quinoa traveled outside the Andes via informal collaborations between scientists, a common and legal practice prior to the development of more stringent international biodiversity governance treaties in the late 1980s and 1990s set limits on such exchanges. The first international treaty on PGR governance, the Union for the Protection of New Varieties of Plants (UPOV), initiated by industrialized countries was in 1961, framed PGRs as part of the "heritage of mankind." This framing supported the free exchange of PGR between nations and researchers based on a "global commons" approach (Halewood et al. 2012), implying that resources should be freely available and accessible to all and hence outside the reach of intellectual property rights (Bilder 1980; Rosendal 2006). As Shiva (1997) points out, throughout much of the twentieth century, genetic resources from the Global South were considered "heritage of mankind" and thus open-access resources. In contrast, the "improved" varieties mostly developed in industrialized nations from these "raw materials" are considered private goods and are sold back to farmers in the Global South for profit. More recently, the World Trade Organization's Trade-Related Aspects of Intellectual Property Agreement (TRIPS) introduced the notion of national sovereignty to legal frameworks managing PGRs. This agreement asserts that individual nation-states should oversee and control their nation's genetic resources. Hayden (2003) maintains that these competing ideas about who "owns" PGRs is about a politics of defining a "public," and national sovereignty does not necessarily guarantee a better outcome for farming communities that free global exchange. While PGRs are increasingly regarded as public resources that should be freely exchangeable, intellectual property rights regimes now protect varieties developed by breeders through plant breeder's rights and patents for genetically modified varieties (Kameri-Mbote and Cullet 1999).

During the 1960s and 1970s, quinoa was disseminated to laboratories and seed banks across Europe. By the mid-1970s, Chilean quinoa was cultivated in experimental plots in Colorado. A collaboration between the *Universidad de Concepción* in Chile and Cambridge University facilitated quinoa's introduction to England and France. Galwey and Risi collected quinoas from across the Andes, which came to form the basis of Cambridge's collection, a collection that was subsequently sent to Finland, Denmark, Holland and other European countries for further experimentation (Galwey and Risi 1984; Carmen 1984; Risi and Galwey 1991). At present, quinoa is in seed banks across the world: 16,422 quinoa accessions are conserved in 30 countries in 59 germplasm banks; 88% of these accessions are in Andean countries (Rojas et al. 2014).

By the early 1990s, breeders in Denmark and Holland had created new quinoa varieties: a variety designed for European climates named "Carmen," and a variety with a thinner saponin layer, and thus more amenable to commercialization, named "Atlas." In 1993, the EU organized the "European Quinoa Trial" which funded field trials of various quinoa varieties in England, Denmark and Italy. In 1996, collaboration between the International Potato Center (CIP) in Peru, the Danish International Development Agency (DANIDA), the FAO, and the *Universidad Nacional del Altiplano-Puno* led to field trials of quinoa in many new countries including Switzerland, Poland, Austria, Germany, Namibia, Australia, and Pakistan (Mujica et al. 1998; Mujica Sanchez et al. 2001).

As consumer demand for quinoa outside the Andes remained stagnant during the 1990s, most of these field trials remained simply trials and uptake of quinoa by farmers outside the Andes was rare. Yet the price began surging in the early 2000s, farmers around the world became interested in commercial quinoa production. New quinoas adapted to different latitudes and climates were released, rendering quinoa no longer exclusive to the Andes. The small-scale commercial quinoa production in Colorado that began in the mid-1980s increased as did production in other parts of the US. Canadian farmers in Saskatchewan and Ontario implemented commercial quinoa production, and by 2013, an estimated 10% of global quinoa production took place in Canada (Quinoa Italia 2015; Bazile and Baudron 2013). The countries growing quinoa increased from 8 in 1980, to 40 in 2010, and to 56 in 2014. Another 20 countries introduced it for the first time in 2015 (Bazile et al. 2016).

Peruvian industrial-scale agriculture along the coast began replacing rice and other commodity crops with monocultures of "improved" quinoas starting in 2014. Large-scale farms in Arequipa reported success with two quinoa crops a year in the same field, with yields of over 6,000 kg per hectare, a feat unattainable in the Altiplano climate and soil conditions. The FAO deemed 2013 the "International Year of Quinoa" in an effort to expand quinoa production globally in the name of global food security, a project that further accelerated the introduction of quinoa into new production zones (FAO 2013; McDonell 2015).

The "improved" quinoas adapted to these regions were also designed to exhibit commercial characteristics, such as high productivity, large grain size, and thin saponin layers, characteristics often lacking in highland quinoas. Early in the quinoa boom, highland smallholders had the competitive advantage of being the only farmers with experience in quinoa production and established quinoa infrastructure. Yet, as quinoa made its way into industrial farming operations with greater "economic efficiency" and into less harsh climates where quinoa exhibits double and triple the yields it exhibits in the Altiplano, local farmers began losing their competitive advantage and their stake in the growing global quinoa market.

Quinoa prices dropped abruptly in 2014 due in part to widespread production along the coast and in other emerging production regions. Peruvian quinoa exporters refocused their operations away from the highland farmers who produced small quantities of quinoa and toward the large and economically "efficient" coastal quinoa operations. Highland smallholders struggled to compete in the market they had established. By early 2015, highland farmers' hold on the market was eroding as quinoa production outside the highlands increased and prices dropped.

Conclusion

The politics of agrobiodiversity in quinoa commercialization was presented as a case study to identify broader issues with traditional crop commercialization to rural development. By examining the ways PGR governance and seed procurement institutions in smallholder quinoa producing communities change with quinoa's commercialization and the movement of quinoa PGR at the global level with the onset of the quinoa boom, this research shows that quinoa producers are losing control of the management of PGRs and seeds. As a result, their ability to benefit (in this case profit) from commercializing a traditional food appears to be short-lived. As farmers are pressured to purchase "improved" seeds to comply with increasingly stringent quality standards and reduce farm-level agrobiodiversity and intra- and inter-species levels to cater to markets, they sacrifice control over their means of (re)production and their ability to make decisions about managing their crops. As international agrobiodiversity governance institutions do not provide producers with mechanisms to control the PGRs of traditional foods, quinoa production was quickly adopted by industrial-scale agricultural producers, both in traditional quinoa-producing nations and worldwide. With increasing production outside the Andes and the corresponding price drop, Peru's quinoa exports have fallen 24% so far in 2015 (Quinua.pe 2015). At present, institutions do not exist to guarantee (or even increase the likelihood) that original producers of a traditional crop will continue to benefit once the crop is commercialized.

As a hub of global agrobiodiversity, the Andean highlands are full of crops with commercialization potential. Meanwhile, the "gastronomic boom" in Peru, and to a lesser extent in Bolivia, is actively creating markets for previously uncommercialized or locally traded crops through the appropriation of indigenous foods into haute cuisine and the exportation of this cuisine to upscale Peruvian (i.e., *Novo Andino*) restaurants across the globe (García 2013). While the gastronomic boom and the attendant the commercialization of Andean traditional foods have been pitched as a rural development opportunity,[13] the results of quinoa's commercialization for smallholder quinoa farmers in the Peruvian highlands brings into question the impacts of traditional crop commercialization on rural development. In fact, many buyers point to complaints of chefs and agroindustrial engineers about the different cooking times and gastronomic characteristics as a key reason they support the homogenization of Puno's quinoa.

In the past year, Puno's quinoa producers have embarked on a project to acquire a collective brand (*marca colectiva*) for their region's quinoa, trademark-like protection that would protect quinoa from Puno as a distinct product that could only be produced in the region and that is considered the intellectual property of the region's agriculturalists. The project seeks to revive Puno's struggling quinoa export industry through establishing its quinoa as the "most traditional" quinoa and creating a niche market for their quinoa in consumer markets. While the *Quinua Puneño* GI project is in its early stages, the GI framework seems to be a possible mechanism to ensure the original producers of a traditional crop continue to benefit once the crop is commercialized. While such a measure may not be sufficient to ensure that traditional producers benefit from traditional crop commercialization, yet such counter-institutions may help combat some of the loss of control of a traditional crop evidenced in the quinoa boom.

End Notes

1. Peru only recently surpassed Bolivia in quinoa production. Up until 2014, Bolivia produced the most quinoa.
2. I put "traditional" in scare quotes to highlight the construction and relational nature of what counts as "traditional." As Eric Hobsbawn (1983) has argued, traditions are invented— "traditional" foods do not exist but must be actively deemed traditional and constructed as such. I do this to highlight the work that goes into ensuring quinoa's traditional nature.
3. Commercializing traditional foods is often situated in a discourse of multicultural development that celebrates cultural difference and sees utility in cultural difference for achieving development goals, an approach that contrasts with the vast majority of development programs that have envisioned "culture" as a barrier to "progress." Commercializing traditional crops is also

often discussed in agricultural development literature through concepts of "neglected and under-utilized species" (NUS) such that under-utilization is code for under-commercialization.

4. By "original cultivators," I refer to those who cultivated a crop prior to a commercialization effort, and thus those who commercialization efforts are generally ostensibly intended to benefit.
5. It's worth noting that Puno's status as the heart of quinoa production is threatened with surging production in regions that previously did not cultivate quinoa. While Puno commanded approximately 80% of Peru's quinoa production in 2010, by 2013, it was only producing 56% of the nation's quinoa as other regions, many with higher productivity due to lower altitudes, increased productive capacity.
6. While many scholarly and popular sources alike claim that the Spanish prohibited quinoa production and consumption (Cusack 1984; Murphy and Matanguihan 2015; Broadening Boundaries in Agriculture: Impact on Health, Habitat and Hunger 2002; Native Crops of Peru—Quinoa ca. 2016), I did not find any original colonial source that verifies this claim, and they did not cite original sources. While it does seem likely that quinoa was banned due to its significance to Inca ethnic identity, it is clear that the Spanish colonizers' did employ culinary colonization: and imposed a culinary regimen based on cereals and sought to displace and devalorize native foods like quinoa (Bazile et al. 2013). The disparagement of quinoa production and consumption little impact on quinoa production and consumption as it came to be a highly stigmatized food. Moreover, the introduction of barley, wheat, oats, beans, and peas, all of which survive in high altitudes.
7. It's important to note that these figures are highly contested and many Bolivians would dispute Peru's claim to be the chief quinoa exporter.
8. Anthropologist Catherine Allen's (2002) work on coca leaf rituals in the Peruvian highlands has argued that in the Andes the communion between humans and non-human beings (e.g., Apus, Pachamama) is enacted in large part through consumption practices (both coca and food products). Likewise, quinoa consumption for many people is considered more than an affirmation of collective identity, but is also a way of establishing and attending to relationships with animating essences or the spiritual domain.
9. The grain of the quinoa that is consumed is also its seed, so the semelleristas are basically selling unprocessed quinoa of extra high quality to use as seed.
10. Semillas de calidad is a general term used to refer to seeds generally but not always improved through scientific breeding programs and that are procured and dispensed by certified distributors who control for uniformity and are evaluated by the National Agricultural Innovation Institute (INIA). The term is often intended as a contrast to "native seeds" which are exchanged through local seed networks and ferias.
11. While farmers are also involved in plant breeding activities, the term 'plant breeder' generally refers to persons or entities involved in commercial plant breeding activities.

12. In this definition, impurities include anything that detracts from a given quantity's uniformity such as grains of different varieties as indexed by difference in size, color, as well as broken grains.
13. Maria Elena García (2013) provides an acute critical analysis of the contradictions in the discourse surrounding Peru's gastronomic boom.

References

Allen C (2002) The hold life has: coca and cultural identity in an Andean community. Smithsonian Institution Press, Washington, DC

Aoki K (1998) Neocolonialism, anticommons property, and biopiracy in the (not-so-brave) new world order of international intellectual property protection. Indiana Journal of Global Legal Studies 6(1):11–58

Appadurai A (ed) (1984) The social life of things. Cambridge University Press, Cambridge

Bazile D, Baudron F (2013) The dynamics of the global expansion of quinoa growing in view of its high biodiversity. In: Bazile D (ed) State of the art report on Quinoa around the world in 2013. FAO, Rome, pp 42–55

Bazile D, Chevarria-Lazo M, Dessauw D (2013) Los sistemas que regulan el intercambio de los recursos genéticos: importancia para el acceso, la circulación y la innovación en el caso de la quinua. In: Bazile D (ed) State of the art report on Quinoa around the World in 2013. FAO, Rome, pp 105–123

Bazile D, Jacobsen S-E, Verniau A (2016) The global expansion of quinoa: trends and limits. Front Plant Sci 7(622)

Bilder RB (1980) International law and natural resources policies. Nat Resour J 20:451

Bosselmann K (1996) Plants and politics: the international legal regime concerning biotechnology and biodiversity. Colorado J Int Law 111(7):111–148

International Potato Center (2002) Broadening boundaries in agriculture: impact on health. Habitat and Hunger, International Potato Center, Lima

Brondizio ES (2008) Amazonian Caboclo and the Acai Palm: forest farmers in the global market. New York Botanical Garden Press, New York

Bruno MC (2006) A morphological approach to documenting the domestication of Chenopodium in the Andes. In: Zeder M (ed) Documenting domestication: new genetic and archaeological paradigms. University of California Press, Berkeley, CA, pp 32–45

Bruno MC, Whitehead WT (2003) Chenopodium cultivation and formative period agriculture at Chiripa, Bolivia. Latin Am Antiq 14:339–355

Brush SB (2013)Review of agrobiodiversity and the law: regulating genetic resources, food security and cultural diversity by Juliana Santilli. J Peasant Stud 40(2)

Carmen ML (1984) Acclimization of Chenopodium quinoa and Chenopodium pallidicaule to Finland. Annales Agriculturae Fenniae 23:135–144

Christensen SA, Pratt DB, Pratt C, et al (2007)Assessment of genetic diversity in the USDA and CIP-FAO international nursery collections of quinoa (Chenopodium Quinoa Willd.) using microsatellite markers. Plant Gen Resour: Charact Utili 5(2):82–95

Collins WW, Qualset CO (1998) Biodiversity in Agroecosystems. CRC Press, New York

Cronon W (1991) Nature's metropolis. W. W. Norton & Company, New York

Cusack D (1984) Quinoa: grain of the Incas. The Ecologist 14(1):21–23

del Carpio L, Yupa VereauF (2014) Diagnóstico Preliminar Para Elaborar un Seguro Agropecuario En La Región Puno. La Positiva, Lima. https://www.apeseg.org.pe/proyecto_seguro_agropecuario/Presentacion_SEGURO_PUNO_14_02_11.pdf

Ehrlich PR, Ehrlich AH (1992) The value of biodiversity. Ambio 21(3):219–226

Eisentraut PJ (1998)Macrobotanical remains from Southern Peru: a comparison of late archaic-early formative period sites from the Puna and Suni zones of the Western Titicaca Basin. Ph.D. Dissertation, Department of Anthropology, University of California, Santa Barbara

FAO 2013 Launch of the international year of Quinoa. https://www.fao.org/quinoa-2013/press-room/news/detail/en/

FAOSTAT (2014) Quinoa production in Peru 1961–2012. FAO, Rome

Galwey NW, Risi J (1984) Development of the Andean grain crop Quinoa for production in Britain. University of Cambridge Annual Report, Cambridge

García ME (2013) The taste of conquest: colonialism, cosmo-politics, and the dark side of Peru's gastronomic boom. J Latin Am Carib Stud 18(3):505–524

Gonsalves J (2010) Economic botany and ethnobotany. Mittal Publications, New Delhi

Halewood M, Noriega IL, Louafi S (2012) Crop genetic resources as a global commons: challenges in international law and governance. Routledge, New York

Hayden C (2003) When nature goes public: the making and unmaking of bioprospecting in Mexico. Princeton University Press, Princeton, NJ

Hellin J, Higman S (2003) Feeding the market: South American farmers, trade and globalization. Kumarian Press, Bloomfield, CT

Hernández Bermejo JE, León J (1994) Neglected crops 1492 from a different perspective. FAO, Rome. https://www.fao.org/docrep/t0646e/T0646E00.htm#Contents

Hobsbawm E (1983) Introduction: inventing traditions. In: Hobsbawm E, Ranger T (eds) The invention of tradition. Cambridge University Press, New York, pp 1–14

INEI (2014a) En el Año Internacional de la Quinua. Instituto Nacional de Estadística e Informatica, Lima

INEI (2014b) Encuesta de Hogares. Insituto Nacional de Estadística e Información, Lima

Jarvis DE, Ho YS, Lightfoot DJ et al (2017) The genome of Chenopodium quinoa. Nature 542(7641):307–312

Kameri-Mbote AP, Cullet P (1999) Agro-biodiversity and international law - a conceptual framework. J Environ Law 11(2):257–279

Kloppenburg J (1988) First the seed: the political economy of plant biotechnology, 1492–2000. Cambridge University Press, New York

Laguna P (2003) Feasability study for fair-trade labeling of Quinoa in Ecuador Peru and Bolivia. German Technical Cooperation, Bonn, Germany

Larson B (1995) Ethnicity, markets, and migration in the Andes: at the crossroads of history and anthropology. Duke University Press, Durham

National Research Council (1989) Lost crop of the incas: little-known plants of the Andes with promise for worldwide cultivation. National Research Council (NRC), Washington, DC

Marca S, Yucra H, Sucari J, Cano I (2009) Producción de Semilla de Quinua. Proyecto Quinua Regional, Gobierno Regional de Puno—Dirección Regional Agraria, Puno, Perú

Marx K (1867) Capital, vol I. L.W. Schmidt, New York

Mayer E (2002) The articulated peasant. Westview Press, Cambridge, MA

McDonell E (2015) Miracle foods: quinoa, curative metaphors, and the depoliticization of global hunger politics. Gastronomica 15(4):70–85

McDonell E (2016) Nutrition politics in the quinoa boom: connecting consumer and producer nutrition in the commercialization of traditional foods. Int J Food Nutr Sci (December)

Ministerio de Agricultura (2014) Quinua: Estadísticas de Producción y Exportación. Ministerio de Agricultura, Lima

Ministerio de Agricultura y Riego ca. (2016) Variedades de Quinua. https://minagri.gob.pe/portal/objetivos/444-granos-andinos/9377-variedades-de-quinua. Accessed 3 Aug 2016

Ministerio de Comercio Exterior y Turismo (MINCETUR) (2013) Reporte Mensual de Exportaciones. Estadísticas Comerciales Exportaciones. Lima: Ministerio de Comercio Exterior y Turismo

Mugabe J (1996) Managing access to genetic resources: towards strategies for benefit-sharing. ACTS, Nairobi

Mujica A, Jacobsen S-E, Izquierdo J, Marathee J (1998) Field Bbook of the American and European test of Quinoa. CIP, UNA-Puno, FAO
Mujica Sanchez A, Jacobsen S-E, Izquierdo J, Marathee JP (2001) Resultados de La Prueba Americana y Europea de La Quinua. FAO, UNA-Puno, CIP
Muller MR (2006) Background study 3: farmers’ rights in Peru—a case study. FNI Report 5/2006. The farmers’ rights project. The Fridtj of Nansen Institute, Lysaker, Norway. www.fni.no/farmers/main.html
Murphy KS, Matanguihan J (2015) Quinoa: Improvement and sustainable production. Wiley, New York
Murra JV (1956) The economic organization of the Inca state. University of Chicago Press, Chicago
Murra JV (1972) El “Control Vertical” de Un Máximo de Pisos Ecológicos El La Economía de Las Sociedades Andinas. Universidad Nacional Hermilio Valdizán, Huánuco, Peru
Native Crops of Peru—Quinoa ca. 2016 Discover Peru. https://www.discover-peru.org/native-crops-of-peru-quinoa/. Accessed 3 Aug 2016
Quinoa Italia (2015) Our Products. https://quinoaitalia.com/
Quinua.pe (2015) Exportación de granos andinos cayó En 24% en lo que va del año. Quinua Peru. https://quinua.pe/exportacion-de-granos-andinos-cayo-en-24-en-lo-que-va-del-ano/. Accessed 29 Nov 2015
Rénique JL (2004) La Batalla Por Puno: Conflicto Agrario y Nación En Los Andes Peruanos 1866–1995. IEP, Lima
Repo-Carrasco R (1991) Contenido de amino ácidos en algunos granos andinos. Avances En Alimentos Y Nutricion Humana, Programa de Alimentos Enriquecidos, Lima
Repo-Carrasco R, Espinoza C, Jacobsen S (2003) Nutritional value and use of the andean crops quinoa (Chenopodium quinoa) and kañiwa (Chenopodium pallidicaule). Food Rev Int 19(1/2):179
Risi J, Galwey NW (1991) Genotype X environment interaction in the andean grain crop quinua (C. quinua) in temperate environments. Plant Breeding 107:141–147
Rojas W, Pinto M, Alanoca C, et al (2014) Estado de la conservación ex situ de los recursos genéticos de quinua. In: Bazile D (ed) State of the art report on quinoa around the World in 2013. FAO, Rome, pp 65–94
Rosendal KG (2006) The convention on biological diversity: tensions with the WTO TRIPS agreement over access to genetic resources and the sharing of benefits. In: Gehring T (ed) Institutional interaction in global environmental governance: synergy and conflict among international and EU policies. MIT Press, Cambridge, MA, pp 79–102
Salomon F (1986) Native lords of Quito in the age of the incas: the political economy of North Andean Chiefdoms. Cambridge University Press, Cambridge
Schiebinger L (2004) Plants and empire: colonial bioprospecting in the Atlantic World. Harvard University Press, Cambridge
Shand H (1997) Human nature—agricultural biodiversity and farm-based food security. RAFI, Ottawa
Shiva V (1997) Biopiracy: The plunder of nature and knowledge. South End Press, Cambridge
Soto J, Valdivia E, Valdivia R, Cuadros A, Bravo R (2012) Descripción de sistemas de rotación de cultivos en parcelas de producción de quinua en cuatro zonas (Siete Distritos) del altiplano peruano. CienciAgro 2(3):391–402
Soto Lavaega G (2009) Maria Elena peasants, national projects, and the making of The Pill. Duke University Press, Durham
Sunder Rajan K (2006) Biocapital. Duke University Press, Durham
Sutton DE (2001) Remembrance of Repasts. Berg, New York
Tapia ME (1979) Historia y distribución geográfica. In: Tapia ME, Gandarillas H, Alandia S, et al (eds) Quinua y Kaniwa. Cultivos Andinos. Centro Internacional para el Desarrollo, Bogotá, Colombia, pp 11–19
Tapia ME (1983) La quinua y kañiwa: cultivos de los sistemas agricolas andinos. In: Tapia ME (ed) Granos Andinos. Bolivia, La Paz, pp 1–6

Tapia ME (1997) Cultivos Andinos Subexplotados y su Aporte a la Alimentación. Oficina Regional de la FAO para América Latina y el Caribe, Santiago, Chile
Tapia ME (2012) La historia de la quinua: Un Proceso de Miles de Años. Paper presented at the Grupo de Investigación e Innovación en Biocomercio: Año Internacional de la Quinua, 2013, Lima, Peru
Tapia ME (2014) El largo camino de la quinoa: ¿quiénes escribieron su historia? In: Bazile D (ed) State of the art report on Quinoa around the world in 2013. FAO, Rome, pp 3–10
Thier D (2010) The story of a cursed crop. The Atlantic, January 20
Van Esterik P (2006) From hunger foods to heritage foods: challenges to food localization in LAO PDR. In: Wilk R, Caldwell M (eds) Fast food/stow food: the cultural economy of the global food system. Altamira Press, Lanham, MD
Zimmerer K (1996) Changing fortunes: biodiversity and peasant livelihood in the Peruvian Andes. University of California Press, Berkeley, CA
Zimmerer K (2003) Geographies of seed networks for food plants (Potato, Ulluco) and approaches to agrobiodiversity conservation in the Andean Countries. Soc Nat Resour 16(7):583–601
Zimmerer K (2011) Vertical environments. In: Dym J, Offen K (eds) Mapping Latin America: a cartographic reader. University of Chicago Press, Chicago, pp 263–268

Chapter 16
Pachamanca-A Celebration of Food and the Earth

Matthew P. Sayre and Silvana A. Rosenfeld

Abstract *Pachamanca* is a traditional form of earth oven baking practiced in the Central Andean Region that translates from the indigenous Quechua language to mean "Earth Pot." This cooking practice is now generally reserved for special events and feasts. We present information from current Pachamanca practices in the town of Chavín de Huántar in Peru. The construction of the earth oven and the preparation of the foods is a group practice that reveals elements of cultural standing, as well as gendered activities. The selection of ingredients, the cooking of the sauces and flavoring agents, the making of the *humitas*, and the preparation and serving of the food is traditionally done by women. The construction of the pit for the oven, the selection of the stones to be used for heating, preparing the fire, the placement of food in the oven, the covering of the oven, and final removal of the food is done traditionally by men. The tasks performed by males and females also reflect the age and relative rank or hierarchy within the community. This traditional form of feast food emphasizes the importance of indigenous ingredients, with notable exceptions that provide insight into the changing nature of food practices under colonialism.

Introduction

The food we eat, to paraphrase Walt Whitman (1959), contains multitudes. Food connects us to the broader world through production, exchange, and history (Douglas 1971; Kogel 2011; Mintz 1996; Weismantel 1988). Many human factors come together in the moment of eating and food traditions are some of the most overt examples of cultural variation. Peruvian cuisine was raised to the rank of National Heritage in 2007 and it is now recognized around the world (Matta 2013). Contemporary food

M. P. Sayre (✉) · S. A. Rosenfeld
One University Parkway, High Point, NC, USA
e-mail: msayre1@highpoint.edu

S. A. Rosenfeld
e-mail: srosenfe@highpoint.edu

J. E. Staller (ed.), *Andean Foodways*, The Latin American Studies Book Series,
https://doi.org/10.1007/978-3-030-51629-1_16

Pachamancas traditionally rely on communal labor. The group effort is a form of labor exchange, and food can be provided in return for labor. However, labor is more commonly exchanged in return for commitments to participate in an exchange of labor and food in the future. Such symbiotic exchange of food and labor is ancient in the Andes and historically has formed tighter social bonds that create future obligations for interaction.

Today, people in the central Andean highlands typically prepare *pachamancas* to celebrate special occasions: harvest time, worker's days, religious festivities (*fiestas patronales*), local celebrations (*aniversarios de pueblos*), etc. (Castillo Posada 2014:266). In the context of archaeological fieldwork, local workers often organize a *pachamanca* to celebrate the end of a field season, or another celebratory event related to the archaeological work. When this feast is prepared to celebrate the end of an archaeological field season, the *pachamanca* is viewed as a local tradition whose preparation requires ingredients primarily procured from local markets or from traditional farmers.

Key components of the *pachamanca* feasts are highlighted by how different individuals perform their roles in the preparation, cooking, and serving of the food. Many of these activities reveal status and role differences that vary by age, gender, and class. As outsiders at these events, we have generally been treated as bystanders who offer little useful labor to the festivities. As such, we usually have been relegated to taking pictures, receiving instructions, and preparing foods that children are capable of producing. In this outsider role the perspective of the participants has to be discerned through observation, conversation, and participation.

Traditionally there is a separation of labor among the male and females in key activities of the *pachamanca*, although gender lines can sometimes be crossed. The women of a family or community are generally responsible for the preparation of sauces, marinades, and *humitas* which are different foods wrapped in ground corn meal. These variable recipes differ depending on the season and the availability of ingredients in distinct parts of the annual cycle. The use of the *huacatay* herb (Peruvian black mint) is marked in certain regions of Peru, whereas banana leaves are more common in more tropical regions. The sauces and marinades are key to the feast, and if they are not well prepared then the entire event can be made less enjoyable.

The *pachamanca* is an event that emphasizes community participation and cohesion. As such, the people who will participate in the meal traditionally provide the majority of the labor for the preparation. The communitarian nature of the feast limits the need to hire people or participate in the cash economy. However, there is at least one part of the preparation that can commonly diverge from this cultural norm—the killing of animals to be consumed as part of the feast. For the meat portion of a *pachamanca* it is common to hire someone from the community who has knowledge regarding killing and butchering animals. Some of the emphasis on specialization arises out of a particular concern to avoid the suffering of animals, and to ensure that the meat is properly butchered into the best cuts for the eventual feast. There are certainly other areas of food preparation that generally necessitate the use of cash to procure all of the necessary ingredients, but it is also possible to avoid these costs if an extended family has sufficient access to raw materials. The butchering of animals often times stands out as one of the few marked activities that requires a specialist.

The construction, maintenance, and monitoring of the earth oven are an activity traditionally managed by males. As Levi-Strauss (1966) stated, cooking with fire is commonly a masculine realm. However, the *pachamanca* does not neatly fit into the culinary triangle of boiling, roasting, or smoking (*ibid.*). The feast does require some roasting, but it is not over an open flame. The heating of stones upon which food will be cooked is in some ways a combination of these three types of cooking. Thus, the *pachamanca* cooks meat in a slightly inefficient manner, as some of the meat is lost, but it is not as inefficient as roasting over an open flame. Older men manage younger men in the selection of proper stones, the construction of the stone oven, and the tending of the fire. They also manage the oven over the hours prior to the collapse of the stone structure and the placement of food upon the heated stones. In somewhat less formal *pachamancas*, or if the need requires, women have been known to manage this activity. It is clearly an area in which women have expertise as many of them prepare daily meals over fire. However, the preparation and tending of the oven is often a time when older men can assert their knowledge over younger men, oftentimes in the form of joking about the relative lack of experience and knowledge that younger men have regarding in the preparation of this important communal feast.

Preparation of *Pachamancas*

The preparation of the *pachamanca* is generally a multi-day affair that begins with a number of concrete steps. The initial phase that begins prior to the cooking is the preparation of sauces, marinades, and *humitas*, as well as the killing and butchering of animals to be consumed, and the procurement of needed supplies such as wood and stones for the pit, and drinks to accompany the meal.

As a feast that brings together communities, a common part of the *pachamanca* process is making sure that many members of the community participate in the preparation of the food and of the oven. This multi-step process is divided into activities prior to cooking and then those directly associated with the cooking event. The days before a *pachamanca* it is common for the preparation activities to be managed by women, who are also frequently in charge of financial matters related to the event. The early days are devoted to harvesting or buying the ingredients, the preparation of sauces and marinades, the gathering of firewood and appropriate stones, and the preparation of labor-intensive foods that are included in the feast.

The preparation of the pit oven generally begins the day of the feast. Building the pit oven involves the cleaning of the stones, excavation of the *manka* in which the cooking will occur, the stacking and heating of the stones, the lining of the *manka* walls, the care/flavoring of the stones with brushes and salt water, the monitoring of the cooking food, and serving the food (Gutarra and Valderrama 2001:52–53). All of these activities provide insights into the roles and status of the people participating in the event (Fig. 16.1). Often *pachamancas* include other performative behaviors

Fig. 16.1 Residents of Chavín de Huántar, Peru constructing *pachamanca* ovens. (Photo by M. Sayre)

in addition to food preparation. For example, musicians are often present who play instruments and are in charge of encouraging people to sing and dance (Castillo Posadas 2014).

Additionally, there are some regional differences in the terms used to describe this method of cooking. The word *pachamanca* is not used across all of Ecuador, Peru, and Bolivia even though Quechua (or Quichua) is spoken by many indigenous groups in these countries; rather, the term *huatia* may be used instead of *pachamanca.* As practiced in the Central Andes of Peru, the *huatia* is a comparatively less elaborate form of cooking. It is constructed of hardened earth and is a type of oven that is particularly useful for cooking potatoes and other tubers. This emphasis on roasting local potato varieties is indicative of the pride and value placed in the flavor of indigenous varieties of tubers. While the *pachamanca* almost always requires the construction of an underground cooking area, the *huatia* tradition does not necessitate this step and can be constructed in a shorter period of time. However, in other regions the term *huatia* can connote a feast or meal quite similar to the *pachamanca*, as described above.

Pachamanca Ingredients

Food Plants

Potatoes (*Solanum tuberosum*) are the iconic staple of Andean cuisine. They are called *papas* in both Quechua and Latin American Spanish. Interdisciplinary lines of evidence indicate that potatoes were domesticated in the highlands of southern Peru over 8,000 years ago (Smith 1999; Spooner et al. 2005). More than four thousand varieties of potato are currently grown in Peru (FAO 2008). This tremendous diversity of sizes, colors, shapes, and textures of these potatoes is valued by local people. Additionally, the potato is commonly described in human terms with its skin being compared to human skin and it is given anthropomorphic attributes such as eyes in ancient effigy vessels (Kogel 2011).

Potatoes were of such importance in pre-conquest Peru that the chronicler Fr. Bernabe Cobo stated that time was measured in the increment that it took to boil them (1997,1653). Their role as a staple in the Andean diet, and common inclusion both in soups and as a side dish also reflects their tremendous nutritional value (Weismantel 1988). In the *pachamanca*, potatoes are commonly added to the earthen pit directly after the placement of the meat on the heated stones. This starchy staple will ensure that all people leave the feast full. And while potatoes are commonly consumed without additional flavorings they can also be eaten with sauces, which typically vary from region to region.

Pachamancas include a great variety of tubers in addition to potatoes. *Oca* (*Oxalis tuberosa*), *olluco* (*Ullucus tuberosus*), and *mashua* (*Tropaeolum tuberosum*) are some of the commonly consumed tubers that until recently, were rarely grown outside of the Andes. These tubers provide taste differences to the potatoes and add to the color presentation of the cooked meal (Fig. 16.2). These tubers are commonly placed near the potatoes in the earthen oven, as they require high and steady heat in order to cook.

Maize, a key ingredient in *pachamancas,* is consumed in both liquid and solid form at these events. Maize (*Zea mays*), or *sara* in Quechua, has long been a high-status food in the central Andes (Hastorf 1990; Staller 2006). This is evident by the emphasis on maize production by the Inka Empire, much of which was grown for use as *chicha*, or corn beer (Cobo 1996, 1653). *Chicha,* which can vary in alcohol strength (Bolin 1998; Isbell 1978), is still a staple of Andean feasting events today. Long fermentation produces high alcohol content beverages, which are commonly enriched among present day Andean cultures with the addition of hard liquor. Maize on the cob (*choclo* in Quechua and South American Spanish[i]) is also commonly consumed during *pachamancas*. *Choclo* is added late to the earthen oven as the steam from the cooking is sufficient to cook the corn. There are other means by which corn is used as an ingredient in *pachamancas*, most commonly as ground corn flour in *humitas.*

Humitas are wrapped corn specialties that can be prepared in sweet or savory versions. This dish is made from ground corn flour wrapped in a corn-husk shells

Fig. 16.2 Retrieving the bounty from a Pachamanca. Particular emphasis is placed on the consumption of local vegetable goods. (Photo by M. Sayre)

and filled with a variety of fillings. *Humitas* are a fairly versatile dish that can be prepared either as a side dish or as sweet dish to be consumed after the main meal. It is also one of the most labor-intensive dishes and must be prepared hours or days ahead of the main cooking event. *Humitas* are also a late addition to the earth oven as they are not cooked directly on heated stones, but rather, are cooked indirectly by the heat and steam from the heated stones.

Habas, or fava beans (*Vicia faba*), are native to the Old World and brought to the Americas by Spanish conquistadors. They are one of the few vegetable ingredients that are not native to the Andes. However, these beans are so well adapted to the area that they have become culturally part of the local cuisine and have almost reached the status of indigenous foods. However, some beans have evolved phytotoxins as an adaptation against predators at these high altitudes (see Johns 1999). Such bean species may have replaced a part of the feast previously occupied by *tarwi* (*Lupinus mutabilis*), a lupine. However, lupines are commonly leached of certain alkaloids in order to be consumed. Such leaching primarily occurs in rivers of moving water. The lupines are later washed again prior to serving. This leaching process requires prior processing before being added to the earthen oven.

A central component of the *pachamanca* feast is the sauce, or sauces, that accompany the food. While sauces are not cooked in the earth oven, they are considered a necessary part of the meal. Many sauces feature chili peppers, or *aji* (in Spanish and *uchu* in Quechua), as a central ingredient. This spice enhances the flavors of certain dishes, and the main ingredients of the *pachamanca.* Spices, particularly hot peppers are a key component of Andean food for millennia (Chiou et al. 2014). Sauces can be made from a variety of chili peppers and combined with numerous ingredients, such as salt, garlic, peanuts, crackers, and vinegar. The inclusion of sauces and condiments is ubiquitous across the Andean region (Gutarra and Valderrama 2001:63) but these sauces vary from region to region.

In this section we have discussed the plant ingredients that are key parts of *pachamancas.* These foods stand in contrast to other daily staples, such as rice, which is a relatively common food of many highland Central Andean households. Imported rice may be considered a "white" food (Weismantel 1988) but the relatively high status of this coastal food does not mean that it is included as a side ingredient in traditional feasts. Rather, its incongruity with the cooking style of *pachamancas* (i.e., requires boiling and is not amenable to cooking in an earthen oven), as well as the *pachamanca* preparators emphasis on highland ingredients relegates rice as an "outside" (sensu Mintz 1996) food to other meals. While bowls of wheat may occasionally be served alongside traditional *pachamanca* fare, this non-native plant is commonly grown in the highlands. Thus, wheat has been incorporated into many local traditions, much along the lines of *habas,* which have become installed firmly in several traditional dishes. Therefore, many of these local, "inside" foods are native to the Andes yet some of the ingredients were introduced from Eurasia after the conquest. Despite the inclusion of some introduced Eurasian foodstuffs, almost all of the ingredients for *pachamancas* are procured locally and consistently are incorporated into daily food consumption practices.

Meat

The meat potions that people prepare and consume during a *pachamanca* are distinctly different from the vegetables in one significant factor: the majority of animal products are from species, primarily terrestrial mammals, not native to the Americas. The meats consumed in the region are generally not those that would have been consumed in the pre-Columbian past, yet they are from Eurasian-introduced animals that are currently part of the highland diet. The meat is produced locally from non-native animals that have acclimated to the highlands.

In the pre-colonial central Andes, the main sources of animal proteins and animal fat were camelids-domestic and wild-, guinea pigs, and other wild game such as deer, *viscachas,* and partridges (deFrance 2014; Rosenfeld 2008). According to ethnohistorical sources, meat was consumed primarily on special occasions among the Inkas. *In Inca Religion and Customs* Cobo (1997, [s]:198) states that "The plebeian ate very little meat, and when they did it was at festivals and banquets." While ethnoarchaeological research carried out among South Andean herders indicates more frequent meat consumption (e.g., Göbel 2001; Yacobaccio 2007), these data are drawn from herders subsisting more on their animals than agricultural produce through much of the year. In general, camelid meat was probably consumed sporadically. The frequency and quantity of meat consumed in the pre-colonial past by the common people was probably also dependent on their distance from the herder communities. Guinea pigs may have been eaten more often given that they can be raised inside houses and they can give birth year-round. It is most probable that people reserved large quantities of camelid meat for special events.

The Spanish initially labeled camelids as *ovejas de las Indias*. The "sheep of the Indies/Americas" is a relevant term here as lamb is a meat commonly consumed in *pachamancas*. These animals produce a relatively fatty meat (see below) that does well when cooked over heated stones. This is similar to another non-native source of meat, pork, which is also a common ingredient in *pachamancas*. Pork is commonly marinated in sauces and then placed directly on the stones and is also a common ingredient in *humitas*.

The tradition of barbecuing beef is rare in the Andean highlands. The amount of grass needed to sustain cattle commonly leads highland farmers to raise a small number of cattle, and these cattle are traditionally not as large as cattle bred in other areas of the world. Thus, relatively lean meat is produced from these animals, which probably limits their desirability as an ingredient in *pachamancas* as trim meats dry out quickly on heated stones. This lack of beef in *pachamancas* does not mean that beef is never consumed by people in the highlands. Today one of the most commonly prepared foods in street markets are *anticuchos*, or grilled beef heart on skewers.

Chicken, (pollo in Spanish or *wallpa* in Quechua), rarely is prepared in *pachamancas*. While chickens were originally domesticated in Asia, there is some debate about whether or not they arrived in the Americas prior to European colonization (Lawler 2014; Thomson et al. 2014). Chickens have become so widely consumed in the Andes that they are considered essentially to be indigenous, a situation similar to fava beans (*habas*). This debate is relevant here as it sheds light on the divide between vegetable and meat ingredients in the *pachamanca*. However, if the chicken did arrive to South America prior to the Spanish, it was most likely present for a relatively short period of time prior to colonization. During the colonial era, chickens rose in prominence and are now one of the more commonly consumed meats in the highlands. The animal's rapid conversion of feed into meat, its production of eggs, and its quick reproduction make chickens a desirable animal to raise in farms and rural backyards. While chicken meat is a staple of modern highland diets, it is not a common ingredient in *pachamancas*. The relatively lean meat of the chicken may be less desirable in comparison to pork or lamb for this dish.

The contrast between the tuber/grain ingredients and the meat ingredients that are at the center of a *pachamanca* is striking. While the majority of the vegetable ingredients are indigenous to the Americas the opposite is true of the meat selection. Meat selection may be based on fat content and/or lack of access to indigenous animals. The animals mostly used in the pre-Hispanic indigenous diet, camelids, and guinea pigs, are lean and have low-fat content (Rosenfeld 2008). Llamas have a weight-based fat value of 8–25% (Mengoni Gonalons 1996) and guinea pigs have a weight-based fat value of 7.8% (Bolton 1979). However, the introduced European animals used in *pachamancas* at present, lamb and pork, have a higher fat content:

lamb is 28% fat and pork is 32% fat of carcass composition (FAO 1991). Both lamb and pork are also cheaper in most regions and have become the preferred type of meat to cook in this manner. Thus, they maintain the "inside" practice of avoiding the purchase of non-local goods.

The placement of the food in the earth oven is a precisely orchestrated event that requires speed and knowledge. When the wood has been removed and the stones collapsed there is precious little time for the placement of the food before its eventual covering. If the food is placed slowly in the earthen oven, too much heat is lost and then there will not be sufficient heat to cook all the food. Meat and tubers are the ingredients that require the most intense heat to cook; therefore, they are placed first in the oven with other ingredients following. This portion of the cooking can be considered the roasting part of the feast. The additional ingredients, such as beans, *habas*, *humitas*, *choclo*, etc., are indirectly cooked by the heat of the stones in a fashion more akin to steaming. After these ingredients are placed in the oven it is generally covered with reeds, cloth, or other material, and then covered in earth. The food then roasts or bakes for an hour or two before being retrieved in the reverse order in which it was placed in the oven.

The food is removed from the *pachamanca* oven and placed in communal food containers that are generally organized by food type. Then the hosts will prepare plates for all of the guests. These plates usually contain portions of all of the cooked foods, so that each guest is able to taste and consume all of the richness and variety of the feast. The serving of the food is generally not arbitrary. Esteemed guests and prominent participants in the *pachamanca* are served prior to distant relatives who may be invited to the feast. In this instance, status is reflected by order and speed rather than the quality or quantity of food served; although, there is significant regional and situational variation in such practices.

We have learned gradually over time how to politely manage the enormous quantities of food served at *pachamancas*. While it is a cross-cultural norm that guests should eat the food that is served to them, it is also true that Andean feasts can lead to portions that are nearly impossible for a single individual to consume. Thus, there are culturally appropriate ways of dealing with this abundance of food. For example, we have seen people pack up uneaten food to be taken home and shared with family members. This tradition allows the participant to not reject any of the graciousness of the hosts while also avoiding the unpleasant feelings that results from eating far more food than any ordinary meal.

Pachamanca During Pre-hispanic Times

One means of documenting the food traditions at the time of conquest is by connecting the writings of early chroniclers with the archaeological evidence (see also Covey, deFrance, Lambaré et al., and Staller, this volume for similar studies combining historical documentation with archaeological analyses). The chronicler Fray Bernabe Cobo (1997 [1653]:198–199) provides some of the clearest descriptions of Andean cooking traditions. However, his account is largely incomplete and biased because he frequently depicts Andean cuisine in a poor light due to its "rustic and crude" nature and its lack of comparability to the cuisine of Spain.

Archaeology can provide insights into past Andean cuisines and food preparation methods (e.g. Hastorf 2003; Rosenfeld 2008; Sayre et al. 2012). While archaeology can reveal a great deal about the foods consumed in the past and where they came from, excavations cannot always uncover evidence of techniques or recipes that were followed to create particular dishes. This is certainly true of *pachamancas*; there is evidence for the use of stones in cooking and for the construction of underground cooking areas (Gutarra and Valderrama 2001; Lavallee 1995; Valdez 2019); yet, the precise recipes and traditions are more difficult to discern on the basis of archaeological evidence. For example, Valdez and Huamani (2019) found evidence of *pachamanca* ovens at the site of Tambo Viejo in the southern coast of Peru. They uncovered a circular pit aligned with rocks and burnt soil inside. Achira leaves were found on the pit surface, perhaps used to cover and protect the food (Valdez and Huamani 2019: 13). Recent and more sophisticated techniques may help researchers demonstrate the existence of pachamanca cooking during pre-Hispanic times by identifying the food residues on the cooking stones. The recovery of concentrations of large smooth stones that are not prone to cracking in fire could possibly be tested for starch grain residue. Proteonomic analysis might also recover evidence of the meat and vegetables that were cooked on these stones.

Conclusions

We all have to eat, yet food is more than mere nutrition; it tells us a great deal about our history and culture. Thus, some foods are emphasized in traditional feasts while other modern staples may be ignored. The *pachamanca* can be viewed as an example of the maintenance of tradition. As we have described, the meal emphasizes the importance of acknowledging the gifts of the earth and giving thanks to the *Pachamama.* It is also a culinary event that emphasizes traditional Andean foods and ancient techniques. This "inside" food, places continued emphasis on local preparation and communal consumption, yet, aspects of the tradition are gradually becoming commodified and marketed in urban settings (Fig. 16.3). In 2003, the *pachamanca* was registered as National Cultural Heritage in Peru. And since 2015, the pachamanca has its own day, the first Sunday in February (*El Comercio* 11-26-2015). The Peruvian Ministry of Agriculture initiated the *Pachamanca* Day, because "it is an emblematic dish and a traditional food preparation in Peru, mostly in the Andean region, with deep sociocultural meaning among the *campesinas* (farming) societies" (our translation[ii], El Comercio 11-26-2015). The Lima newspaper *El Comercio* comments that *Pachamanca Day* will promote the commercialization of Peruvian traditional ingredients involved in the preparation of the *pachamanca.* Thus, it is conceivable that this "inside" food practice may eventually evolve into an "outside" food event that primarily exists to be marketed to tourists. Already in Lima, *pachamanqueros*

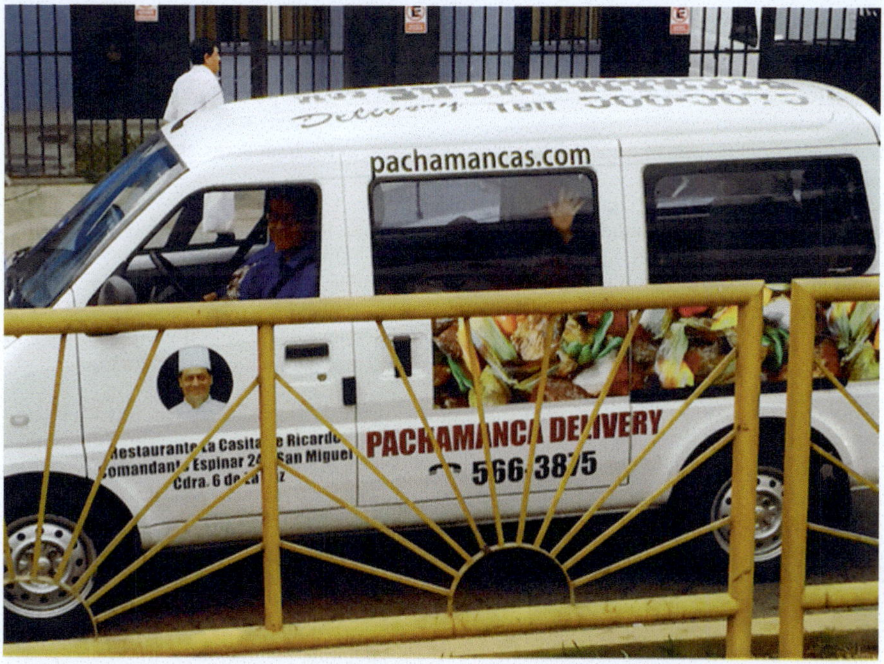

Fig. 16.3 Modern Pachamanca delivery service in Lima, Peru. (Photo by M. Sayre)

from the highlands are hired to prepare the traditional Andean dish in the backyard of urban upper-class houses following more *sanitized* rules (Castillo Posadas 2014:264). However, currently in the highlands the continued emphasis on communal labor allows communities to come together to celebrate food, place, and time. As an important regional tradition, it seems probable that the *pachamanca* will continue to serve as valuable practice for preserving traditional Andean foodways and their association with highland beliefs and ritual feasting traditions.

Notes

i. While *choclo* technically refers to a large-kernel variety of field corn originated in the Peruvian Andes, the term is used throughout South America to refer to cobs of corn in general.

ii. 'Entre los considerandos de la norma se señala que la **Pachamanca** "es un plato emblemático y a la vez una forma tradicional de preparación de alimentos en el Perú, la cual se realiza principalmente en la región andina y tiene un profundo significado sociocultural en las sociedades campesinas". Diario El Comercio, November 26, 2015. https://elcomercio.pe/peru/Pachamanca-dia-primer-domingo-febrero-246807.

References

Alberti G, Mayer E (1974) Reciprocidad y intercambio en los Andes peruanos. Instituto de Estudios Peruanos, Lima

Bolin, Inge 1998 Rituals of respect. The secret of survival in the high Peruvian Andes. University of Texas Press, Texas

Bolton R (1979) Guinea pigs, protein, and ritual. Ethnology 18:229–252

Bray TL (2003) To dine splendidly. In Bray T (ed) The archaeology and politics of food and feasting in early states and empires. Springer, Boston, MA, pp 93–142

Castillo Posadas A (2014) Pachamanca: la cosmovisión andina a través de la alimentación. *Lienzo.* Revista de la Universidad de Lima 35:253–268

Chiou K, Hastorf C, Bonavia D, Dillehay T (2014) Documenting cultural selection pressure changes on Chile Pepper (Capsicum baccatum L.) Seed size through time in Coastal Peru (7,600 B.P.–Present). Econ Bot 68(2):190–202

Clarke MJ (2001) Akha feasting: an ethnoarchaeological perspective. In: Dietler M, Hayden B (eds) Feasts: archaeological and ethnographic perspectives on food, politics, and power. Smithsonian Institution Press, Washington D.C., pp 144–167

Cobo BF (1996 [1653]) History of the Inca Empire. Translated and edited by Hamilton R. Foreword by Rowe J (5th ed). University of Texas Press, Austin

Cobo BF (1997) [1653]) Inca religion and customs. University of Texas Press, Austin

deFrance SD (2014) The luxury of variety: animals and social distinction at The Wari Site of Cerro Baúl, Southern Peru. In: Arbuckle BS, McCarty SA (eds) Animals and inequality in the ancient world. University Press of Colorado, Boulder, CO, pp 63–84

Dietler M (2001) Theorizing the feast: rituals of consumption, commensal politics, and power in African contexts. In: Dietler M, Hayden B (eds) Feasts: archaeological and ethnographic perspectives on food, politics, and power. Smithsonian Institution Press, Washington, D.C., pp 65–114

Douglas M (1971) Deciphering a meal. In: Myth, symbol, and culture, edited by Clifford Geertz. New American Library, NY, pp 74–101

FAO (Food and Agricultural Organization of the United Nations) (1991) Guidelines for slaughtering, meat cutting and further processing. In: FAO animal production and health paper (91). Rome, Italy. FAO Corporate Document Repository. https://www.fao.org/docrep/004/t0279e/t0279e05.htm

FAO (Food and Agricultural Organization of the United Nations) (2008) International year of the potato: new light on a hidden treasure. Italy, Rome

Göbel B (2001) El ciclo anual de la producción pastoril en Huancar (Jujuy, Argentina). In: Goñalons GM, Olivera D, Yacobaccio H (eds) El Uso de los Camélidos a Través del Tiempo. GZC/Ediciones del Tridente, Buenos Aires, pp 91–115

Gutarra Carhuamaca J, Valderrama Leon M (2001) Pachamanca: El Festín Terrenal. Universidad de San Martin de Porres, Lima

Hastorf CA (1990) The effect of the Inka state on Sausa agricultural production and crop consumption. American antiquity. 262–290

Hastorf CA (2003) Andean luxury foods: special food for the ancestors, deities and the elite. Antiquity 77:545–554

Hayden B (2001) Fabulous feast: a prolegomenon to the importance of feasting. In: Dietler M, Hayden B (eds) Feasts: archaeological and ethnographic perspectives on food, politics, and power. Smithsonian Institutional Press, Washington D.C., pp 23–64

Hobsbawm E, Ranger T (1983) Introduction: inventing traditions. In: Hobsbawm E, Ranger T (eds) The invention of tradition. Cambridge University Press, Cambridge, UK, pp 1–14

Isbell BJ (1978) To defend ourselves. Ecology and ritual in an Andean Village. Waveland Press, Inc., Illinois

Johns T (1999) The chemical ecology of human ingestive behaviors. Annu Rev Anthropol 28:27–50

Kogel A (2011) Food, power, and resistance in the Andes. Lexington Books, NY

Lavalle D (1995) Telermachay: Cazadores y Pastores de Prehistoricos de los Andes. Instituto Frances de Estudios Andinos, Lima
Lawler A (2014) Why did the Chicken Cross the World: The Epic Saga of the Bird that powers civilization. Atria Books, NY
Levi-Strauss C (1966) The culinary triangle. Peter Brooks, trans. Partis Rev Partis Rev 33(4):586–595
Matta R (2013) Valuing native eating: the modern roots of Peruvian Heritage. Anthropol Food (online), 8:1–22. https://journals.openedition.org/aof/7361
Mauss M (1967) The gift: forms and functions of exchange in archaic societies. Norton, NY
Mengoni Goñalons G (1996) La domesticacion de los camelidos sudamericanos y su anatomia economica. *Zooarqueologia de Camelidos*. Perspectivas Teoricas y Metodologicas 2:23–35
Mintz S (1996) Tasting food, tasting freedom: excursions into eating, culture, and the past. Beacon Press, Boston, MA
Rosenfeld SA (2008) Delicious guinea pigs: seasonality studies and the use of fat in the pre-Columbian Andean diet. Quatern Int 180:127–134
Sayre M, Goldstein D, Whitehead W, Ryan Williams P (2012) A marked preference: *Chicha* de *Molle* and Huari State consumption practices. Ñawpa Pacha J Andean Archaeol 32:231–282
Smith B (1999) The emergence of agriculture. Scientific American Series. W.H. Freeman & Co, New York
Spooner DM, McLean K, Ramsay G, Waugh R, Bryan GJ (2005) A single domestication for potato based on multi-locus amplified fragment length polymorphism genotyping. Proc Natl Acad Sci 102(41):14694–14699
Staller JE (2006) The social, symbolic, and economic significance of Zea mays L. in the Late Horizon period. In: Staller J, Tykot R, Benz B (eds) Histories of maize. Academic Press, San Diego, pp 449–467
Thomson VA, Lebrasseur O, Austin JJ, Hunt TL, Burney DA, Denham T, Rawlence NJ, Wood JR, Gongora J (2014) Linus Girdland Flink, Anna Linderholm, Keith Dobney, Greger Larson, and Alan Cooper (2014) Using ancient DNA to study the origins and dispersal of ancestral Polynesian chickens across the Pacific. Proc Natl Acad Sci 111(13):4826–4831. https://doi.org/10.1073/pnas.1320412111
Valdez L, Huamani M (2019) Investigaciones arqueologicas en Tambo Viejo, Acari, costa sur del Peru. Revista Huacaypata 14:6–28
Valdez LM, Marcelino Y, Huamaní N (2019) Investigaciones arqueológicas en Tambo Viejo, Acarí, costa sur del Perú. Revista Haucaypata. Investigaciones arqueológicas del Tahuantinsuyo, Nro. 14: 6-28. Lima.
Weismantel M (1988) Food, gender, and poverty in the Ecuadorian Andes. Waveland Press Inc., Illinois
Whitman W (1959) Leaves of grass: the first, 1855th edn. Viking Press, NY
Yacobaccio HD (2007) Andean camelid herding in the South Andes: Ethnoarchaeological models for archaeozoological research. Anthropozoologica 42(2):143–154

Chapter 17
Ethnicity and Ritual in the *Atacameños* Andes: Water, Mountains, and Irrigation Channels in Socaire (Atacama, Chile)

América Valenzuela and Ricardo Moyano

Abstract The water in the indigenous community of Socaire (Antofagasta Region, Chile), as well as in other communities in the Atacama Desert, is a very important resource for the development of agriculture and livestock. Water has historically played a major role in terms of its symbolic manifestations and associated rites and ritual contexts. After the beginning of the agricultural year, in late October, the local *atacameños* celebrate the cleansing of irrigation canals with a ceremony, which renews their commitment to the mountains (*mallkus*), and the tutelary spirits of the community. These spirits provide the necessary water for life. From this Andean ritual, we introduce ethnographic and archaeoastronomical evidence focused upon the symbolic, political, and cultural complex, and how these have reference to the agricultural cycle and its relationship with the surrounding landscape, i.e., water and mountains, as a possible Inca *ceque* system. We avoid primordialist approaches of the culture, and we show that the symbolic elements related to water and irrigation channels are part of the contemporary system of inter-ethnic relations.

Keywords Archaeoastronomy · Ethnicity · Mountain worship · Atacama · Chile

The ecology, environment, climate, and surrounding landscape of San Pedro de Atacama, Chile is challenging. It is situated at 2,407 masl or around 7,000 feet, and the climate is extremely dry and mild. In order to cultivate the food crops necessary to sustain indigenous populations in this region of the Chilean highlands, communities have developed complex system of labor organization, ritual practices, and feasting related to the maintenance of the irrigation networks and canals fed by rivers from the Bolivian Altiplano. Cultural practices have particularly focused upon cleaning

A. Valenzuela (✉)
Departamento de Antropología, Facultad de Ciencias Sociales, Universidad Alberto Hurtado, Santiago, Chile
e-mail: avalenzuel@uahurtado.cl

R. Moyano
Sociedad Interamericana de Astronomía En La Cultura, Santiago, Chile
e-mail: astronomiaintercultural@gmail.com

J. E. Staller (ed.), *Andean Foodways*, The Latin American Studies Book Series,
https://doi.org/10.1007/978-3-030-51629-1_17

irrigation canals to ensure a successful harvest and widespread agricultural cultivation. Such cultural practices are central to fostering and maintaining the ethnic identity of various indigenous communities in the modern world. These contemporary indigenous ritual practices are related to a devotion to traditional ancient Andean customs, and represent a conscious response to a long history of marginalization and discrimination by external forces beyond the control of local communities.

The Community of Socaire

Socaire is one of the nine *atacameña* or *Lican Antay* communities associated with the municipality of San Pedro de Atacama. It is located 160 km from Calama City, southeast of the Atacama Salar in Northern Chile (23°35′28.68″S 67°52′36.10″W/3274 masl). The *atacameña* organization is formed by smaller territorial units called *ayllus*: Quítor, Conde Duque, Solcor, Yaye, Tchecar, Séquitor, Solor, Cúcuter, Poconche, Béter, Coyo, and Tulor. These *ayllus* were added to the named *atacameñas* communities that were composed mostly of indigenous Andean populations[1] (Fig. 17.1). During Inca times (c. A.D. 1470–1532), this region was situated in Collasuyu or the southern quarter of Tawantinsuyu or land of the four quarters. This Quechua term had direct reference to the Inca Empire because of its quadripartite organization. Atacama was of great interest to Spanish conquistadores because of its rich copper and silver ore deposits. Among indigenous populations, this is an important region for raising camelids and agriculture, particularly potatoes, quinoa, and maize which are all critical food crops to indigenous agriculture. The people supplied and maintained a section of the Inca highway, *Camino Real* or *QhapaqÑan* between San Pedro de Atacama and the valley of Copiapo (28°S. Latitude), as well as the Circumpunean Area which includes southwestern Bolivia and northwestern Argentina (Aldunate and Castro 1981; Núñez 1991, 1993; Sanhueza 1992).[2]

The Atacama region is situated east of the Atacama Desert, the driest desert on earth. This is a region where there has never been any recorded rainfall. Although the higher altitude regions to the east of the desert are characterized as arid, precipitation does occur during the astral winter from January to March, as well as around its high mountains and volcanoes. Mountain peaks in the sometimes exceed 5,000 masl, e.g., Mt. Lincancabur (5,920 m), Chiliques (5,770 m), Socompa (6,051 m), and Llullaillaco (6,739 m). Mt. Llullaillaco has the highest documented archaeological site in the world, a Inca Capac Hucha burial on its summit (CIADAM 1985, 2001; Reinhard 1993, 2006). The environmental conditions have broader implications for effecting

[1] The last National Population Census of 2002 indicates that the atacameños number 21,015 people, of which 13,874 live in the region of Antofagasta, that is 66% of its total population. The atacameños are distributed as follows: Antofagasta 1,203; Mejillones 25;Sierra Gorda 37; Taltal 35; Ollagüe 4; San Pedro de Atacama 2,862; Tocopilla 25; María Elena 25 and Calama 9.658 (8,747 urban and 911 rural), which reveals that the current Atacama residence has an orientation toward the urban (INE 2005, http://www.ine.cl/).

[2] The Tropic of Capricorn in 1500 A.D. was around latitude 23°30′S, only 10 km north of Socaire.

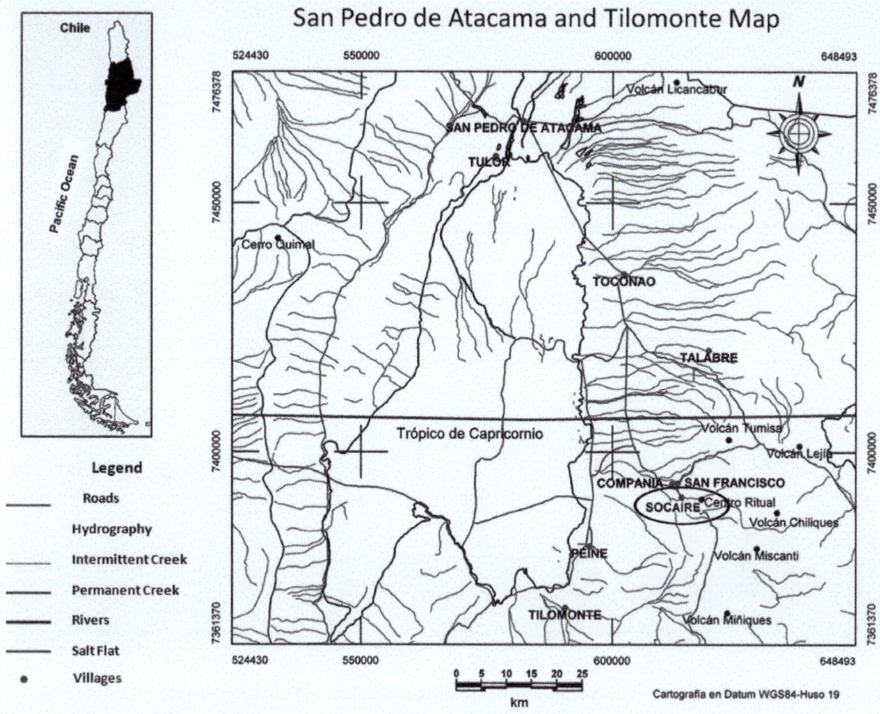

Fig. 17.1 Map of general location of Socaire in northern Chile (Courtesy of Moyano)

the ecology typical of a high desert ecosystem or Puna. Communities like Socaire, located in the highlands (above 3,000 masl), face a complex survival relationship with the environment, which leads to the development of a social and political economy based on the interaction of a diversity of resources from different ecological zones, what is generally termed Andean complementarity (see, e.g., Castro 1998; Murra 1975, 1978) (Fig. 17.2).

The Atacama region has had both widespread agricultural cultivation and breeding of livestock (camelids) dating back to pre-Columbian times. These are related to the Andean tradition of ancestral forms of irrigation and methods of community labor (*minga*) as well as European monocrop agriculture which was introduced in early colonial times. The integration of Old World domesticated crops and animals such as alfalfa, wheat, sheep, goats, and donkeys transformed the local ecology (Gundermann and González 1995). However, Socaire's geographical characteristics are disadvantageous with respect to communication between various sectors of the Atacama Desert, particularly the lowland urban centers such as Antofagasta and Calama. In this context, several authors noted the impact of copper mining and non-metallic minerals in the San Pedro de Atacama area. Mining of copper and other metals such as silver displaced agricultural activities and ushered in problems regarding water management and sulfide and arsenic contamination. In turn, this

Fig. 17.2 General view of Socaire and the Miscanti/Miñiques lagoons (Courtesy of Moyano)

fostered indigenous mobility to larger towns, brought about not only a loss of ethnic identity and ancestral veneration, but also widespread poverty and overcrowding (Gunderman and González 1995; Núñez 1992).

Despite restrictions to agricultural cultivation, present-day social life in Socaire is partially centered on agricultural activities. The community determines the rituals relating to documenting the annual cycle or "*sistema de cargos*" and civil-religious hierarchy: These include what are referred to as water judges (September 30–October 1), water distributor (March 31–April 1), and the *minga* who maintained and cleaned the irrigation canals in this region (October 24–26), and were also central to caring for livestock, planting, and harvesting of food crops between August and May. Community members also oversaw the major Christian festivities and feasting associated with San Bartolome (August 24), St. Barbara (December 4), Carnival (February–March), Holy Cross (May 3), San Juan (June 24), and Santiago (July 25) (Fig. 17.3). June 24 is associated with the winter solstice and the livestock branding feast, which coincides with the cultural perception of the "short days becoming longer" (Barthel 1986; Hidalgo 1992; Moyano 2011, 2015).

The most important date for the *socaireños*, possibly of pre-Hispanic origin, is the cleansing of the canal channel in late October. On this date (24–26 October), people in the community invoke the spiritual mountain essences from a ceremonial center located in the Nacimiento ravine, 5 km east of Socaire (Fig. 17.4). The water ritual is performed by a specialist, called the *cantal mayor*. This individual sings and

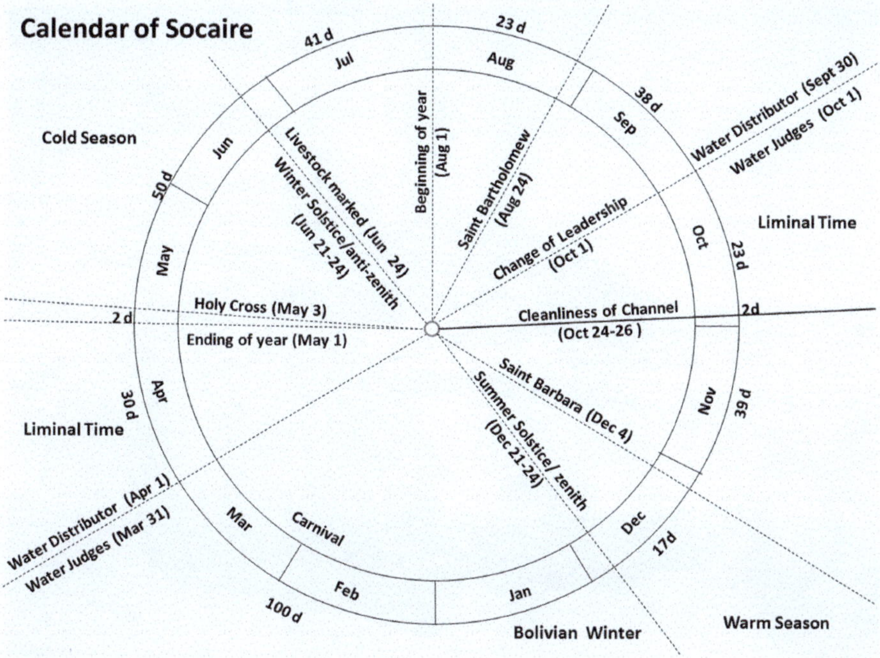

Fig. 17.3 The seasonal calendar of Socaire (Courtesy of Moyano)

dances, (a performance called *talatur*) in a circle around a bottle of *aloja* -alcoholic beverage made of Algarrobo (*Prosopis chilensis*)-inside, inside of a stone circle representing the landscape (*merendadero*), which includes hierarchical names of the surrounding mountains from the east in a clockwise direction. This ritual is interpreted by archaeoastronomy as a "3D Cosmogram," that includes the following categories: Above/East—Down/West, Right/Day—Left/Night, and Northern/Southern groups of mountains, in a system of imaginary lines or local *ceques* (Moyano 2011: 94, 99; 2015: 926; following the ideas of Barthel 1986; Grebe 1996; Grebe and Hidalgo 1988; Mariscotti de Görlitz 1978; Núñez 1993; Tichy 1983).

The system of radiating lines to the cardinal directions called *ceques*" correspond to 41 directions or *ceques* (*zeq'e,* any kind of line in Quechua) that had geographic reference to 328 *huacas* or sacred places (stones, springs, hills, and palaces) in the surrounding landscape of Cuzco, capital of the Inca Empire (Bauer 1998; Farrington 2013, 2014; Zuidema 1995 [1964], 2011).[3] In Cuzco, these lines were centered on the Temple of the Sun or Coricancha (*Qorikancha*) or Golden Enclosure. The Coricancha functioned as the center for establishing kinship, political hierarchies, tribute, and major holidays associated with a lunisolar and sidereal calendrical scheme. This

[3]In calendrical terms, number of 328 is equal to 12 sydereal lunar months of 27.3 days. The remaining 37 days to computing the 365 solar year, are completed with the period of invisibility of Pleiades, between May 3 and June 9, date that coincide with the beginning of the ceque calendar in Cuzco (Zuidema 2011).

Fig. 17.4 The ceremonial center of Socaire (Courtesy of Moyano)

system was related to *Collana, Payan,* and *Cayao* categories, which was connected to *panacas* and *ayllus* (social groups), within a dual, ternary, and quadripartite organization that governed the endo and exogamic relationship between *hanan* (upper) and *hurin* (lower) Cuzco (Zuidema 2011). In the Andes, these lines are oriented in the direction of certain mountains, recognized as sacred or *huacas* (*tata-mallkus*)[4] (Moyano 2011). These mountains are culturally perceived as the source of rain, thunder and lightning, the wind and literally associated mineral or deposits related to the metallurgical arts (Staller and Stross 2013). They are also astronomical markers during the annual cycle for agricultural cultivation and harvesting and breeding livestock (Del Río 2005; Farrington et al. 2015; Mariscotti de Görlitz 1978; Pino 2005; Sanhueza 2004, 2005). Cuzco had at least two *ushnus* (pillar, gnomon, and hole): one at Machu Picchu and the other the Sunturhuasi building in Cuzco, which allowed astronomical observations of the solar cycle on October 30 and February 13 (zenith passages), and nadir passages on August 18 and April 26, the beginning and ending of the agricultural cycle (Zuidema 1980: 320–321, 2011: 164).

[4]The *mallku* or *tata mallku* is associated among the Atacameños as spirits and mythical ancestors, that inhabiting spaces or sacred places like mountains, waterfalls, rivers, and even stones. They are recognized for having great powers and abilities for sicken and punish to humans, but also bestow protection, health, and good climate for agricultural production and livestock care. However, in return must be a "pago" or offering in coca leaves and alcohol (Valenzuela 2000).

Recent research has focused on understanding how these imaginary lines to the east are mental projection of a "left human hand," delineated by the volcanoes: Tumisa, Lausa, Chiliques, Ipira (Miscanti), and Miñiques (Moyano 2011; Moyano and Uribe 2012). The Miñiques volcano was called "the hand of thunder god or Tunupa" (Moyano and Bustamante 2015), and later associated by the Inca with lightning (*Illapa*) (Staller and Stross 2013). Ethnographic descriptions according to the socaireños refer to the Miñiques volcano as the lowest mountain, because in relation to the other four volcanoes or "fingers" it conveys this perception (Moyano 2011: 95, 102–103). This phenomenon could be explained by the Pareidolia, a psychological response that identifies the images from chaotic elements. This implicit image allows human beings, from memorization and pattern recognition, to identify and distinguish known forms from a surrounding background, for example, a set of patches that resemble animals, people, or parts of them, like a Rorschach test (Bustamante 2007).

The Cleaning of the Irrigation Channels

Before the cleaning of canal channel ceremony (October 24–26), villagers gather in an assembly in order to elect two people called captains (*capitanes*) to be responsible for coordinated the work. They are in charge of organizing and overseeing the distribution of the labor. Captains make a hierarchical distinction in their function as leaders: the senior captain assigns each task, while the assistant captain is responsible for reviewing, refining, and approving the pending tasks. Ethnographic evidence suggests there are distinctions between the roles of the senior and assistant captain (Barthel 1986 [1957]). The "capitán" has a higher status and control associated with male gender and the assistant captain related to female gender and the word "*capitana.*" Likewise, they are associated with the musical instruments associated with the festival: the *capitan* with the "clarín" (a type of trumpet) and the *capitana* with the "putu" (bull horn) to invoke the "sound of water that comes from the mountains" (Barthel 1986: 152).

Once elected, the captains are accompanied by a person designated as "toreador," responsible for signaling the channel for the labor of cleaning. Then the captains, the judge, and the *tareador* of water, already elected, are directed to the starting point (water well), in order to turn off the water and let the channel dry and begin the *minga* or communal work. Then on this last weekend of October, all villagers gather in the center of Socaire and begin the cleaning. Captains take a list of community members and the number of crop hectares, to assign sections of the channel equal to the hectares of each family or lineage group. The villagers begin their work in the vicinity of the old chapel (Socaire's colonial church)[5] (Fig. 17.5), and from there go

[5] Archaeo-astronomical data demonstrates that the orientation of the Socaire's church it is defined by the location of Miñiques volcano (5,927 m AMSL), identified as a condor or an human face (Moyano 2010, 2011, 2015).

Fig. 17.5 The sixteenth-century Church in Socaire (Courtesy of Moyano)

up progressively to the water well on the following afternoon. The first 100 meters are cleaned without any special measure to a sector known as *El Chorro*, where every third year people are required to clear sediments. The villagers move quickly because they have already cleared the hardest part of the channel. Our data clearly indicates that the irrigation canals are culturally conceptualized as a line projected in the geography connecting Socaire with the water well in the Nacimiento ravine (5.25 km away). A *ceque* or ceremonial path linked human activities, mainly in the October festival, with sacred sources of water and mountains in the local landscape (Moyano 2011, 2015; cf. Tilley 1994). In other words, a system of imaginary lines responded to the myths and local needs of the Atacama and was related to water rights, political division of the territory, agricultural rituals, and kinship.

After completing each labor the Socaireños await the review of the assistant captain, before continuing with the next task. There are times when the captain does not accept the job, because it is not satisfactory, and in this case people need to remove the mud and weeds accumulated in the channel wall again. Once the work is approved, the workers then go to senior captain to be assigned to another task, in another section of the channel (Fig. 17.6). This is done in rigorous order, because it is not possible to give a task in advance or clean the same section of the channel again. The work is characterized by an atmosphere of collaboration and celebration. Throughout the day people drink alcohol, including wine, beer, and *aloja*. The latter is a fermentation of the fruit of algarrobo (*Prosopis chilensis*), a species of Andean tree

Fig. 17.6 Communal work on the canal in Socaire (Courtesy of Azócar)

from the family of *Fabaceae*, that is prepared specially for this occasion. Toasted flour and coca leaves are shared with the sound of the *clarín* and *putu*, musical instruments that provide a background for the feast. The most enthusiastic villagers, provide shouts of approval at the very break of singing or playing. At noon, when the work is relatively advanced, guests and family start arriving with a variety of meats. The best food goes to the workers, because their work was particularly intense on this important holiday.[6] After lunch work resumes until sunset, then the villagers stop at the place where, year after year, an evaluation of the day is done. They then choose the "chacheros" responsible for collecting mountain grass that produces intense charring and smoke for the "cantales" ceremony the next day.

During the second day, dancing and meetings take place among local families in Socaire. According to Barthel (1986: 152), such festivities were for pleasure rather than ceremonial. Nevertheless, our fieldwork confirms that the social function of drinking and eating among workers, and drinking and eating with the Mother Earth or *Pachamama,* are important parts of communal work (Valenzuela 2006; Moyano 2010). The food and drink are particularly important in bringing people together. During this day, villagers clean a canal and celebrate by dancing of *talatur* in honor of the channel. They perform a water petition completed by the *cantales,* "or those that sign" (a master and his assistant) (Barthel 1986: 156–157). Around noon community

[6]The rest is defined by a circle of stones similar to the ceremonial center, described in the ceremony of rains (Moyano 2010: 53).

Fig. 17.7 Celebration of the awakening of the grandparents (Courtesy of Azócar)

members, families, and guests approach a nearby water well called "merendadero" or ceremonial center of Socaire (CCS), and there the *cantales* perform a ceremony known as "awakening of the grandparents"[7] (Fig. 17.7).

In the *meredadero*, the *cantales* offer food and alcohol to the *Tata Mallkus* (ancestor) or sacred mountains to the south and north in a hierarchical order, always starting by the geographical east direction (Mariscotti de Görlitz 1978: 79–80) (Fig. 17.8).

There are different versions concerning the number and name of the mountains (see Table 17.1); however, the consensus is that requests for the coming rains by the master and his assistant moves clockwise from the southern mountains Litinque and Chiliques first and then to the northern group, and then following the Lausa volcano it reverts to following a counterclockwise movement (see Fig. 17.1).

After the day's labor, exhausted *cantales* enter slowly and solemnly to the *merendadero* to present their greetings and give thanks for the workday. *Cantales* then present greetings and invite people to join them to drink wine and chew coca leaves. After the greetings, work is resumed and people go in the direction of the water well,

[7]The CCS is formed by (1) a plane surface surrounded by a circle of rocks (5.5 × 4.3 m) called *merendadero*; (2) a vertical rock known as *cerro grande*, at north of *meredadero*; (3) a rock of 1.0 × 1.5 m that symbolized the Chiliques volcano; (4) an area called *covero*, south of *merendadero*, where the *cantales*burn wood, coca leaves, and flamingo feathers; and (5) a system of seating formed by rocks, at north of *merendadero* (Barthel 1986; Hidalgo 1992; in Moyano 2011: 98).

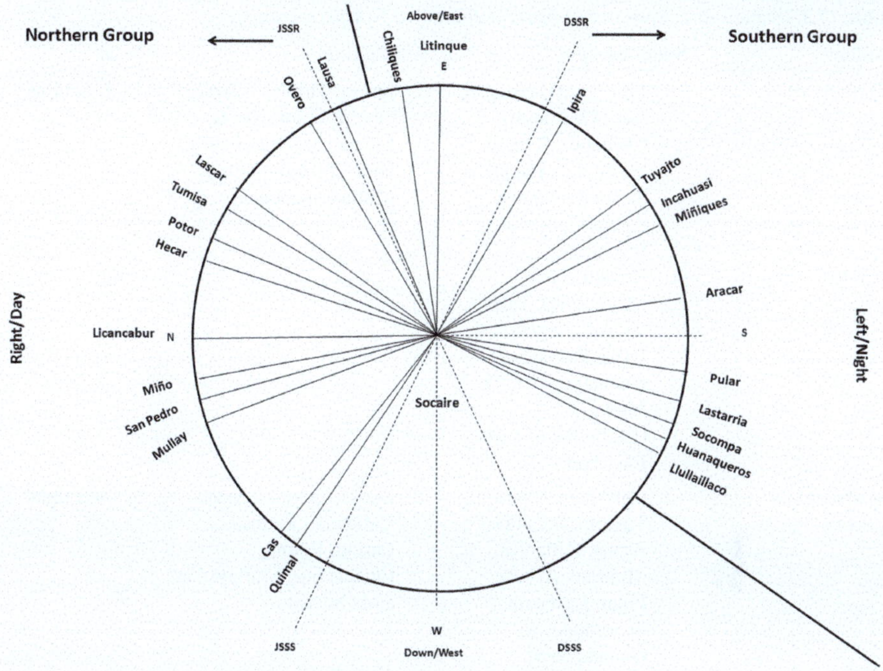

Fig. 17.8 The ceques surrounding Socaire (Courtesy of Moyano)

where they take another break, and enjoy music and dance (thanks to the *clarin* and *putu*). Such festivities also include Socaire women and children, and also include consumption of more *aloja* and wine. People ask for the *talatur* and request the presence of the *maestra* (women master) to lead them in songs to *huacas*, sacred springs, and hills, with little bells called *chomorrón*. The *maestra* uses a *chomorron,* a musical instrument formed by three pairs of little bronze bells, that ring the chords to *talatur.* People embrace and form a big circle first listening to the first verses of *talatur* in Kunza, an endangered language, moving from one direction to another. The *cantales* encourage the dancers by banging rods on their buttocks while playing the *clarin* and *putu* and shouting "¡*más lindo, más lindo*!" (¡cutest, cutest!). While the community of Socaire enjoys the *talatur*, the *cantales* privately make the last payments or *convidos* in the *merendadero*, where the *convidos* o *pagos* of ritual mountain offerings are located (Fig. 17.9).

During the ceremony of the "awakening of the grandparents," the *cantales* make a hole in the *covero* and deliver the first payment of the day and wait for an offering that is donated by the families of the community. Then the principal *cantal* requests offerings of *kajtcher* at the water well from the people. The *kajtcher* is a bottle of *aloja*, from which hangs a *tuftuca* (a piece of *llama* fat), typically accompanied by a *kipe* or bag containing small amounts of seeds, corn, quinoa, wheat, beans, and coca leaves. This set is tied with a red ribbon, and with flamingo feathers symbolizing

Table 17.1 Sacred directions in Socaire (Southern and Northern groups)

Author	South group	North group	Total
Barthel (1986)	15: *Litintique* (Litinque), *Ipira* (Miscanti), Chiliques, Laguna Verde, Las Fuentes de los Miñiques, Aguas Calientes, Incahuasi, Huanaqueros (Argnetine), Talaus, Arácar (Argentina), Pular, Socompa, Huanaqueros (¿?), *Llullaco* (Llullaillaco) and Lastarria	12: Lausa, Tumisa, Chasca, Overo, Potor, Hécar, Licancabur, San Pedro, Niño (Miño), Quimal, Mullay and Cas	27 directions
Mariscotti de Görlitz (1978)	12: Chiliques, Miscanti, Tuyaito, Incahuasi o Aguas Calientes, Miñique, Aracar, Pular, Lastarria, Socompa, Guanaqueros, Llullaillaco and Guanaqueros	10: Lausa, Covero, Aguas Calientes, Tumisa, Lascar, Potor, Licancabur, Mullay, Cas and Quimal	22 directions
Tichy (1983)[a]	16: Chiliques, Litintique,Miscanti, Tuyajto, Incahuasi, Miñique, Aracar, Capur, Pular, Lastarria, Socompa, Huanaqueros, Llullaillaco, Huanaqueros, Tilomento and Peine	13: Lagusa, Overo, Tumisa, Lascar, Potor, Hecar, Macon, Licancabur, Mullay, (Soncor), Cas, Quimal and Cerros Negros	29 directions (27?)
Reinhard (1983)[b]	No information	No information	More than 20 directions
Grebe and Hidalgo (1988)	14: Chiliki, Likintiki, Ipira, Laguna Verde, Miñiki, Kosor, Kulámar, Arakar, Pular, Salín, Iyaco, Lastarria, Puntas Negras and Tulan	16: Lausa, Tumisa, Chascal, Cerro Overo, Agua Caliente, Yoyoque, Pilire, Lascar, Ekar, Aritas, Likankabur, San Pedro, Niño, Kimal, Muyay and Gusyka Kas	30 directions

(continued)

Table 17.1 (continued)

Author	South group	North group	Total
Hidalgo (1992)	20: Chiliques, Ipira, Laguna Verde, Miñiques, Laco, Tuyajto, Incahuasi, Huanaqueros, Cósor, Culámar, Arakar, Púlar, Salín, Llullaillaco, Lastarrias, Socompa, Arizal, Puntas Negras, Pajonales and Tulan	20: Laúsa, Tumisa, Pibor, Chasca, Overo, Aguas Calientes, Laskar, Pilir, Hékar, Pótor, Llolloque, Aritas, Licancabur, Cajones, Chajchar, San Pedro, Miño, Quimal, Mullai and Cas	40 directions
Zuidema (1989, 2011), based on Barthel (1986), and Mariscotti de Görlitz (1978)	15: …?	12: …?	27 directions
Valenzuela (2000)	20: Lipira o Miscanti, Laguna Verde o Miñiques, Aguas Calientes, Laco, Incahuasi, Tuyacto, Huanaqueros, Cunatar, Talaus, Cosor, Aracar, Salin, Pular, Silla, Llullaillaco, Pajonales, Puntas Negras, Lastarria, Tilomonte and Tulan	16: Chiliques, Tumisa, Overo, Patos, Aguas Calientes, Hécar, Láscar, Tumbre, Laguna Verde, Licancabur, San Pedro, Niño, Moto, Kimal, Mullay and Cas	36 directions

[a]Tichy (1983) notes that Socaire is located very close to the Tropic of Capricorn, which means that each December solstice, the Sun reaches the zenith at noon. Zuidema (1995 [1964]) proposed a radial system based on imaginary lines projected from Socaire to mountains and astronomical points, as solstices and equinoxes, based on division of the circle of 360° in 40 sections of 9° each (Tichy 1983: 71)

[b]Reinhard resumed work of Barthel (1986) and described archaeological sites in almost all mountains invoked during the October ceremony except in Lastarria and Lascar, due to volcanic activity at Chiliques, Pular, Quimal, and Licancabur volcanoes, there are records of roads, structures, platforms, and offerings from Inca time (CIADAM 1985, 2001)

each member of Socaire's families. These are representative of age and gender categories. Black feathers are associated with adult males, while pink and red feathers are associated with adult females, and smaller white feathers with children. The *kajtcher* delivery is a time of great formality and the protocol requires a salute to the *cantales* who in turn receive the bottle and drink (*convidar*) when invited to the channel. Then *cantales* serve a glass of *aloja* and raise their right hand toward the mountains and the earth, *Pachamama.* They invite people to smoke, drink, and chew coca

Fig. 17.9 Talatur of Socaire (Courtesy of Azócar, October 2014)

leaves, then kneel and pray the "Our Father" while viewing the mountains. At this time the *controls* might offer the first payment. Everyone then follows the officials to the beginning of the *cerro grande*, represented by a large stone at *merendero's* shore, and then continue on with the other sacred elements of the landscape. Each *convidos* to the mountains is syncretized with Christian prayers and with salutes to the Mother Earth (*Pachamama*), the channel, and the ancestors (*abuelos*). Ultimately they remove the flamingo (*Phoenicopterus andinus*) feathers and slop *kajcher*, the *llama* fat, into a *covero* fire repeatedly, once for each of the mountains and springs located between Chile, Bolivia, and Argentina, while they chant the following verse:

> "(…) reciban todos esta hojita y alojita todos los tata cerros, más alegre y más content porque hoy es tu cumpleaños Tata putarajni, para que dé más agüita el tata putarajni, en el nombre del Padre, del hijo y del espíritu santo, amén" (cantal mayor, Socaire, October 2010). Translation: "(…) receive all this leaflet and lodge all the hills, more joyful and more content because today Tata putarajni, is your birthday, give more water to the tata putarajni, in the name of the Father, the son and the holy spirit, amen".

This specialized knowledge of the landscape could be interpreted as a magical relationship with the mountains, through their symbolic contact with them:

> "(…) cuando invitan a tomar a todas las divinidades y mientras van derramando gota a gota el trago por el suelo, (…) recogen nombre por nombre todos los lugares de su territorio y los insertan en un espacio más vasto y lleno de fuerzas, mediante la recitación casi interminable de cada lugar, cada recoveco donde se reconoce un poder especial. No es una simple

> enumeración del espacio, sino que van llamando a estos lugares para que desempeñen cada uno una function propia" (Harris y Bouysse-Cassagne 1988 p. 259). Translation: "(…) when they invite all the gods to take and as they pour drop by drop on the floor, (…) they take name by name all the places of their territory and insert them in a more vast and full of forces, through the almost endless recitation of each place, each corner where a special power is recognized. It is not a simple enumeration of the space, but they are calling to these places so that they each play their own function" (Harris y Bouysse-Cassagne 1988 p. 259).

After the ceremony of the *cantales*, workers and community are told to come back to the *merendadero.* The *socaireños* invite one by one to deliver greetings to the *controls* and then listen to the final speech. The major *cantal* delivers an emotional speech emphasizing in the high and low points of the celebration. The *cantal also* celebrates the respect that the community has had by following established traditions, performing good work, and providing space for the communal lunch. Finally, people come together to share food and drinks, and after that, the *cantales*, *capitanes*, and farmers go to the water well to get water. The opening of the channel gates is done with aplomb and enthusiasm. But when *cantales* give *convidos* to the mountains, this moment is mixed with great anxiety and excitement. People often scream when the water rushes into the canal which is a fast-moving flow filled with sediment. After the giving of the water, people return to the village for a ceremonial channel cleaning accompanied with a social dance at the local headquarters. Some *socaireños*, particularly those most engaged in the preceding work, accompany the water down, drinking *algae*, flirting, dancing, and singing the *talatur*. The water is received by singing and shouting, amid a feast on the banks of the canal at sunset. Lower down the hill, youths burn patches of grass to illuminate the slope of the hill and to signal that the water is going down. The night people accompany the water with shouts and applause in the middle of the main square singing the *talatur* in the middle of the main square well into the night. The sound of the *cumbias* is the signal that the ceremony in October is over.

Contemporary Power and Ethnicity in Socaire

There is no doubt that the ceremony of channel cleaning is the most important social event in Socaire. Cleaning the canal invokes the mountains, which are the source of the water, is part of the syncretism of ideas that combine aspects of indigenous and Christian traditions.

Hidalgo has indicated that in October is possible to understand the community as

> "(…) en acción, ante la eventualidad de concretar una obra o realizar una tarea en beneficio del colectivo, en el cual los hombres y mujeres que participan se movilizan aportando su esfuerzo personal. El trabajo comunitario es un rasgo del grupo que refuerza su espíritu de cuerpo, la identidad y pertenencia a una comunidad determinada" (Hidalgo 1992 p. 259). Translation: "(…) in action, before the eventuality of concretizing a work or performing a task for the benefit of the group, in which the men and women who participate, mobilize contributing their personal effort. Community work is a feature of the group and it reinforces their spirit of body, identity and belonging to a particular community" (Hidalgo 1992 p. 259).

Although the agricultural production decreases each year there by systematically decreasing the *campesino's* need of water for irrigation, the ceremony of cleaning of the channel affirms the atacameños claims for land, particularly since the adoption of indigenous rights in Chile (N°19.253).[8] In this sense, the cleaning of the channel not only benefits farmers that need water for agricultural production, it has relevance for the entire community, even those who have moved to the cities of Calama and Antofagasta. Thus, the participation in the October ceremony has more social implications than agricultural, it plays a large part in the process of ethnic vindication and identity in this Atacama region.

Historically the *atacameños* has been characterized by a sparse population and inhabit spaces of little or no interest to state officials. This is a marginal frontier region on the edge of a desert, unsuitable for life. This is the impression federal authorities have projected of the *atamameños* in the national mentality. It is often reversed, particularly when these "indigenous" become the subject of a policy based on the recognition of cultural difference. As Bolados has indicated "*ahora y desde una perspectiva de discriminación positiva, pasaron a constituirse en ciudadanos de primera*" (Bolados 2010: 82). The author notes that the process of a claim coincides with other social movements ushered in during the 1990s, after previous military dictatorships, and the beginning of a democracy in Chile. This resulted in a social platform for indigenous societies, previously stigmatized and this platform became a path of ethnic struggle and positive recognition. This is apparent when we consider that several anthropology and archaeology departments have courses on this topic and the speeches of local elites are imbued with the historical continuity and cultural traditions of the Atacama region.

With regard to the archaeological discourse and ethnic consciousness in San Pedro de Atacama, Ayala (2007, following Uhle 1903) noted that the Atacama Desert is a cultural area with pre-historic origins as evidenced by landscape modifications and ancient archaeological sites throughout the region some dating back to 600 B.C. (see, e.g., Staller 2005). These ideas were later taken up in the 1930s by Oyarzún and Latcham, who then used archaeological excavations to define the "Cultura Atacameña" (Atacama Culture) (Ayala 2007: 46). The archaeological and ethnohistorical records continued to influence the scientific paradigm, governed by an idea of cultural continuity within the anthropological discipline in Chile (Castro 1998: 9). Therefore, Mostny's ethnographic research in a little town located south of Socaire (1954) called "Peine, un pueblo atacameño," became a classic of the Chilean ethnography by describing important aspects of the indigenous lives of the Atacama oasis. Something similar happened when the Jesuit Gustavo Le Paige, who became pastor in *San Pedro de Atacama* in the 1950s, began archaeological research and constructed an archaeological museum (R.P. Gustavo Le Paige Archaeological Museum). Le Paige inaugurated a tradition that is now known as "estudios atacameños" (*atacameños* studies), characterized by collections of cultural artifacts and the excavation of human remains. In the first National Congress of Chilean Archaeology, which took place in San Pedro de Atacama in 1963, Le Paige defended

[8]http://www.conadi.gob.cl/documentos/LeyIndigena2010t.pdf (accessed on February 3, 2016).

his thesis of the pre-Hispanic origin of the *atacameños*. However, the conference attendees could not agree on their ethnic associations with other surrounding pre-Hispanic cultures. They want to substitute the denomination of "Cultura Atacameña" by "Complejo Cultural San Pedro" (Ayala 2007: 47). Despite this clarification, the academics remain unclear with reference to the ancient inhabitants of San Pedro de Atacama particularly with the discovery of ancient ceramic fragments, textiles, and funerary complexes (Castro 1998: 10).

Archaeological research has challenged research primarily ascribed to Murra (1972 [1975]) that the "Andean verticality" or complementarity was the basis of Andean political economies. Such cultural practices have enriched discussion surrounding indigenous ethnic diversity and their mobility (coastal, valley, desert, and mountain areas), and have incorporated an considerable variation into Andean political economy. This paradigm shift in Chilean archaeology, lead to the conclusion that cultural composition in the Atacama region is a mix of various populations and traditions, moving across the landscape from the *Altiplano* to the desert and the coast. Therefore, the idea of cultural continuity in archaeology is unsustainable. In the last two decades, more researchers are contributing to our understanding of indigenous ethnicity and in the process revitalizing certain indigenous traditions. Until the 1990s, the indigenous ceremonies surrounding the cleaning of irrigation channels carried out by communities were of little interest to anthropological and archaeological research. In this sense, the *atacameña's* concept of ethnicity is linked to resistance, cultural continuity, and ethnic affections and identities (Castro 1998; Muñoz 1993).

A second group of studies frames the *atacameños* in ethnic instrumentalism (Bartolomé 2006), which is that the indigenous are interested in participating in the national society because of their ability to manage favorable positions within the policies of the government of the center-left political parties called "*Nueva Mayoría*" (New Majority). Various experts on the indigenous issue such as Boccara (2007, Boccara y Bolados 2011), Bolados (2010), and Gundermann (1995, 1997, 2004) have agreed with the primacy of the ethnic instrumentalism in Atacama as related to the necessity of the new Chilean democracy after dictatorship (March 1990). The result was the *atacameños* leaders became empowered thanks to the management of symbolic and material capital in the area of *San Pedro de Atacama*. A closer review of local historical processes in reference to ethnography (Morales 2010, 2013; Núñez 1992; Rivera 2005, 2006), proposed that ethnicity in Atacama is not a product of cultural continuity or by the visibility of the state, but rather a contemporary social construction powered by the social pact between the state, pastors, farmers, and workers in Atacama. On the other hand, indigenous people have been integrated into the Chilean state since the late XIX. Specifically, Morales (2010, 2013) has shown that ethnicity in San Pedro de Atacama today, is the result of the appearance of a twenty-year scenario linking the resolution of issues between indigenous and non-indigenous. But the social field we call "ethnicity" is useful in mitigating the differences between indigenous, federal, private corporations, and national society, and in adjusting the differences and internal fragmentations of the *atacameños*. Indigenous leaders traditionally used the archaeology and agricultural practices in their speeches to legitimize cultural narratives as in the ceremony of cleaning the channel of San Pedro de Atacama, Alto Loa, and Calama areas reinforcing their ethnic and self-identity with this territory and region and culturally linked by a common history. In other words, it provides an understanding for the cultural diversity and historical autonomy of Atacama indigenous communities (Valenzuela 2006: 210).

This statement makes sense when we consider ethnicity in the *atacameña* narrative and its reference to the strength and vitality of traditions among various indigenous cultures. Especially high-status indigenous individuals, who have struggled with federal authorities regarding their natural heritage (sites and archaeological objects), intangible heritage (traditions, customs, and activities), as well as the state and private investment related to archaeological and natural parks, tourism, and agreements with mining, energy, and astronomy projects. However, the "common people" do not necessarily share the benefits of such investments. Morales (2010) has noted that this group is far removed from the level of discourse and narrative of the *atacameña* elite, causing Vázquez, to note that "the notions of indigenous, ethnicity and nation" should not only be understood and important to educated people who are able to address such issues to government politicians. Concepts are also needed which define their ethnic domination (Vázquez 2010: 11).

Finally, in a study about mobility systems of grazing llamas in the Atacama highlands, Morales (1997) has documented strong relationships between the archaeological sites and local customs. Shepherds from Talabre, Camar, and Socaire, as well as for communities south of San Pedro de Atacama, places with archaeological remains, tombs, or rock art, are perceived as sacred or spaces of "*gentiles*" (ancestors). This speech is mainly handled by elders and adults who express fear to visit these places, because the *gentilares* are inhabited by the "abuelos" or "tata abuelos" (grandfathers). These correspond to the spirit power of the natural world, which links mythical ancestors that are recognized as powerful and capable of generating diseases, punishing people, promoting rains and water, and caring for livestock and crops to particular ethnic groups. Therefore, the people need to offer these ancestors coca leaves and *chicha* to foster the regeneration of natural food resources. Such animistic conceptions of the environment reinforce a hierarchical status within the *atacameña* communities, and in the words of Morales (1997), allows the legitimization of the political differences of ethnicity. Therefore, it should not be treated as cultural continuity, but instead as a result of struggles and local ethno-political processes of the Atacama communities (Bolados 2010: 93).

The cleaning of channels in Socaire represents one of the most important ritual activities in the social life of the community. Not only because of its importance to the cultivation of food crops necessary to sustaining the community, but also because it is central to the reproduction and continuity of the agricultural cycle. The cleaning of the irrigation canals provides a cohesion between the community and its various ethnic groups, to the water and mountains. When referring to the importance of the cleaning of channels in terms of ethnicity, it should not only be perceived in terms of a set of elements with some historical continuity. The channel cleaning has strong connections to ceremony, rituals, and feasting providing various ethnic groups with a sense of belonging to the *atacameño*. At the end of the 1990s, when the ethnic claiming processes among the *atacameños* became positive and appealing, the cleaning of channels began to acquire a political dimension in defining what is known as "genuine" and as indigenous. This perception of the differences among the Atacama people, established by the contact and interaction with others (Barth 1976), is the result of the relationship between the indigenous and the Chilean state, i.e., the relationship between the ethnic and their stereotypes (Gundermann 1995).

These types of rituals mediate the processes of transformation of indigenous identity to an art of cultural heritage in the Antofagasta province, in turn generating a sense of belonging among the *atacameños* as possessors of a heritage rooted in time. The most noticeable effect of these processes in the *atacameña* communities has been the generation of a capacity for organization and institutionalization of aspects of indigenous social life. The ceremony of cleaning of channels is a prime example of this process. In this way, through the ritual is legitimized giving meaning, like never before, to the ethnic discourse in Atacama: this is obvious in the origin myths, belief in certain spirits, and the cult of water and ancestors. Therefore, to the *socaireños* the underlying importance of the cleaning of channels is the knowledge and experience transmitted by generations, not in the sense of historical and cultural continuity, but rather as a process of reinterpretation and self-identification against the others, i.e., the local cultures and the state. It is also presented with an unstable *atacameño* subject undefined by the anthropological evidence, which equated culture with a territory or a language. The individuals who organize and authorize the cleaning of the canals in the October ceremony are a prime example, consisting of not only the *socaireños*, but also relatives, friends, and people from other communities. Anthropologists and archaeologists are sometimes not well received in such ceremonies. In particular, they restricted access to the feast and the ceremonial center. This is an example of symbolic power that *atacameños* have over the "other" and also involves the recreation of important ceremonial traditions, or as commoners from Socaire say "doing things as the grandparents" (personal communication 2010).

It is important to emphasize the *socaireños* have had a long cultural history, whether pristine or unrelated to cultural syncretism, and in the ceremony of cleaning of channels, the members of ethnic groups in the community have raised a sense of ethnicity against the Chilean state that has systematically discriminated against them. The October ceremony, not only allows the *socaireños* projection against the label of "otherness," but also generates a sense of identity within the policies of modern Latin American neoliberalism. The Chileans refer to the *socaireños* as "Bolivians, drunks, blacks, animals, and dirties"—these names are, unfortunately, still are used in the North of Chile today to refer to the culturally distinct communities of the deserts and mountains.

Final Remarks

Our research on the October ceremony of channel cleaning channel in Socaire indicates that it represents the most important ritual and political activity among the *atacameños*. Its importance provides the community with sustenance through the cultivation and harvesting of food crops essential to their survival and to provide resources from other regions of the Andes.

First, the ceremony recognizes the deep knowledge *socaireños* have of their natural environment in this case a highland desert, through the hydrology and irrigation of the landscape, and also manages the highly complex agricultural cultivation of numerous food crops and this is a tradition that has lasted for generations. Our ethnographic evidence indicates water and the management of the natural landscape have a profound significance to the social, material, and productivity of community life. Another dimension is the ritual and cultural importance of the October ceremony itself; the cleaning of the channels is strongly linked to ethnic identity and a cultural recognition of the symbolic associations and spiritual relationships ethnic groups have to the mountains that channel the water for their cultivation. The food crops consist of domesticated food crops associated with cuisines which have strong ethic associations to various ethnic groups in the community. These cultures are responsible for tracking meteorological cycles for their human survival. When participating in maintaining the *cantales*, they also make *pagos* or offerings and *convidos* to venerate *huacas* or sacred places in nature, particularly those associated with calculations of the annual cycles, i.e., local *ceques*. Socaire's *ceques* system is then delineated by the existence of lines projected out to visible and non-visible points in the surrounding landcape, to certain mountain peaks and other features associated with the cardinal directions. These sets of directions have two centers. One is located in the ceremonial center or *merendadero*, in the vicinity of the water well in the Nacimiento ravine, while the second center is probably located in the courtyard of the sixteenth century Socaire's church. Our ethnographic research suggests these lines are perceived as an "area of protection for *convidos*" or lines of ritual offerings to the mountains. These lines serve not only to establish a direct relationship between the *cantales* (ritual specialist) and sacred landscape, but also link different *mallkus* or sacred mountains not visible from Socaire that define the territorial limits of the community. Finally, the validity and centrality that the cleaning of channels has to *socaireños* relates to the ethnic claim processes. In the last two decades, the Chilean state authorities must now recognize a moral debt to these indigenous communities and provide them opportunities for dialogue, tolerance, and reconciliation of cultural differences. Although this context has shown its own paradoxes, we must recognize that it opened the way for dialogue between the central government and the *atacameños*. The old stigma of the indigenous is currently positive, creating the possibility that *atacameños* can maintain their cultural tradition, i.e., the ceremony of cleaning of the channel. It has come to be revitalized and perceived in terms of power and political autonomy within multi and pluri-ethnic processes. At the same time, it provides anthropology and cultural studies a new dynamic for our understanding of so-called ethnicities in the Andean region.

Acknowledgments We would like to thank the indigenous community of Socaire, especially to Josefa Cruz and her family. Also to Aliro Plaza, Yanet y Gustavo Cruz, Elsa and Silvestre Varas, Dina Tejerina, Berta y Gladys Cruz and the *cantal* Laureano Tejerina (R.I.P), and the field photography of Rodrigo Azócar, anthropologist of University of Chile. Special thanks to the council of the *atacameños* people and all the communities of *San Pedro de Atacama* in their important struggle for indigenous rights.

References

Aldunate C, Castro V (1981) Las chullpas de Toconce y su relación con el poblamiento en el Loa Superior. PeriodoTardío, Ediciones Kultrún, Santiago

Ayala P (2007) Relaciones entre atacameños, arqueólogos y estado en Atacama (norte de Chile). Estudios Atacameños 33:133–157

Barth F (1976) Los grupos étnicos y sus fronteras. La organización social de las diferencias culturales. Fondo de Cultura Económica, México

Barthel T (1986 [1957]) El agua y el festival de primavera entre los atacameños. Allpanchis 28 (XVIII):147–184

Bartolomé M (2006) Los laberintos de la identidad. Procesos identitarios en las poblaciones indígenas. Avá 9:28–48

Bauer B (1998) The Sacred Landscape of the Inca. University of Texas Press, Austin

Boccara G (2007) Etnogubernamentalidad: La formación del campo de la salud intercultural en Chile. Chungara 39(2):185–207

Boccara G, Bolados P (2011) ¿Qué es el multiculturalismo? Política étnica y pueblos indígenas en el Chile neoliberal. Revista de Indias 250 (LXX):651–690

Bolados P (2010) Neoliberalismo multicultural en el Chile democrático: Gubernamentalizando la salud atacameña a través de la participación y el etnodesarrollo. PhD. Thesis in Anthropology, IIAM-UCN, Chile

CIADAM (Centro de Investigaciones Arqueológicas de Alta Montaña) (1985) El enigma de los santuarios indígenas de alta montaña. Revista del CIADAM 5. San Juan, Argentina

CIADAM (Centro de Investigaciones Arqueológicas de Alta Montaña) (2001) Revista del CIADAM 6 (Años 1987–1999). San Juan, Argentina

Castro V (1998) La dinámica de las identidades en la subregión del río Salado, Provincia de El Loa, II region. Paper presented at the Encuentro Nacional Interinstitucional de Investigadores de Identidades Culturales. Universidad de Chile, Santiago, pp 5–50

Bustamante P (2007) Pareidolia y apofenia como fenómenos extendidos y como herramienta de análisis de obras rupestres y sitios arqueológicos pertenecientes a diversas culturas. Rupestreweb: http://www.rupestreweb.info/pareidolia2.htm. Accessed 4 July 2018

Del Río M (2005) Etnicidad, territorialidad y colonialismo en los Andes: Tradición y cambio entre los Soras de los siglos XVI y XVII. IEB-IFEA, Bolivia

Farrington I (2013) Cusco. Urbanism and Archaeology in the Inca World. University Press of Florida

Farrington I (2014) The centre of the world and the Cusco Ushnu complexes. In: Meddens F, Willis K, McEwan C, Branch N (eds) Inca Sacred Space Landscape, Site and Symbol in the Andes. Archetype Publications, London, pp 197–207

Farrington I, Moyano R, Gustavo Díaz M (2015) El paisaje ritual en El Shincal de Quimivil. La importancia de los estudios arqueoastronómicos. In Una Capital Inca al Sur del Kollasuyu: El Shincal de Quimivil, edited by Rodolfo Raffino, Lidia Lácona, Reinaldo Moralejo, Diego Gobbo and Ma. Guillermina Couso. Dirección Provincial de Antropología, Provincia de Catamarca—Museo de la Plata—Fundación de Historia Natural Félix de Azara, Argentina, pp 41–61

Grebe ME (1996) Patrones de continuidad en el mundo surandino: creencias y cultos vinculados a los astros y espíritus de la naturaleza. In Cosmovisión andina. Centro de Cultura, Arquitectura y ArteTaipinquiri. La Paz, Bolivia, pp 205–220

Grebe M, Hidalgo y B (1988) Simbolismo atacameño: Un aporte etnológico a la comprensión de significados culturales. Revista Chilena de Antropología 7:75–97

Gundermann H (1995) Comunidad indígena y ciudadanía. La experiencia aymara en el norte de Chile. Allpanchis 46:91–130

Gundermann H (1997) Etnicidad, identidad étnica y ciudadanía en los países andinos y el norte de Chile. Los términos de la discusión y algunas hipótesis de investigación. EstudiosAtacameños 13:9–26

Gundermann H (2004) Inicios de siglo en San Pedro de Atacama: Procesos, actores e imaginarios en una localidad andina. Chungara 36(1):221–239

Gundermann H, González H (1995) Tierra, agua y sociedad atacameña, un scenario cambiante. In Agua, ocupación y economía campesina en la region atacameña, edited by Pierre Pourrut and LautaroNúñez. Universidad Católica del Norte, Antofagasta, pp 78–106

Harris O, Bouysse-Cassagne T (1988) Pacha: entorno al pensamiento aymara. In: Albó X (ed) Raíces de América: El mundo aymara. Alianza Editorial, Madrid, pp 217–274

Hidalgo B (1992) Organización social, tradición y aculturación en Socaire. Una aldea Atacameña. Undergraduate thesis, University of Chile

Instituto Nacional de Estadísticas (INE) (2005) Estadísticas sociales de los pueblos indígenas en Chile-Censo 2002 Chile.

Mariscotti de Görlitz AM (1978) Pachamama Santa Tierra. In Indiana 8. Gebr. Mann, Berlin

Morales H (1997) Pastores trashumantes al fin del mundo. Un enfoque cultural de la tecnología: enunacomunidadandina de pastores. Undergraduate Thesis of Anthropology, University of Chile

Morales H (2010) Etnopolítica en Atacama. Laberintos de la etnicidad atacameña en Chile. Inaugural-Dissertation zur Erlangung der Doktorwürde Am Fachbereich Geschichts- und Kulturwissenschaften Lateinamerika-Institut (LAI) Altamerikanistik der Freien Universität Berlin

Morales H (2013) Construcción social de la etnicidad: Ego y alter en Atacama. EstudiosAtacameños 46:145–164

Mostny G (1954) Peine un pueblo atacameño. Publicación N°4 del Instituto de Geografía, Facultad de Filosofía, Universidad de Chile

Moyano R (2010) La mano de Dios en Socaire: Estudio de un calendario agrícola en Atacama, norte de Chile. Master Thesis, ENAH, México

Moyano R (2011) Sub-tropical Astronomy in Southern Andes: The Ceque System in Socaire, Atacama, Northern Chile. In Ruggles CLN (ed) IAUS 278 Archaeoastronomy and Ethnoastronomy: Building bridges between Cultures. Cambridge University Press, pp 93–105

Moyano R (2015) Landscape, mountain worshiping and astronomy in Socaire. In: Ruggles CLN (ed) Handbook of archaeoastronomy and ethnoastronomy. Springer Science + Business Media, New York, pp 921–929

Moyano R, Uribe C (2012) El volcán Chiliques y el "morar-en-el-mundo" de una comunidad atacameña del norte de Chile. Estudios Atacameños 43:187–208

Moyano R, Bustamante P (2015) Fenómenos de pareidolia en Socaire, norte de Chile. Paper presented at the III Jornada Interamericana de Astronomía Cultural, Río de Janeiro, Brasil

Muñoz B (1993) Procesos de cambios socialesen el área de San Pedro de Atacama. Pérdida y recuperación de la identidad étnica. Una contribución antropológica para el desarrollo. Universität en Tübirgen, Holanda

Murra JV (1972 [1975]) Formaciones económicas y políticas del mundo andino. Instituto de Estudios Peruanos

Murra JV (1978) La organización económica del estado Inca. Siglo Veintiuno Editores, México

Núñez P (1991) Sobre economía prehispánica de Socaire, Norte de Chile. Actas del XI Congreso de Arqueología Chilena. Museo Nacional de Historia Natural, Sociedad Chilena de Arqueología, Santiago de Chile, pp 201–210

Núñez P (1993) Un canal de regadío incaico: Socaire—Salar de Atacama. Actas del XII Congreso Nacional de Arqueología Chilena. Boletín 4, Museo de la Araucanía. Chile, pp 145–149

Núñez L (1992) Cultura y conflicto en los oasis de San Pedro de Atacama. Editorial Universitaria, Santiago

Pino J (2005) El ushnu y la organización espacial astronómica en la sierra central del Chinchaysuyu. Estudios Atacameños 29:143–161

Reinhard J (1983) Las montañas sagradas: un estudio etnoarqueológico de ruinas en las altas cumbres andinas. Cuadernos de Historia 3:27–62
Reinhard J (1993) Llullaillaco: an investigation of the world's highest archaeological site. Latin Am Indian Lit J 9(1):31–65
Reinhard J (2006) The Ice Maiden: Inca Mummies, Mountain Gods, and Sacred Sites in the Andes. National Geographic Books
Rivera F (2005) Anti-flexibilización con identidad/alteridad. La cuestión étnica atacameña contemporánea y su contexto neoliberal-flexible". Revista de Antropología Experimental 4:1–17
Rivera F (2006) Entorno liberal y la alteridad étnica anti-flexible de los atacameños contemporáneos. Revista Chilena de Antropología 18:59–89
Sanhueza C (1992) Tráfico caravanero y arriería colonial en el siglo XVI. Estudios Atacameños 10:173–187
Sanhueza C (2004) Medir, amojonar, repartir: territorialidades y prácticas demarcatorias en el camino incaico de Atacama (II Región, Chile). Chungara 36(2):481–492
Sanhueza C (2005) Espacio y tiempo en los límites del mundo, los incas en el despoblado de Atacama. Boletín del Museo Chileno de Arte Precolombino 10(2):51–77)
Staller JE (2005) La agricultura de la irrigación en el Período Formativo del Desierto de Atacama. In: Rivera MA (ed) Arqueología del Desierto de Atacama, Chile: La Etapa Formativa en el Area de Ramaditas/Guatecondo. Santiago: Ediciones Universidad Bolivarana, pp 89–102
Staller JE, Stross B (2013) Lightning in the Andes and Mesoamerica. Pre-Columbian, Colonial, and Contemporary Perspectives. Oxford University Press
Tichy F (1983) El patrón de asentamientos con sistema radial en la meseta central de México: ¿Sistemas ceque en Mesoamérica? Jahrbuch f¨ur Geschichte von Staat, Wirtschaft und Gesellschaft Lateinamerikas 20:61–84
Tilley C (1994) A phenomenology of landscape: places. Paths and Monuments, Berg, Oxford
Valenzuela A (2000) Socaire: Contexto, Problemas y Transformaciones en la Agricultura de un Pueblo Atacameño. Unpublished manuscript, Universidad de Chile, Santiago
Valenzuela A (2006) Atacameños de Calama. Diversidad, transitoriedad y fragmentación0nen las organizaciones atacameñas urbanas y surelación con el estado chileno. Master Thesis, CIESAS, Guadalajara
Vázquez L (2010) Multitudy distopía. Ensayos sobre la nueva condición étnica en Michoacán. UNAM, México
Zuidema RT (1980) El ushnu. Revista de la Universidad Complutense 28:317–362
Zuidema RT (1989) Reyes y guerreros: Ensayos de la cultura andina. Fomciencias, Lima
Zuidema RT (1990) Ceques and Chapas: an Andean Pattern of Land Partition in the Modern Valley of Cuzco. In: Circumpacifica Festschrift für Thomas S. Barthel, edited by Bruno Illius and Matthias Laubscher. Peter Lang, Frankfurt, pp 627–643
Zuidema RT (1995 [1964]) El sistema de ceques del Cuzco. Translated by Ernesto Salazar. Pontificia Universidad Católica del Perú, Lima
Zuidema RT (2011) El calendario Inca. Tiempo y espacio en la organización ritual del Cusco, la idea del pasado. Fondo Editorial del Congreso del Perú, Lima

The manufacturer's authorised representative in the EU is Springer
Nature Customer Service Centre GmbH, Europaplatz 3, 69115 Heidelberg,
Germany. If you have any concerns regarding our products, please
contact ProductSafety@springernature.com

Printed and bound by CPI Group (UK) Ltd, Croydon, CR0 4YY
30/09/2026
02225247-0016